As Pastoralists Settle

Social, Health, and Economic Consequences
of Pastoral Sedentarization in
Marsabit District, Kenya

STUDIES IN HUMAN ECOLOGY AND ADAPTATION

AS PASTORALISTS SETTLE
Social, Health, and Economic Consequences of Pastoral Sedentarization in Marsabit
District, Kenya
Edited by Elliot Fratkin and Eric Abella Roth

A Continuation Order Plan is available for this series. A continuation order will bring delivery of each new volume
immediately upon publication. Volumes are billed only upon actual shipment. For further information please
contact the publisher.

As Pastoralists Settle

*Social, Health, and Economic Consequences
of Pastoral Sedentarization in
Marsabit District, Kenya*

Edited by

Elliot Fratkin

Smith College, Northampton, Massachusetts

and

Eric Abella Roth

University of Victoria, Victoria, British Columbia, Canada

Kluwer Academic Publishers • New York and London
New York, Boston, Dordrecht, London, Moscow

Library of Congress Cataloging-in-Publication Data

As pastoralists settle : social, health, and economic consequences of the pastoral
    sedentarization in Marsabit District, Kenya / edited by Elliot Fratkin, Eric Abella Roth.
      p. cm. — (Studies in human ecology)
    Includes bibliographical references and index.
    ISBN 0-306-48594-X (hb.) — ISBN 0-306-48596-6 (eBook)
      1. Marsabit District (Kenya) — Economic conditions. 2. Marsabit District (Kenya) — Social
    conditions. 3. Nomads — Kenya — Marsabit District — Sedentarisation.   I. Fratkin, Elliot M.
    II. Roth, Eric Abella.   III. Series.

HC865.Z7M373 2004
306.3′49′0967624—dc22

                                                                            2004042174

ISBN HB: 0-306-48594-X
    PB: 0-306-48595-8
    e-Book: 0-306-48596-6

© 2005 Kluwer Academic / Plenum Publishers, New York
233 Spring Street, New York, New York 10013

http//www.kluweronline.com

10 9 8 7 6 5 4 3 2 1

A C.I.P. record for this book is available from the Library of Congress

Contributors

Wario R. Adano, M.Sc., Amsterdam Research Institute for Global Issues and Development Studies, University of Amsterdam, Nieuwe Prinsengracht 130,1018 VZ Amsterdam, The Netherlands (w.adano@frw.uva.nl)

Elliot Fratkin, Professor, Department of Anthropology, Smith College, Northampton, Massachusetts, 01063 (efratkin@smith.edu)

Masako Fujita, M.A., Department of Anthropology, Box 353100, University of Washington, Seattle, Washington 98195-3100 (masakof@u.washington.edu)

John G. Galaty, Associate Professor, Department of Anthropology, McGill University, Montreal, PQ, Canada (john.galaty@mcgill.ca)

Joyce Giles, M.A., Department of Anthropology, University of Victoria, P.O. Box 3050, Victoria, British Columbia, Canada V8W 3P5

Joan Harris, R.N., Hornby Island, British Columbia, Canada, bigtreetralee@hot.mail.com

Peter D. Little, Professor, Department of Anthropology, University of Kentucky, 211 Lafferty Hall, University of Kentucky, Lexington, Kentucky 40506-0024 (Pdlitt1@uky.edu)

John G. McPeak, Assistant Professor, Department of Public Administration, Syracuse University, Syracuse, New York 13244 (jomcpeak@maxwell.syr.edu)

Leunita Auko Muruli, Ph.D., Institute of African Studies, University Of Nairobi, P. O. Box, 30197, Nairobi, Kenya (laterem@yahoo.com)

Martha A. Nathan, M.D., Assistant Professor of Medicine, Tufts University School of Medicine, Boston MA Massachusetts 02111, and Brightwood Health Center, 380 Plainfield Street, Springfield Massachusetts 01107 (martygif@comcast.net)

Elizabeth N. Ngugi, R.N. Ph.D., Strengthening STDs/HIV Control Unit, Department of Community Health, University of Nairobi, P. O. Box 19676, Nairobi, Kenya (stdhiv@arcc.or.ke, Engugi@ratn.org)

Walter Obungu Obiero, Ph.D., Family Health International, 2101 Wilson Boulevard, Suite 700, Arlington, VA 22201 (wobiero@fhi.org)

Eric Abella Roth, Professor, Department of Anthropology, University of Victoria, P.O. Box 3050, Victoria, British Columbia, Canada V8W 3P5 (ericroth@uvic.ca)

H. Jürgen Schwartz, Professor, Humboldt University of Berlin, Philippstrasse 13, Haus 7, Berlin-Mitte; Berlin D-10115; Germany horst.juergen.Schwartz@rz.hu-berlin.de

Bettina Shell-Duncan, Associate Professor, Department of Anthropology, University of Washington, Box 353100, Seattle, WA 98195 (bsd@u.washington.edu)

Kevin Smith, Ph.D., United States Agency for International Development, Nairobi, Kenya (kevsmith@ usaid.gov)

David Wiseman, M.D., Hornby Island, British Colombia, Canada, bigtreetralee@ hotmail.com

Karen Witsenburg, M.Sc., Amsterdam Research Institute for Global Issues and Development Studies, University of Amsterdam, Nieuwe Prinsengracht 130,1018 VZ Amsterdam, The Netherlands (k.witsenburg@frw.uva.nl)

Studies in Human Ecology and Adaptation: A Note From The Series Editors

This volume, *As Pastoralists Settle*, edited by Elliot Fratkin and Eric Abella Roth, is the first in a new series launched by Kluwer Academic Publishers which will present studies in human ecology and adaptation under our general editorship. The objective of this series is to publish cutting-edge work on the bio-social processes of adaptation and human-environmental dynamics. We are committed to three initial volumes, including both archaeological and field research on contemporary populations, and hope that academic interest will permit us to add more, thereby making the series a major and continuing resource in the field of human ecology. Generally, scientific publications have the largest impact when they can be utilized in the classroom or by non-academic practitioners in addition to stimulating further research by specialists. Key to making these studies accessible to a broad readershp is that they are clearly written and tightly focused. We have quite deliberately taken the journal *Human Ecology* as a model since it focuses on empirically rooted original research addressing a wide interdisciplinary readership. As a consequence, the journal has achieved considerable recognition among academic publications worldwide which deal with environmental issues. As with contributions to *Human Ecology*, potential manuscripts in this series are subject to peer review.

We hope that readers will agree that *As Pastoralists Settle* is both an outstanding contribution to the field and a splendid way to inaugurate this new book series. Following the detailed introduction by the very able editors, we have studies here by some of the foremost scholars of East African pastoralism, many of whom are contributors to *Human Ecology* in different capacities. Some 240 million, or approximately half of the world's agro-pastoralists live in Africa, of whom twenty-five million live in East Africa. Clearly the sedentarization of pastoralists is a major concern for those working in areas of national development, touching as it does on problems of inter-group competition including possible conflict, health and the delivery of human services, gender roles, and the economics of food production and distribution. While there are a number of books which treat the settlement of pastoralists in different parts of the world, this is the first interdisciplinary collaboration involving anthropologists, health specialists, agronomists, experts in public administration and development, and others all focusing on the long-term causes and consequences of pastoral settlement in one relatively delineated geographic area—Marsabit district, Kenya. This geographic focus gives the book a unique perspective that looks beyond events, numbers, and trends to examine the complex interplay of bio-social processes including health, & nutrition, child development, economic activities, and perceptions of risk and reward, as well as the unfortunate but ever-present potential for conflict. The international nature of the collaborative effort here is also unusual but extremely welcome and rewarding in the diversity of views it allows.

We are indebted to the scholars whose work is presented here and which cumulatively represents many lifetimes of meticulous and often arduous research. The necessarily anonymous reviewers have made a significant contribution to the publication process, which we gratefully acknowledge here. We would also like to express our gratitude to our far-sighted, perhaps even fearless, editors at Kluwer Academic Publishers Myriam Poort and Teresa Krauss.

Daniel G. Bates
Ludomir R. Lozny

Hunter College, C.U.N.Y.

Contents

Chapter 1

Introduction

The Social, Health, and Economic Consequences of Pastoral Sedentarization in Marsabit District, Northern Kenya

ERIC ABELLA ROTH AND ELLIOT FRATKIN

1. INTRODUCTION

Formerly nomadic livestock-keeping pastoralists have settled in many regions of the world in the past century. Some groups, including those in the former Soviet Union, Iran, and Israel, have settled in response to state-enforced measures; others including Saami in Norway or Bedouins in Saudi Arabia, in response to changing economic opportunities. East Africa, home to many cattle- and camel-keeping pastoral societies, has been among the most recent to change. The shift to sedentism by East African pastoralists increased dramatically in the late 20th century as a result of sharp economic, political, demographic, and environmental changes. Prolonged drought, population growth, increased reliance on agriculture, and political insecurities including civil war and ethnic conflict have all affected the ability of pastoralists to keep their herds. Still, the majority of pastoralist households in Kenya, Ethiopia, Somalia, and Tanzania remain committed to raising livestock, even as they adapt to farming or urban residence. Pastoral production remains a major economic focus in the savannas and scrub deserts of Africa, due to both its ecological adaptability and the economic incentive to market livestock and their products (Fratkin, 2001).

Pastoralists settle for a variety of reasons, some in response to 'pushes' away from the pastoral economy, others to the 'pulls' of urban or agricultural life. Maasai people in southern Kenya, for example, lost grazing lands due to the growth of agricultural and

ERIC ABELLA ROTH • Department of Anthropology, University of Victoria, Victoria, British Columbia, Canada V8W 3P5.
ELLIOT FRATKIN • Department of Anthropology, Smith College, Northampton, Massachusetts, 01063.

1

pastoral populations, the privatization of land for commercial farms and ranches, and the expansion of tourist game parks, leading many pastoralists to combine sedentary maize cultivation with livestock keeping (Campbell, 1999; Galaty, 1992; McCabe, 2003). In the more arid and less densely populated areas of northern Kenya, pastoralist families have settled in response to a different set of factors. These include population growth, increasing drought and famine, the need for physical security and the economic attractions of peri-urban residence. Towns offer opportunities to sell livestock and agricultural productions, proximity to schools and health clinics, and prospects for finding wage-paying employment.

International development agencies, religious organizations, conservation groups, and national governments have frequently advocated the settling of nomadic or semi-nomadic pastoralists in Africa. Many of these organizations view nomadic pastoralism as wasteful, unproductive, or 'primitive,' and advocate permanent settlement as a means to benefit the local and national economy, assimilate marginal populations thereby forging national identity, and provide health and education to these isolated peoples (Dyson-Hudson, 1991; Fratkin, 1997; Kituyi, 1993). In the 1960s and 1970s, international donors including the World Bank and USAID encouraged the privatization of formerly communal rangelands among the Maasai and the establishment of individual ranches in Kenya and Tanzania, a situation which led to widespread land titling and loss of grazing resources (Galaty, 1994; Hodgson, 2001; Jacobs, 1980; World Bank, 1984). In northern Kenya, religious organizations involved in famine relief work including the Catholic Relief Services, encouraged poor pastoralists to settle permanently at famine relief points to deliver food and social services, but also to disengage pastoral populations from their nomadic lifestyle (Fratkin, 1991; Hogg, 1982). National governments in Africa and elsewhere have long been concerned with sedentarizing pastoralists as a means of control, taxation, and inhibiting cross-border migrations (Bayer and Waters-Bayer, 1994; Salzman, 1996).

Photo 1. Catholic Mission Sister distributing famine relief foods to Rendille at Korr, 1985.

More recently, international development agencies have accepted findings from ecologists and anthropologists that point to the adaptive aspects of pastoralism, showing that pastoralists live in unstable, arid, and sparse environments described as being "in disequilibria" (Scoones, 1994). Pastoralists pursue a variety of risk-aversion strategies that enable them to cope with periodic drought, conflict, or resource loss; these strategies include settling near towns, engaging in farming, seeking wage jobs, and moving to new locations, as they continue to herd their animals (Little et al., 2001; World Bank, 1993, 1997).

Recent studies on pastoral sedentarization have pointed to a variety of costs and benefits to people changing their lifestyle. Several studies point to problems of impoverishment and destitution for pastoralists who settle (Hogg, 1986; Little, 1985; Talle, 1999), which may particularly affect women (Mitchell, 1999; Talle, 1988). Others point to increased marketing benefits (Ensminger, 1992; Sato, 1997; Zaal and Dietz, 1999), particularly to women selling milk and agricultural products (Fratkin and Smith, 1995; Little, 1994a; Smith, 1999; Waters-Bayer, 1988). There have been only a few studies on the health and nutritional consequences of pastoral sedentarization, most of which have reported negative effects: poorer nutrition, inadequate housing, lack of clean drinking water, and higher rates of certain infectious diseases (malaria, anthrax, bilharzia), despite better access of settled populations to formal education and health care (Duba et al., 2001; Galvin et al., 1994; Hill, 1985; Klepp et al., 1994; Nathan et al., 1996; Sellen, 1996).

For the past fifteen years, the authors, cultural anthropologist Elliot Fratkin and demographic anthropologist Eric Abella Roth, with several collaborators including medical doctor Martha A. Nathan, have conducted multidisciplinary and longitudinal research examining the bio-social concomitants of sedentism for Ariaal and Rendille pastoralists of Marsabit District, northern Kenya. Throughout this research, we have looked at changes in economy (Fratkin, 1989, 1991; Roth, 1990), fertility and population change (Roth, 1993, 1994), education and HIV awareness (Roth, 1991; Roth et al., 1999; Roth et al., 2001), women's roles (Fratkin and Smith, 1995), and changes in health and nutrition among formerly pastoral and now settled populations (Fratkin et al., 1999; Nathan et al., 1996). Between 1994 and 1997, with support from the US National Science Foundation, we focused our research on measuring health and nutritional consequences of pastoral sedentarization for women and children in five Ariaal and Rendille communities which pursued different economic strategies, including full pastoralism, agro-pastoralism, irrigation agriculture, and famine-relief dependency. Selecting forty women from each community and their under 6-year-old children (588 individuals), we measured health and nutritional outcomes every two months over this three-year period. The time period included a year of above-average rainfall year (1995) and a 'drought' year of below average rainfall (1996) (our study ended before the destructive El Niño rains of October–November 1997). The results of these studies, reported in Chapters 8, 9, and 10, present strong evidence that women and children living in nomadic pastoralist communities had significantly lower levels of malnutrition and morbidity than any of the four settled communities studied, and lent evidence to the debate about the value of livestock pastoralism in arid lands.

In the course of presenting our research among Ariaal and Rendille in this volume, we wanted to incorporate related research on pastoral sedentarization that other biological and social scientists has undertaken in Marsabit District, Kenya. This volume consequently presents a variety of studies dealing with problems of pastoral sedentarization in northern Kenya among Rendille, Gabra, Boran, Samburu, and LChamus. These researchers include veterinary biologist H. Jürgen Schwartz; geographers Wario Adano and Karen Witsenburg; social anthropologists John Galaty, Peter Little, and Kevin Smith; economist John McPeak,

Photo 2. Marty Nathan and AnnaMarie Aliyaro weigh child in Ngrunit.

biological anthropologists Bettina Shell-Duncan and Masako Fujita; Kenyan public health researchers Elizabeth Ngugi, Leunita Auko Muruli, and Walter Obungu Obiero; and health practitioners David Wiseman MD and Joan Harris RN.

The chapters presented here offer an interdisciplinary approach enabling us to understand pastoral sedentarization and its social, ecological, economic, and health consequences. Studies in this volume analyze voluntary sedentarization in northern Kenya in terms of economic diversification, environmental consequences of agriculture, agro-pastoralism, and town life, degrees of formal education, employment, and market integration, levels of maternal and child health and nutrition, and levels of political security in an environment of increased ethnic tensions. In so doing, we attempt to present a comprehensive study of pastoral sedentarization in one particular region of Africa—Marsabit District, northern Kenya.

2. WHY PASTORALISTS SETTLE

Africa contains one half of the world's pastoral peoples with over 20 million pastoralists and 240 million agro-pastoralists (Blench, 2001: 10). Twenty five million pastoral and agro-pastoral people live in the East African countries of Kenya, Tanzania, Uganda, Ethiopia, Eritrea, Sudan, and Somalia (Morton, 2003: 6). Kenya in particular is home to a variety of pastoral societies including Nilotic-speaking groups originally from Sudan (Maasai, Samburu, Chamus, Pokot, Turkana) and Afro-Asiatic speakers from Ethiopia and Somalia (Boran, Gabra, Rendille, Sakuye, and Somali; see Figure 1). Although these pastoralists inhabit 70% of Kenya's land, they number less than 2 million of Kenya's total

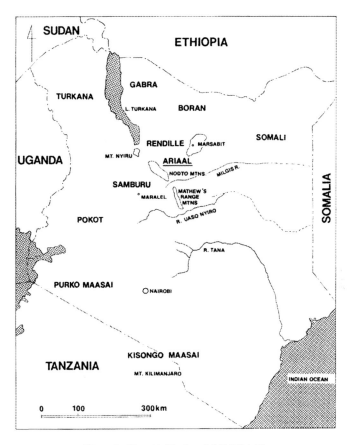

Figure 1. Marsabit District rainfall, 1994–97.

population of 30 million people. Maasai reside in the south in Kajiado and Narok Districts, Pokot and Turkana in the northwest; Samburu in the north center (Samburu District); Boran, Gabra, Rendille (including Ariaal), and Sakuye in the north (Marsabit District); Boran and Somali in Isiolo District, and Somali groups including Ogaden, Gurreh, Orma, Hawiyeh; and Arjun in the northeast (Wajir, Garissa, and Madera Districts).

Pastoralists are people whose livelihood depends on the raising of domestic animals including cattle, camels, goats, sheep, and donkeys for milk, meat, transport, and trade (Swift 1988 broadens his definition of pastoralists to include those relying on at least 50% of their livelihood from domestic livestock). A pastoral herd owner's fundamental goal is to maintain enough animals to provide food (e.g., milk, meat, and blood) and material goods (livestock, hides, and wool which can be exchanged for grains, tea, household goods) to the household group. The pastoralist must have a sufficiently large household labor force to provide pasture, water, and security to the herds. East African herders raise more female than male animals to produce milk for both humans and nursing livestock, and to replace periodic loss through reproduction. Male animals are kept for transport, meat, and trade, and to satisfy social obligations including marriage payments (Dyson-Hudson and Dyson-Hudson, 1982; Swift, 1988).

Photo 3. Ariaal man watering cattle at Ngrunit.

Pastoralists occupy savannas, semi-arid, or arid deserts where rain-fed agriculture is precarious. They are distinguished from livestock ranchers by their practice of taking herds to pasture and water, rather than having fodder brought to the animals. Consequently pastoral populations have historically relied on mobility to graze their herds over wide areas. Pastoralists have typically occupied communally shared land, utilizing kinship ties of marriage and descent for mutual herding and defense. Their herds are often large, in poor condition, but hardy enough to survive periodic drought and sparse vegetation.

East Africa's pastoralists have increasingly settled in the past forty years in response to the problem of shrinking rangeland caused by population growth, the expansion of cultivation in pastoral land, and to the impact of increased commercialization and privati zation. The major factors leading to pastoral sedentarization include:

Population Growth—Kenya, Tanzania, and Uganda have among the world's highest popu-lation growth rates (2.1%, 2.9%, and 2.9% annual increase, respectively in 2000), attrib-uted to high but declining total fertility rates (4.7 in Kenya, 5.6 in Tanzania, and 6.9 in Uganda) coupled with declines in child mortality. However, HIV/AIDS is spreading rapidly, resulting in high mortality among young adults and declining life expectancies (48 years for males in Kenya, 52 in Tanzania, and 42 in Uganda) (Population Reference Bureau, 2000). Rapid population growth has affected rural and urban areas, where farmers increasingly move onto less productive lands to raise their crops and families. Furthermore, pastoralists have increased farm cultivation, leading to a loss of pasture and water resources for pastoral production. In the more arid north of Kenya where agriculture is possible only in isolated highlands, population growth in both herds and humans has led to increasing competition with pastoral neighbors for pasture and water, leading to recent armed attacks between Turkana and Pokot, Boran and Rendille, Turkana and Samburu, and Somali and Boran.

The growth of human and livestock populations in East Africa's pastoral regions, although modest, has direct consequences for land management and resource use in these arid regions. This is particularly manifest around permanent water and dry season grazing resources, which are located in more populated highlands and which are attracting sedentarizing populations. The concentration of populations directly contributes to economic transformation, environmental degradation, and political conflict in these regions.

Drought and Famine—Drought is the periodic absence or decline in rainfall, while famine is the widespread disruption of the food supply leading to starvation and emigration. Drought is a regular feature of the African climate, but it occurred with greater frequency in the second half of the 20th century compared to the first, reported in East Africa in 1960–61, 1968–69, 1974–76, 1979–81, 1991–93, 1996, and 2000. Severe famine occurred in 1968–73 in the Sahelian countries of West Africa (Burkina Faso, Chad, Mali, Mauritania, Niger, Nigeria, Senegal, and Sudan), and in 1982–84 in the Horn of Africa (Djibouti, Ethiopia, Kenya, Somalia, and Sudan). Pastoralists have historically adapted to conditions of drought or low and erratic rainfall by physical mobility, dispersion of their herds and people, and seeking different food sources through fishing, hunting, gathering, and agriculture. Today pastoralists have added new options including migration to famine relief centers, urban migration for wage labor, and the widespread adoption of agriculture.

Loss of Common Property Resources—Where livestock among most East African pastoralists constitutes individual or family property, access to land (for pasture, water, minerals, and security) is usually held in common as a communal resource (i.e., shared by territorial or kinship groups) or as common property open to all. Since Independence, Kenya has moved away from recognizing communal land tenure in favor of individual tenure rights. In pastoral Maasai regions the government established "group ranches" in the 1960s and subsequently promoted private and individual land titles in the 1980s, leading to a scramble for land similar to the American West in the 19th century (Galaty, 1994). Kenyan and Tanzanian pastoralists have also lost former grazing lands to national game parks including Amboseli, Mara Masai, Tsavo, and Samburu Parks in Kenya and the Serengeti, Ngorongoro Crater, and Mkomazi in Tanzania (Brockington, 1999; Homewood, 1995). While pastoralists in northern Kenya's Marsabit District live in more arid and less populated conditions, they too are experiencing land crowding and, in highland locations including Marsabit Mountain, are beginning to privatize and title farm plots (see Adano and Witsenburg; Smith, this volume).

Commoditization, Sedentarization, and Urban Migration—Pastoralists have increasingly shifted their economy from subsistence production (producing mainly milk for the household consumption) to commercial production (beef and dairy products for sale both to domestic and export markets). The sale or trade of livestock is not new to pastoralists— northern Kenyan pastoralists were trading livestock for grains with southern Ethiopian farmers of Ethiopia in the 19th century and in colonial-era markets like Isiolo and Nanyuki in Kenya (Adano and Witsenburg, this volume; Falkenstein, 1995; Sobania, 1991). However, both the demands and opportunities for market sales of livestock in northern Kenya have increased substantially in the past twenty-five years, as have opportunities for wage labor (McPeak and Little, this volume). The education of children is seen as an investment in creating wage earners, and the possibility of gaining education for their children is a motivating factor in pastoral sedentarization (see McPeak and Little;

Roth et al., this volume). However, increased commoditization of the livestock economy has benefited those with large livestock herds, allowing them to remain in the pastoral economy, while those without sufficient herds often migrate out of the pastoral economy and seek jobs in towns or livelihood on farms (Fratkin and Roth, 1990).

Political Turmoil, Civil War, and State Intervention—Although not experiencing the devastating civil wars of neighboring Sudan, Ethiopia, and Somalia, northern Kenya has seen its share of violence, mainly from banditry and inter-ethnic livestock raiding in an area not well policed. During their colonial rule, the British established tribal grazing areas in northern Kenya which, although unpopular, reduced armed conflict between competing pastoral groups. However, following independence in 1963, the impact of drought and livestock loss, coupled with the massive influx of automatic weapons from neighboring countries' conflicts, have led to more raiding and casualties, and have increased the general level of insecurity. During the past forty years, Somali, Boran, and Turkana have periodically raided (and been counter-raided by) Samburu, Ariaal, and Rendille, while well-armed Pokot have extensively raided Turkana in the 1990s, and Gabra, normally peaceful, raided surrounding Dasenech, Rendille, and Boran neighbors in Marsabit District in 1992. National police occasionally bring order to these regions, but government forces are often incapable of tracking down cattle rustlers or policing broad and inhospitable areas. Pastoralists have also been subject to state interventions, including the creation of game parks in Maasai and Samburu lands, and the creation of international refugee camps in Turkana, Marsabit, and Wajir Districts.

Sedentarization is the process of individuals, households, or entire communities of formerly nomadic populations settling into sedentary, non-mobile, and permanent communities. Sedentism is neither a recent event nor a unidirectional process, and has occurred in many regions of the world at different points in history. Fulani pastoralists in West Africa long have had ties to sedentary agricultural villages and mercantile towns, trading or selling livestock, leather, and meat for grains and other commodities (Bayer and Waters-Bayer, 1994). In East Africa, pastoralists obtained necessary grains by trading regularly with agricultural neighbors (e.g., Maasai with Kikuyu in the 19th century (Waller, 1993)), or taken up agriculture themselves, as did the Arusha or 'agricultural' Maasai (Spear, 1997). Pastoral sedentarization in the 19th and early 20th century in East Africa was prompted largely by new market opportunities rather than the population pressure and ecological decline which characterize the later 20th century. Maasai settled near roads and urban areas including Nairobi for access to cattle markets, while in northern Kenya, Boran cattle herders were encouraged by the British to settle on Marsabit Mountain to provide beef and milk to police posts and road crews during the 1930s (Adano and Witsenberg, this volume).

In the past thirty years formerly pastoralist livestock herders settled in response to land crowding, population growth, and competition with other groups. Maize cultivation is becoming increasingly important for Maasai pastoralists living in the Ngorongoro conservation area in Tanzania (McCabe et al., 1992), while cultivation became an important source of livelihood for LChamus people in the early 20th century (Little, 1992). In addition to cultivation, pastoralists have also settled near urban areas to market milk, meat, and livestock, forming sedentary communities in peri-urban communities (Fratkin and Smith, 1995; Little, 1994a; Salih and Baker, 1995).

Photo 4. Settled Ariaal at Kitaruni near Karare, Marsabit Mountain—Lugi Lengesen and family.

2.1. Features of Sedentarization

We summarize several features of pastoral sedentarization that derive from our research in Marsabit District, but which may also apply to other regions experiencing pastoral sedentism:

1. *Sedentarization represents an alternative economic strategy as part of a larger set of diversification strategies.* Diversification is an essential component of pastoral decision making to cope with varying and unpredictable resources. Pastoralists practice multi-species herding, enabling them to utilize different herding environments. Similarly, as chapters by McPeak and Little, Adano and Witsenburg, and Smith show, settled life in towns and farms represents additional resources to utilize, where one can take up, permanently or temporarily, farming, wage labor, or entrepreneurial activities including shop keeping, livestock marketing, charcoal or beer production, etc.

2. *Sedentarization does not result in a sharp break with the pastoral community or economy.* Former pastoralists living in towns or farms often own livestock which are herded by kinsmen or friends in the pastoral economy, or divide up their households with some members farming and others herding. Social ties are maintained by marriage, age-set rituals, and exchanges and serve to keep the pastoral and agricultural/town communities integrated. Sedentarization is a process that operates along a continuum from highly mobile pastoral households to permanently settled households, of which individuals may move from one domain to the other.

3. *Sedentarization does not imply one type of lifestyle or economic activity, but includes a range of economic choices.* Some communities, including Ariaal and Boran on Marsabit

Mountain, are sedentary livestock keepers who graze and water their cattle from permanent homes on the mountain, while taking advantage of schools, dispensaries, and markets in Marsabit town. Other communities may be exclusively agricultural, including the Rendille living in Songa or Boran at Badessa on Marsabit Mountain. In the lowlands, pastoralists live in close proximity with their animals to towns, including Gabra near Maikona or Rendille near Korr, while impoverished pastoralists may live in town depending on famine relief foods or odd jobs for a living.

4. *Sedentarization is usually accompanied by larger socio-cultural changes.* Despite ties to the pastoral communities, settled townspeople and farmers often undergo dramatic changes in customs and relationships, including a departure from communal and kin-based relations in the pastoral communities to individualized identities in the towns and farms, including the adoption of capitalist concepts of private property and individual gain (Meir, 1997). For settled Rendille in the farming community of Songa, former age and gender roles break down, including the collective power of male elders to control younger men (Smith, 1999). Several chapters in this volume point to a breakdown in the "moral economy" of redistribution, where women living on isolated farming plots no longer share food with others as they did in the pastoral setting. But some women benefit from new opportunities in the market economy, selling milk or vegetables, opportunities denied them in male-dominated livestock economy (Fratkin and Smith, 1995). But other women may be forced out of poverty to depend on beer brewing and prostitution to survive in the urban economy, increasing their exposure to HIV/AIDS and other sexually transmitted diseases (Klepp et al., 1994).

5. *Sedentarization entails costs and benefits.* Former pastoralists living in settled communities often have increased access to health care, formal education, and markets, but they may also incur losses in nutritional status and new health hazards, particularly women and children (Nathan et al., 1996; Sellen, 1996). Nathan et al. (Chapter 10 this volume) show that settling results in greater childhood malnutrition through loss of meat and milk protein, although pastoralists suffer seasonal shortages. And although towns provide increased access to health clinics, and vaccinations, there are higher respiratory and diarrheal disease rates among children living in towns, with particular differences between highland and lowland residences. Breastfeeding mothers living in highland farms suffer more malnutrition than women living in lowland pastoralist communities, as protein-poor grains replace milk as a staple food (Fujita et al., Chapter 11 this volume). However, settled life also results in explicit benefits. McPeak and Little (Chapter 5) point to education and wage labor as important strategies to avoid risk, where having one employed child can guarantee food security for an entire household. Roth and Ngugi (Chapter 13) point to the importance of educating females, despite the strong bias for male education, and correlate female's education to knowledge about and reduction in sexually transmitted illnesses.

2.2. Research on Pastoral Sedentarization

Ecological, economic, and sociological research on pastoralism has grown substantially in the past few decades (for reviews see Blench, 2001; Fratkin, 1997; Galaty and Johnson, 1990). Comparative studies on pastoral development, sedentarization, and social change have also grown (Fabietti and Salzman, 1996; Galaty et al., 1981; Meir, 1997; Salzman, 1980; Salzman and Galaty, 1980). There has been relatively less written on biosocial aspects of pastoral sedentarization, particularly about health, nutritional, and demographic changes associated with settled life. Allan Hill's 1985 edited volume *Population, Health, and Nutrition in the Sahel* assembled several independent studies on health, nutrition, and demography among farming and pastoral groups in Mali, although it was not in itself a systematized research program. One study in Hill's volume reported that nomadic groups have higher rates of tuberculosis, brucellosis, syphilis, trachoma, and child mortality (children five and under), and settled agricultural populations had higher rates of bilharzia, intestinal helminths and other parasites, and higher malaria and anemia rates particularly among those groups living in by rivers (Chabasse et al., 1985). Hill's study did not look at sedentarization as a process, nor did it compare health and nutritional outcomes for nomadic versus settled communities of the same ethnic group.

There have been several interdisciplinary research projects on pastoral regions in northern Kenya from which our studies have benefited greatly. These are the UNESCO-IPAL project in Marsabit District (1976–82), the South Turkana Ecosystem Project (STEP) in northwestern Kenya in the 1980s, and the recent Pastoral Risk Management Project (PARIMA) in Samburu, Baringo, and Marsabit Districts, Kenya, and southern Ethiopia. Several researchers from these projects contribute to this volume, including H. Jürgen Schwartz (IPAL), Peter Little, Kevin Smith, and John McPeak (PARIMA), and Bettina Shell-Duncan (STEP).

The Integrated Project in Arid Lands (IPAL) was funded by UNESCO's Man and the Biosphere Program, in conjunction with the newly created United Nations Environment Program's (UNEP)'s 'Conference on Desertification' in Nairobi in 1977. Staffed initially by rangeland ecologists and wildlife scientists, IPAL conducted ecological research in western Marsabit District and provided a strong database on vegetative, water, climate, livestock resources, and pastoral practices (IPAL, 1984; O'Leary, 1990; Sobania, 1979; Schwartz, 1980a, 1980b). Although initially concerned with environmental degradation and "desertification", by 1980 the IPAL project, now funded by German Development Corporation (GTZ) and directed by Kenyan Walter Lusigi, combined research on pastoral land use with practical programs including rural market development, roads, and water sources to reduce grazing pressures among Rendille, Ariaal, and Gabra. Although IPAL officially disbanded in 1985, GTZ continued pastoral and agricultural development work in Marsabit District through the 1990s. IPAL was one of the first collaborative efforts looking at arid lands ecology, and while made up principally of ecologists and biologists, it also produced reports on traditional pastoral practices (O'Leary, 1985; Schwartz, 1980a, 1980b; Sobania, 1979). IPAL did not, however, look at issues of health, nutrition, demography, or indeed the effect the project itself was having on sedentarization among Rendille and Ariaal (Fratkin, 1991; Little, 1994b).

The South Turkana Ecosystem Project was a highly coordinated collaboration of ecologists, biological anthropologists, and cultural anthropologists who carried out research on ecology, health, nutrition, and fertility of nomadic Turkana of Kenya (Little and Leslie, 1999). Several STEP studies examined health and nutrition among nomadic and settled farming Turkana populations. Here researchers found that settled Turkana experienced reduced fertility, increased morbidity (particularly from malaria) and increased child mortality. Settled children under five showed more growth stunting than nomadic children, although settled children over five were heavier, which is attributed to the greater role of carbohydrates in their diets, particularly for children receiving supplemental feeding in schools. Nomadic Turkana women, however, were taller, heavier, and had lower blood pressure than settled women (Brainard, 1990; Campbell et al., 1999; Galvin, 1992).

The Pastoral Risk Management Project (PARIMA), which was funded by USAID's Global Livestock Collaborative Research Support Program, conducted comparative research among Samburu, LChamus, Rendille, Gabra, Boran, in northern Kenya and Borana in southern Ethiopia (Little et al., 2001). While addressing issues of range ecology, PARIMA focuses more on the socio-economic choices of pastoralist decision-makers, particularly in averting risks of drought, land crowding, and insecurity. PARIMA examines alternative production strategies in a variety of pastoral, agro-pastoral, and urban settings, including marketing and wage labor, some of the results of which are presented by John McPeak and Peter Little in this volume.

3. RENDILLE SEDENTARIZATION PROJECT

Research on Ariaal and Rendille sedentarization in Marsabit District was initiated by Elliot Fratkin, Eric Abella Roth, and Martha Nathan MD, who in 1990 and 1992 compared nutritional and health outcomes of women and children in both pastoral and sedentary Rendille and Ariaal communities (Nathan et al., 1996). In 1994, we began a three-year research project, dubbed the Rendille Sedentarization Project, which investigated social, health, and economic consequences by comparing women and their children in five separate Ariaal and Rendille communities, including a large lowland pastoral community (Lewogoso), a lowland 'famine-relief' town (Korr), two highland communities on Marsabit Mountain (agro-pastoral Karare and the agricultural 'scheme' at Songa), and the small town of Ngrunit in the Ndoto Mountains (see Figure 2).

The research surveyed two hundred households every two months over the three-year period, looking at health, nutrition, demographic, social, and economic processes. This research found that pastoral children under six years were three time less likely to be severely malnourished than children living in settled communities, including those relying on famine relief foods or agricultural produce (Fratkin et al., 1999). In this volume, Nathan et al. report morbidity differences between highland and lowland Rendille and Ariaal populations, with higher rates of respiratory illnesses in the highlands and higher rates of malaria in the lowlands, and with pastoral Lewogoso children having the lowest respiratory and diarrheal morbidity of any of the settled communities. Fujita et al. (2004, also this volume) demonstrate that pastoral women are larger and less malnourished than settled women living in Songa agricultural community.

The Rendille Sedentarization Project focused on several research themes central to the study of pastoral sedentarization. These include health and nutritional change, sedentism and commercialization, changing gender roles, and educational opportunities. These themes are briefly introduced below, and discussed in specific context in this volume's chapter contributions.

3.1. Health and Nutrition

One way to evaluate the adaptive success of newly settled populations is by examining whether or not their recent transition to sedentism successfully provides adequate nutrients, thereby maintaining their health, biological function, and productive capacity during the lean season (Huss-Ashmore, 1993: 202, 215).

Pastoral diets generally are characterized as high in protein but low in calories, with marked seasonal variation in both protein and energy content (Galvin, 1985, 1992; Galvin

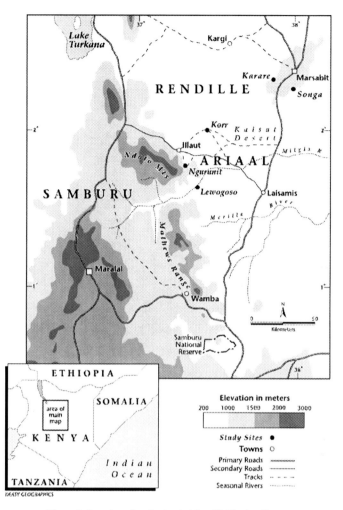

Figure 2. Location of study sites in Marsabit District, Kenya.

and Little, 1999; Little et al., 1993; Nathan et al., 1996; Nestel, 1986; Shell-Duncan, 1995). The lean season for Marsabit pastoralists is at the end of the two dry seasons (November–March and May-August) when both drinking water and livestock pasture becomes scarce, in turn limiting milk availability for human consumption. During dry periods, small stock are increasingly sold to purchase foods particularly grains (maize meal or *posho*) and other carbohydrates (sugar to mix with tea). The milk-based, high-protein diet of pastoralists, nonetheless, appears to contribute positively to their adaptation to a highly seasonal environment with limited resources for dietary energy (Galvin and Little, 1999). The positive ramifications of a pastoralist high-protein diet may be particularly significant for infants, pregnant women, and lactating mothers, who are particularly at risk from poor environments (Panter-Brick, 1997). Since protein is an indispensable nutrient for reproductively active pastoral women and for infants and growing children (Galvin and Little, 1999),

Photo 5. Settled Rendille woman of Songa.

the potential protein loss associated with agricultural sedentism may also have a negative impact on maternal nutritional health.

Market integration of rural producers in Africa may have both positive and negative consequences on child health and nutrition. Sales of agricultural commodities may diminish child nutrition when they lead to substitution of cheaper, poorer foods for high calorie or protein ones (Lappe and Collins, 1977). However, other studies report improved child nutrition associated with commercial agriculture when combined with subsistence production in various production strategies of Taita farmers of Kenya (Fleuret and Fleuret, 1991). Ensminger's (1991) study of the economic transformation of Orma of Kenya found increased residence in market centers and agricultural commercialization associated with improved nutritional markers (weight for height) for adults and male children, but not for female children.

Today in settled communities of former pastoralists, certain families may have established a wider economic resource base by engaging in the commercial livestock economy and taking up cash-crop agriculture. This allows them not only to alleviate seasonal fluctuation of food availability but also to widen the variety of food in their diet. Typically, there are contrasting seasonal patterns of nutritional stress between agriculturists and pastoralists. Critical periods for agriculturists coincide with the food shortage and high labor demand associated with planting and harvesting (Simondon et al., 1993: 166). Families with sufficient agricultural and/or pastoral resources will be able to even out the seasonal stresses associated with each subsistence mode. By contrast, poorer families who rely on

Photo 6. Korea Lekutan of Karare village.

smaller pastoral or agricultural holdings for their subsistence and cash income are more likely to experience seasonal stresses distinct from those of wealthier families.

The Rendille Sedentarization Project examined many of these interrelated factors for five Rendille communities, four settled, and one still nomadic pastoral control. A startling finding of this research was that children in the pastoral community were three times less likely to be malnourished than children in settled communities, a fact attributed to access to camel's milk among the pastoralists versus dependency on maize meal among settled children (see Roth et al., Chapter 9, this volume). However, town life also offered access to health care, particularly vaccinations and interventions for infectious diseases including malaria.

A prevalent theme in our investigation of Rendille sedentarization is that it constitutes an ongoing process consisting of constraints and opportunities, attracting both wealthy and poor members of the pastoral community. Sedentarization has yielded a number of beneficial effects including improved access to drinking water, education, and health care and to a market economy (Fratkin, 2004; Roth, 1991; Smith, 1999). But the process of pastoral sedentarization has also resulted in the widening disparity in wealth distribution and access to food resources, reported among other settling pastoralists (Hogg, 1986; Little, 1985). This has led to a decrease in the nutritional health of children as evidenced by decreased height and weight when compared to similar-aged samples from pastoral populations (Nathan et al., 1996; Roth et al., Chapter 9, this volume). This negative consequence on children's growth patterns is consistent with recent findings that the transition to sedentism was accompanied by a major dietary shift from a protein-rich pastoral diet to more cereal-based diet (Fratkin et al., 1999; Fujita, 2000; Fujita et al., 2004.).

3.2. Sedentism and Commoditization

East African pastoralists have long exchanged livestock for agricultural products with their farming neighbors (Kerven, 1992; Sobania, 1991). Today Marsabit's pastoralists consistently sell 5–10% of their herds every year, with higher offtake occurring in response to higher prices or to drought when buying grains is necessary because milk supplies are low (Little, 1992; O'Leary, 1990). Most pastoralists raise livestock for daily food consumption, and production is subsistence-based rather than market-oriented. Marsabit's pastoralists have increasingly entered the livestock market in the past twenty years. Settled communities produce a variety of agricultural and livestock products for exchange, including seasonally available vegetables (kale, tomatoes), cattle milk, and live cattle and goats.

The effects of sedentarization and commoditization appear varied, reflecting respective access to markets, type of large stock raised, and the effects of drought. Comparing 1995 annual livestock sales data from the lowland Rendille community of Korr (n. households = 145) with those of the Ariaal Rendille highland community of Karare (n. households = 251) for the same year, Fratkin et al. (1999) found the two communities differed markedly. Residents of Korr who keep camels and small stock sold mainly goats for local consumption, while highland Karare residents earned three times as much income as Korr by selling cattle and milk products for both local and national markets. We attribute this to Karare's greater integration in the commercial livestock market than is Korr, deriving from its geographic position on the main road and close to the major market at Marsabit town. Korr is an isolated mission outpost 45 km from the main road.

3.3. Sedentism and Gender Roles

Among pastoral households living near urban centers, women are active players who promote and sell surplus milk or vegetables to customers in town. Those who own sufficiently productive resources can successfully and regularly earn cash, whereas the poor without such resources cannot (Fratkin and Smith, this volume Chapter 8). By implication, wealthier women may be more successful in obtaining stability of both food quantity and nutritional quality. Shell-Duncan and Obiero (2000) documented such cases of intra-community discrepancy among Rendille women in highland settlements, where poorer women faced risk of low dietary intake of milk, a key staple food.

Gender inequality, if not outright oppression of women, has been noted in pastoral African societies as it has in rural agricultural ones. One particularly heated debate has concerned female circumcision, or 'female genital mutilation' as termed by some of its critics (see Shell-Duncan et al., Chapter 12, this volume; and Shell-Duncan and Hernlund, 1999). Female circumcision is found widely in northeast African populations, and is practiced by Rendille, Gabra, Boran, Samburu, Ariaal, Somalis, and other groups in Marsabit District. In this volume Shell-Duncan, Obiero, and Muruli report on recent changes in female circumcision among Rendille, where the practice is becoming medicalized in settled communities, even though it is outlawed by the Kenyan government and medical services in hospitals and clinics are officially denied to these women. This chapter points to important areas of changing attitudes about gender equality (and inequality) and represents a significant development with pastoral sedentarization.

Photo 7. Bettina Shell-Duncan and Rendille families in Korr.

3.4. Formal Education

One potential advantage of town life is increased education for pastoral children. Most growing towns in northern Kenya have an elementary school with burgeoning enrollment. Today education represents a tremendous potential engine for both demographic and social change; with female education one of the most important factors associated with mortality and fertility decline in Kenya (Brass and Jolly, 1993; Dow et al., 1994). Caldwell (1979) first called attention to the role of female education in lowering both mortality and fertility rates. His analysis of Nigerian surveys revealed that mother's education was a far more important determinant of childhood mortality than familial economic markers, including father's occupation. Caldwell (1982) also argued that female education leads to increased autonomy, exhibited in the use of modern contraceptives, a prediction borne out throughout sub-Saharan Africa (National Research Council, 1993).

A more neglected aspect of the education–demographic link is that dealing with parental decisions about which children to enroll in primary school. Attendance at school means that the child is at least partially removed from the household labor pool, to which children in Third World settings make significant contributions from an early age. Culture-specific patterns may also influence parental decisions about schooling. In a study of Rendille childhood education based on survey data collected in Korr in 1987 (Roth, 1991), boys were overwhelmingly chosen more frequently than girls to attend school. Since young women reside away from their natal home following marriage, the benefits of education accrue to the groom's household, while the expense of female schooling, represented by lost labor in addition to school fees, would be borne by her parents. Ten years later, the survey was repeated in the same study site, yielding a 1996 sample of 546 school aged children drawn from 145 households. Findings showed that parents from earlier age-sets were only one-half as likely to send a child to school, versus parents from more recent age-sets.

Photo 8. School girls practice for performance in Karare, Marsabit Mountain, 1999.

This points to a change in parental decision making over time, with far more children going to school now. However, no change is found for gender selection of children, as boys are still two and one-half times more likely to attend school. Thus, one decade after the original study, sedentary Rendille families at Korr are increasingly sending their children to school, yet the potential for demographic and social change from female education still appears unfulfilled.

4. THE CHAPTER CONTRIBUTIONS

This volume examines these issues through in-depth and longitudinal studies of pastoral and settled communities in Marsabit District, Kenya. We begin with descriptions of environment and historical background, and move through the topics of economic transformation, changing roles of women, and the health and nutritional consequences of sedentism. The book concludes with discussions of changing values and practices, particularly among women, including the practice of female circumcision in Marsabit and the role of female education in diminishing the risks of sexually transmitted infections and HIV/AIDS.

Following this introduction, Chapter 2 by Fratkin and Roth describes the geographical and historical setting of pastoral sedentarization in Marsabit District, using the example of the Ariaal and Rendille. John Galaty's Chapter 3, "Time, Terror, and Pastoral Inertia", provides a time depth to understanding ethnic conflict and competition among pastoral groups in Marsabit District, relating the establishment of colonial administrative rule and the incidence of conflict and violence to which settlement and movement patterns accommodate themselves. Galaty shows how earlier conflicts continue to play out in the contemporary period in national politics and competition for farming lands on Marsabit Mountain, and explores the impact of the recent influx of automatic weapons on inter-group violence.

Chapter 4 by H. Jürgen Schwartz presents a longitudinal examination of the ecological and economic consequences of reduced mobility for pastoralists in northern Kenya, and presents a stark picture of what may happen in Marsabit District. Arguing from experiences of pastoralists world-wide, Schwartz predicts that subsistence-oriented, migratory pastoralism will be replaced by commercialization of the livestock economy, which will lead to an increasing economic polarization that separates large and successful pastoralists from poorer subsistence-based herders, who may ultimately be driven out of the system. Schwartz does not expect dryland farming to offer an adequate solution in Marsabit, given the aridity of the area and its inability to support agriculture on a large scale, but expects that poorer pastoralists, concentrating on small stock to trade for grains, will continue to exist on the margins of more successful herders who may, as among Maasai to the south, begin to privatize range and farm land.

Chapter 5, "Cursed If You Do, Cursed If You Don't" by John McPeak and Peter D. Little, compares pastoral risk management among six different communities in northern Kenya, and presents a more varied picture of possible outcomes. Based on their research of specific Ariaal, Boran, Chamus, Gabra, and Samburu communities, they argue that sedentarization presents both new opportunities and new risks, including cultivation and pursuit of wage labor. Sedentarization does not necessarily mean a full-time departure from pastoralism, nor does it necessarily jeopardize pastoral production. They also distinguish the relationship between sedentarization and vulnerability in livestock wealth, and sedentarization and vulnerability in food security, where the relationship between food security and sedentarization varies between communities in complicated ways. Pastoralists living in very arid regions, including Rendille near Kargi, rely most heavily on their herds, deriving a higher share of their income from livestock and livestock products, and also have more milk available for home consumption than do pastoralists in more agriculturally-suited lands. Other communities may derive more benefits from mixed economies. For many communities, investments in education and the pursuit of steady wage paying jobs are an important hedge against risk and livestock loss. However, while agriculture and education provide important alternate strategies, McPeak and Little note that these do not provide households with many benefits during periods of drought. McPeak and Little's study provides a complex picture of different strategies engaged by herders, and point to a greater variety of outcomes than a simple pastoral/non-pastoral framework.

Chapter 6 by geographers Wario Adano and Karen Witsenberg, "Once Nomads Settle: Assessing the Process, Motives, and Welfare Changes of Settlements on Mount Marsabit", discusses the ecological and social changes that have occurred on the District's main highland region, where many Rendille and Boran have settled to take up agriculture, agro-pastoralism, or urban residence. This research compares planned-scheme settlements, where poor pastoralists have settled through development and church mission assistance, from those who settled without development assistance. Adano and Witsenburg found that those children in the planned-schemes obtain higher educational levels, especially among female-headed households, and that scheme households are more likely to rebuild livestock herds, which they note is a favourable trend for the future.

Chapter 7 "From Milk to Maize" by Kevin Smith provides an in-depth look at the Songa community of Rendille, a principle agricultural scheme on Marsabit Mountain. Although made up of predominately poor Rendille who migrated to the scheme beginning in the 1970s, Songa's residents have overwhelmingly positive attitudes toward farming. Smith argues that the economic strategy of agriculture will remain a preferable option, but one that will not be available to all Rendille and Ariaal. Land pressures have increased conflicts, particularly with Boran, and have also led to a popular movement toward land titling and privatization as a legal protection against future land loss. As with Adano and Witsenburg's analysis, Smith finds increased opportunities for women who have opportunities to sell agricultural produce and educate their children. However, Smith suggests (and Fujita et al.'s Chapter 11 confirms) that Songa women suffer greater malnutrition than pastoral Ariaal, experiencing greater seasonal shortages, protein-deficient diets, and greater female workloads associated with agriculture than pastoralism.

The question of changing women's status continues in Fratkin and Smith's Chapter 8, "Women's Changing Economic Roles with Pastoral Sedentarization." This study (originally published in *Human Ecology* in 1995) shows a variety of economic strategies pursued by women in different Ariaal and Rendille communities. Women living in the nomadic settlements in the lowlands have little economic independence, but those living near towns (including Ngrunit or Marsabit) can sell milk to regular customers, much as Songa women sell agricultural produce. This study demonstrates that women are much more likely to spend their earnings on food for their children than are men, and argues that economic integration for women has direct consequences for household health and nutrition. Towns offer other opportunities, including employment for wages or entrepreneurship, although these tend to be at lower socio-economic levels where women find work as servants, casual laborers, prostitutes, or beer brewers.

The book moves into detailed analysis of health and nutritional changes associated with pastoral sedentarization in Chapters 9, 10, and 11. Data for these analyses were collected between 1994–97, when five communities ranging from fully pastoral Lewogoso to agricultural Songa were surveyed every two months for data on morbidity, diet, anthropometric measurements, and economic livelihoods. Eric Roth et al.'s Chapter 9, "The Effects of Pastoral Sedentarization on Children's Growth and Nutrition among Ariaal and Rendille," describes dietary differences for children in the five communities, and shows convincingly that pastoral diets based on milk result in significantly less malnourishment than the settled communities which are based mainly on maize meal. While there are probably seasonal and drought-based calorie deficits, the pastoral diet relying on camel's milk consistently provided more protein than all other communities.

In Chapter 10, Nathan et al. compare illness diaries in the five communities with outpatient records from four dispensaries, two in the highlands and two in the lowlands. Included in the morbidity study are data obtained by physical examinations of the study's mothers and children in the five communities undertaken in 1995. Clinic data of adults and children reveal a higher prevalence of respiratory diseases and a lower prevalence of malaria in the highlands than lowland clinic data. The highland–lowland data was not reproduced in the child morbidity surveys of the five communities, but the nomadic lowland pastoral children of Lewogoso had markedly less days of colds and diarrhea than children from any of the settled communities, although they had more fever days per child in the normal rainfall year. In both the clinic data and the children's morbidity survey, incidence of fever and malaria dropped significantly in the year of low rainfall.

Fujita et al.'s Chapter 11 "Sedentarization and Seasonality: Maternal Dietary and Health Consequences in Ariaal and Rendille Communities" compares data on women's diets and anthropometrics between lowland nomadic Lewogoso and highland sedentary Songa community. This study reveals that although there is a high seasonality in pastoral diets compared to highland Songa, pastoral women have larger AMAs (arm muscle areas) than Songa women, although this difference declines if Songa women have milk animals. Although wealth differences between women in both communities affected their

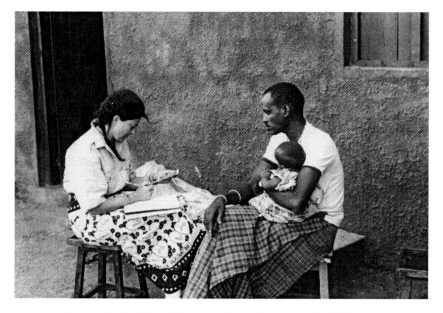

Photo 9. Martha Nathan MD interviews Larian Aliyaro about his child in Korr.

anthropometric indices, these differences were much more pronounced in Songa. Poor women in Songa had significantly decreased nutritional indices, suggesting that agriculture in Marsabit District does not in itself improve the nutritional well-being of women. Significantly, breast-feeding mothers in Songa had smaller AMAs than did their counterparts in Lewogoso, suggesting poorer caloric and protein levels, or higher caloric expenditures which one would expect in agricultural Songa where, as Smith showed in Chapter 7, women engage in more labor intensive activities than in pastoral Lewogoso.

The final two chapters focus on changes in women's status and behavior brought about by sedentarization. Shell-Duncan et al.'s Chapter 12 discusses the changing nature of female "circumcision" (genital cutting) in Rendille society, where even among settled Rendille whose educational and economic opportunities increase, the practice is strongly rooted. As the authors show, women who are not circumcised cannot marry and thus bear children, and face ostracism from their communities. Due to a variety of international and local pressures, Kenya has made the practice of female genital cutting illegal. Rather than dis-appearing, however, the practice continues but increasingly is carried out antiseptically as a medical procedure, albeit still illegally. Shell-Duncan and her Kenyan associates argue that steps in improving the aseptic conditions for excision significantly reduce medical risks for the women involved, and encourage improvement of medical conditions, so women can benefit as much as men by local practices.

The final Chapter 13 by Roth and Ngugi build on previous research by Roth investigating parental decision making in children's education, noting a large bias for males over females in educational opportunities in Marsabit District. Yet these authors also show the strong correlation between women's educational levels and knowledge and prevention of sexually transmitted illnesses, including HIV/AIDS. While other demographic research has demonstrated that female's educational levels are positively correlated with changes in fertility behavior, the relationship between girls' schooling and declines in the spread of

HIV/AIDS is of pronounced importance in Africa. This research also shows the importance of these efforts in pastoral regions, and the increased opportunities for changes in sexual behavior that can accompany sedentarization.

5. CONCLUSION

The contributions to *As Pastoralists Settle* demonstrate aspects of a larger *process* of sedentarization. By adopting a processual perspective concerned with the interaction of different parameters, both physical and cultural, they manage to avoid the debilitating dichotomy of past East African pastoral studies. On one side of this dichotomy is the qualitative tradition emphasizing cultural representation, identity, and social history (cf. Anderson and Broch-Due, 1999; Pálsson, 1990), while the other is the quantitative approach of the natural sciences focusing solely on biological and physical factors associated with pastoral ecology, adaptation, health and nutrition (cf. Little and Leslie, 1999). Some adherents of the latter approach criticize the former's emphasis on "current anthropological theory and criticism, rather than concentrating on the acquisition of data that might have been used to test hypotheses and build the structure of a firm scientific base of information" (Little, 1997: 31). In turn, the former approach criticizes the "thin description" of the physical and biological science approach, which attempt to quantify and reduce matters that are, rooted in particular social histories and forces. As Broch-Due and Anderson (1999: 13–14) argue in their introduction to *The Poor Are Not Us: Poverty and Pastoralism*, "Quantitative measures of poverty, defined most usually by incomes or nutrition, and, for pastoralists particularly, by the counting of livestock, are often both inadequate and misleading when detached from as raw statistics from the context of their cultural meanings."

The chapters in this volume easily and frequently span the chasm between these two approaches. Examples of this holistic approach include Schwartz' principal components analysis of Rendille settlement variables which reveals that less than half the variation in site selection is attributable to ecological factors, exemplified by median rainfall, range condition, vegetation cover index, and availability of permanent water. Cultural factors, including a circumcision ritual and a period of increased banditry are equally important determinants. These findings, derived from quantitative multivariate statistical analyses, parallel those in Galaty's chapter, based on historic records, that inter-ethnic conflict is an integral determinant of both past and present settlement patterns in northern Kenya. Another example is provided by the McPeak and Little's chapter employing quantitative methodologies to examine both ecological (e.g., rainfall), herd mobility and socio-economic variables (e.g., income, household expenditures, and herd offtake rates), while also providing qualitative portraits of individuals in differing settings and circumstances. A third example is represented in the Shell-Duncan, Obiero and Muruli's chapter in which male and female Rendille voices explaining *emic*, or internally conceived, views of female genital mutilation are followed by etic, or externally imposed, analyses employing logistic regression methodology.

This focus on process entails a reliance on data, quantitative *and* qualitative, contemporary or historical, over *time*. This last factor is also a strong theme running through these chapters. Our concern with time takes many forms. It is represented by (1) the repeated surveys collecting dietary, morbidity, and growth data in the chapters on maternal-child health (Roth et al., Nathan et al.; and Fujita et al.), (2) the McPeak and Little's chapters on socio-economic diversification, (3) the Adano and Witsenberg's assessment of livestock holdings at the onset of sedentarization and years later, and (4) Roth et al.'s inquiry into female child education over a fifteen-year span.

A third collective thread linking the chapters in this volume is the view that sedentarization is neither good nor bad, but rather represents a series of opportunities and constraints upon populations. Changing opportunities, particularly for women, include increased involvement in cash economies as revealed in Smith's chapter on sedentary agriculture and the Fujita et al.'s chapter on maternal health, and the potential benefits of female education as documented by the Roth, contribution. Of course along with opportunities there are constraints and costs involved in sedentarization. Nathan et al. showed that a more sedentary lifestyle has not led to improved child health or nutrition, one of the most fundamental and important measures of population adaptation. Similarly, while not as dramatic, the Fujita et al.'s chapter on maternal diet, morbidity and anthropometrics indicates that sedentarization has not significantly improved maternal health.

Finally, while emphasizing cultural, socio-economic, and biological changes arising from sedentarization, many of these chapters reveal that some important factors have not changed much at all, or now constitute a mixture of stability and change. Thus Shell-Duncan, Obiero, and Muruli find no differences in the prevalence of female genital mutilation among sedentary and nomadic sample, but do note an increased trend toward the use of medical assistance in the form of sterile razors, anti-tetanus injection, and prophylactic antibiotics among sedentary communities. Similarly, Galaty's research clearly shows a continuation of the historic pattern of inter-ethnic group violence, only with livestock raiding now combined with more recent disputes over arable land adjoining sedentary agricultural communities. In pointing out that major changes and new adaptations are linked to older, previously established patterns of culture and behavior, these chapters and this book, demonstrate that rather than representing simple dichotomies—e.g., nomadic

Photo 10. Participants of Rendille Sedentarization Project in Karare, 1995. From Left: David Wiseman, Daniel Lemoille (behind Wiseman), Margaret Leala McPeak, Korea Leala (in glasses), Mama Leala, Lugi Lengesen (in hat), Unnamed, Eric Abella Roth, Joan Harris, Kevin Smith, Elliot Fratkin.

versus sedentary, pastoral versus agrarian—sedentarization in northern Kenya is a complex, ongoing process.

ACKNOWLEDGEMENTS. There are many institutions and people we would like to thank who participated in or contributed to this research program on pastoral sedentarization in northern Kenya. We are grateful to the Office of the President, Republic of Kenya for permitting us to undertake research in Marsabit District between 1987–2002 (Research Permits OP/13/001/C 1594 for Elliot Fratkin and OP/13/001/19C 249/34 for Eric Roth). We thank our colleagues at the Institute of African Studies, University of Nairobi, Department of Community Health, University of Nairobi, and the Kenya Agricultural Research Institute, and to physicians and staff at the Marsabit District Hospital and health clinics at Korr, Ngrunit, Songa, and Karare who contributed their time and knowledge for our study. We also appreciate the hospitality shown to us by personnel at the African Inland Church, Food for the Hungry, and the Catholic Diocese of Marsabit.

Funding for the Rendille Sedentarization Project was provided to Elliot Fratkin and Martha Nathan by the United States National Science Foundation (Research Grants SBR-9400145 and SBR-9696088), the Mellon Foundation, the Geisinger Medical Foundation, and Smith College, and to Eric Roth by the Social Sciences and Humanities Research Council of Canada, the Association of Universities and Colleges of Canada, the Canadian International Development Agency, and the University of Victoria, Canada.

We extend our special thanks to our field assistants Anna Marie Aliyaro, Larian Aliyaro, Korea Leala, Daniel Lemoille, Kawab Bulyar, and Patrick Ngoley, and to the researchers who contributed to this volume and to many beneficial conversations and interactions, Wario Adano, Masako Fujita, John Galaty, Joyce Giles, Joan Harris Peter D. Little, John McPeak, Leunita Muruli, Elizabeth Ngugi, Walter Obiero, Bettina Shell-Duncan, Kevin Smith, H. Jürgen Schwartz, David Wiseman, and Karen Witsenburg.

Finally, and most appreciatively, we thank the women, men, and children of Lewogoso Lukumai, Karare, Korr, Ngrunit, and Songa in Marsabit District, Kenya. It is our strongest hope that the results of our studies will benefit the health and well-being of these, and other, pastoral and settled communities.

Unless indicated, photographs throughout the volume are by Elliot Fratkin.

REFERENCES

Anderson, D.M. and V. Broch-Due, 1999 (eds.), *The Poor Are Not Us: Poverty and Pastoralism in Eastern Africa.* Oxford: James Currey.

Bayer, Wolfgang and Ann Waters-Bayer, 1994, Coming to Terms: Interactions between Immigrant Fulani Cattle Keepers and Indigenous Farmers in Nigeria's Subhumid Zone. *Cahiers d'etudes Africaines* XXXIV (1–3): 213–229.

Blench, R., 2001, *'You Can't Go Home Again': Pastoralism in the New Millennium.* London: Overseas Development Institute. www.odi.org.uk/pdn/eps.pdf.

Brainard, J.M., 1990, Nutritional Status and Morbidity on an Irrigation Project in Turkana District, Kenya. *American Journal of Human Biology* 2: 153–163.

Brockington, D., 1999, Conservation, Displacement, and Livelihoods: The Consequences of Eviction for Pastoralists Moved from the Mkomazi Game Reserve, Tanzania. *Nomadic Peoples* (NS) 3(2): 74–96.

Brass, W. and C. Jolly, 1993, *Population Dynamics of Kenya.* Washington, D.C.: National Academy Press.

Caldwell, J., 1979, Education as a Factor in Mortality Decline: An Examination of Nigerian Data. *Population Studies* 33: 395–413.

Caldwell, J., 1982, *Theory of Fertility Decline.* New York: Academic Press.

Campbell, B.C., P.W. Leslie, M.S. Little, J.M. Brainard, and M.A. DeLuca, 1999, Settled Turkana. In *Turkana Herders of the Dry Savanna: Ecology and Biobehavioral Response of Nomads to an Uncertain Environment*, edited by M.A. Little and P.W. Leslie, pp. 333–352. New York: Oxford University Press.

Campbell, D.J., 1999, Response to Drought Among Farmers and Herders in Southern Kajiado District, Kenya: A Comparison of 1972–1976 and 1994–1995. *Human Ecology* 27(3): 377–416.

Chabasse, D., C. Roure, A.G. Rhaly, P. Rangque, and M. Quilici, 1985, The Health of Nomads and Semi-Nomads of the Malian Gourma: An Epidemiological Approach. In *Population, Health and Nutrition in the Sahel*, edited by A.G. Hill, pp. 319–333. London: Routledge and Kegan-Paul.

Dow, T., L.H. Archer, S. Kasiani, and J. Kekovole, 1994, Wealth Flow and Fertility Decline in Rural Kenya, 1981–92. *Population and Development Review* 20: 343–364.

Duba, H.H., I. Mur-Veeman, and A. van Raak, 2001, Pastoralist Health care in Kenya. *International Journal of Integrated Care*, 1 March 2001. www.ijic.org/puclish/articles/000019/index.html.

Dyson-Hudson, N., 1991, Pastoral Production Systems and Livestock Development Projects: An East African Perspective. In *Putting People First: Sociological Variables in Rural Development*, Second Edition, edited by M. Cernea, pp. 157–186. Oxford: Oxford University Press.

Dyson-Hudson, N. and R. Dyson-Hudson, 1982, The Structure of East African herds and the Future of East African Herders. *Development and Change* 13: 213–238.

Ensminger, J., 1991, Structural Transformation and its Consequences for Orma Women Pastoralists. In *Structural Adjustment and African Women Farmers*, edited by C. Gladwin, pp. 281–300. Gainesville: University of Florida Press.

Ensminger, J., 1992, *Making a Market: The Institutional Transformation of an African Society*. New York: Cambridge University Press.

Fabietti, U. and P.C. Salzman, 1996, *The Anthropology of Tribal and Peasant Pastoral Societies*. Pavia: Ibis.

Fleuret, P. and A. Fleuret, 1991, Social Organization, Resource Management, and Child Nutrition in the Taita Hills, Kenya. *American Anthropologist* 93: 91–114.

Fratkin, E., 1989, Household Variation and Gender Inequality in Ariaal Pastoral Production: Results of a Stratified Time Allocation Survey. *American Anthropologist* 91(2): 45–55.

Fratkin, E., 1991, *Surviving Drought and Development: Ariaal Pastoralists of Northern Kenya*. Boulder: Westview Press.

Fratkin, E., 1997, Pastoralism: Governance and Development Issues. *Annual Review of Anthropology* 26: 235–261.

Fratkin, E., 2001, East African Pastoralism in Transition: Maasai, Boran, and Rendille Cases. *African Studies Review* 44(3): 1–25.

Fratkin, E., 2004, *Ariaal Pastoralists of Northern Kenya: Studying Pastoralism, Drought, and Development in Africa's Arid Lands*, Second edition. Needham Heights MA: Allyn and Bacon.

Fratkin, E. and E.A. Roth, 1990, Drought and Economic Differentiation among Ariaal Pastoralists of Kenya. *Human Ecology* 8: 385–402.

Fratkin, E.M., E.A. Roth, and M.A. Nathan, 1999, When Nomads Settle: The Effects of Commoditization, Nutritional Change, and Formal Education on Ariaal and Rendille Pastoralists. *Current Anthropology* 40(5): 720–735.

Fratkin, E. and K. Smith, 1995, Women's Changing Economic Roles with Pastoral Sedentarization: Varying Strategies in Alternate Rendille Communities. *Human Ecology* 23(4): 433–454.

Fujita, M., E.A. Roth, M.A. Nathan, and E. Fratkin, 2004, Sedentism, Seasonality and Economic Status: A Multivariate Analysis of Maternal Dietary and Health Statuses Between Pastoral and Agricultural Ariaal and Rendille Communities in Northern Kenya. *American Journal of Physical Anthropology* 123(3): 277–291.

Galaty, J.G., 1992, 'This Land is Yours': Social and Economic Factors in the Privatization, Subdivision and Sale of Maasai Ranches. *Nomadic Peoples* 30: 26–40.

Galaty, J.G., 1994, Rangeland Tenure and Pastoralism in Africa. In *African Pastoralist Systems*, edited by E. Fratkin, K. Galvin, and E.A. Roth, pp. 185–204. Boulder: Lynne Rienner Publishers.

Galaty, J.G., D. Aronson, P.C. Salzman, and A. Chouinard, 1981, *The Future of Pastoral Societies*. Leiden: E.J. Brill.

Galaty, J.G. and D.L. Johnson, 1990, *The World of Pastoralism*. New York, The Guilford Press.

Galvin, K.A., 1985, *Food Procurement, Diet, Activities and Nutrition of Ngisonyonka Turkana Pastoralists in an Ecological and Societal Context*. Ph.D. Dissertation. State University of New York: Binghamton.

Galvin, K.A., 1992, Nutritional Ecology of Pastoralists in Dry Tropical Africa. *American Journal of Human Biology* 4: 209–221.

Galvin, K.A. and M.A. Little, 1999, Dietary Intake and Nutritional Status. In *Turkana Herders of the Dry Savanna: Ecology and Biobehavioral Response of Nomads to an Uncertain Environment*, edited by M.A. Little and P.W. Leslie, pp. 125–145. New York: Oxford University Press.

Galvin, K.A., D.L. Coppock, and P.W. Leslie, 1994, Diet, Nutrition and the Pastoral Strategy. In *African Pastoral Systems: An Integrated Approach*, edited by E. Fratkin, K. Galvin, and E.A. Roth, pp. 113–132. Boulder: Lynne Rienner Publishers.

Goyder, H. and C. Goyder, 1988, Case Studies of Famine: Ethiopia. In *Preventing Famine: Policies and Prospects for Africa*, edited by D. Curtis, M. Hubbard, and A. Shepherd, London: Routledge.

Hill, A.G., 1985, *Population, Health and Nutrition in the Sahel*. London: Routledge and Kegan Paul.

Hodgson, D., 2001, *Once Intrepid Warriors: Gender, Ethnicity, and the Cultural Politics of Maasai Development*. Bloomington: Indiana University Press.

Hogg, R., 1982, Destitution and Development: The Turkana of Northwest Kenya. *Disasters* 6(3): 164–168.

Hogg, R., 1986, The New Pastoralism: Poverty and Dependency in Northern Kenya. *Africa* 56(3): 319–333.

Homewood, K.M., 1995, Development, Demarcation and Ecological Outcomes in Maasailand. Africa 65(3): 331–350.

Huss-Ashmore, R.A., 1993, Agriculture, Modernization and Seasonality. In *Seasonality and Human Ecology*, edited by S.J. Ulijazek and S.S. Strickland, pp. 202–219. Cambridge: Cambridge University Press.

IPAL, 1984, *Integrated Resource Assessment and Management Plan for Western Marsabit District, Northern Kenya*. Integrated Project in Arid Lands Technical Report No. A-6. Nairobi: UNESCO.

Kerven C., 1992, *Customary Commerce: A Historical Reassessment of Pastoral Livestock Marketing in Africa*. London: Overseas Development Institute.

Kituyi, M., 1990, *Becoming Kenyans: Socio-economic Transformation of the Pastoral Maasai*. Nairobi: ACTS Press.

Klepp, K.I., P.M. Biswalo, and A. Talle, 1994, *Young People at Risk: Fighting AIDS in Northern Tanzania*. Oslo: Scandinavian University Press.

Lappe, F.M. and J. Collins, 1977, *Food First: Beyond the Myth of Scarcity*. Boston: Houghton Mifflin.

Little, M., 1997, Adaptability of African pastoralists. In *Human Adaptability: Past, Present and Future*, edited by S.J. Ulijaszek and R.A. Huss-Ashmore, pp. 61–81. Oxford: Oxford University Press.

Little, M.A., S.J. Gray, and P.W. Leslie, 1993, Growth of Nomadic and Settled Turkana infants of North-west Kenya. *American Journal of Physical Anthropology* 92: 335–344.

Little, M.A. and P.W. Leslie, 1999, *Turkana Herders of the Dry Savanna: Ecology and Biobehavioral Response of Nomads to an Uncertain Environment*. New York: Oxford University Press.

Little, P.D., 1985, Social Differentiation and Pastoralist Sedentarization in Northern Kenya. *Africa* 55(3): 243–261.

Little, P.D., 1992, *The Elusive Granary*. Cambridge: Cambridge University Press.

Little, P.D., 1994a, Maidens and Milk Markets: The Sociology of Dairy Marketing in Southern Somalia. In *African Pastoralist Systems: An Integrated Approach*, edited by E. Fratkin, K. Galvin, and E.A. Roth, pp. 165–184. Boulder: Lynne Rienner Publishers.

Little, P.D., 1994b, The Social Context of Land Degradation ("Desertification') in Dry Regions. In *Population and Environment: Rethinking the Debate*, edited by L. Arizpe, M.P. Stone, and D.C. Major, pp. 209–251. Boulder: Westview Press.

Little, P.D., K. Smith, B.A. Cellarius, D.L. Coppock, and C.B. Barrett, 2001, Avoiding Disaster: Diversification and Risk Management Among East African Herders. *Development and Change* 32: 401–433.

McCabe, J. Terrence, S. Perkin, and Clare Schofield, 1992, Can Conservation and Development Be Coupled Among Pastoral People? An Examination of the Maasai of the Ngorongoro Conservation Area, Tanzania. *Human Organization* 51: 353–366.

McCabe, J.T., 2003, Sustainability and Livelihood Diversification among the Maasai of Northern Tanzania. *Human Organization* 62(2): 100–111.

Meir, A., 1997, *As Nomadism Ends: The Israeli Bedouin of the Negev*. Boulder: Westview Press.

Mitchell, J.D., 1999, Pastoral Women and Sedentism: Milk Marketing in an Ariaal Rendille Community in Northern Kenya. *Nomadic Peoples* NS (1999) 3(2): 147–160.

Morton, J., 2003, *Conceptualizing the Links between HIV/AIDS and Pastoralist Livelihoods*. Paper presented to the Annual Conference of Development Studies Association (UK).

Nathan, M.A., E.M. Fratkin, and E.A. Roth, 1996, Sedentism and Child Health Among Rendille Pastoralists of Northern Kenya. *Social Science and Medicine* 43(4): 503–515.

National Research Council, 1993, *Factors Affecting Contraceptive Use in Sub-Saharan Africa*. Washington, D.C.: National Academy Press.

Nestel, P., 1986, A Society in Transition: Developmental and Seasonal Influences on the Nutrition of Maasai Women and Children. *Food and Nutrition Bulletin* 8: 2–14.

O'Leary, M.F., 1985, *The Economics of Pastoralism in Northern Kenya: The Rendille and the Gabra.* IPAL Technical Report F-3. Nairobi: UNESCO.

O'Leary, M.F., 1990, Drought and Change amongst Northern Kenya Nomadic Pastoralists: The Case of Rendille and Gabra. In *From Water to World-Making: African Models in Arid Lands,* edited by G. Pálsson, pp. 151–174. Uppsala: The Scandinavian Institute of African Studies.

Pálsson, G. (ed.), 1990, *From Water to World Making: African Models and Arid Lands.* Uppsala: Scandinavian Institute of African Studies.

Panter-Brick, C., 1997, Biological Anthropology and Child Health. In *Biosocial Perspectives on Children,* edited by C. Panter-Brick, pp. 66–101. Cambridge: Cambridge University Press.

Population Reference Bureau, 2000, Population Bulletin 2000. Washington, D.C.: Population Reference Bureau.

Roth, E.A., 1990, Modeling Rendille Household Herd Composition. *Human Ecology* 18: 441–455.

Roth, E.A., 1991, Education, Tradition and Household Labour Among Rendille Pastoralists of Northern Kenya. *Human Organization* 50: 136–141.

Roth, E.A., 1993, A Reexamination of Rendille Population Regulation. *American Anthropologist* 95(3): 597–611.

Roth, E.A., 1994, Demographic Systems: Two East African Examples. In *African Pastoralist Systems,* edited by E. Fratkin, K. Galvin, and E.A. Roth, pp. 133–146. Boulder: Lynne Rienner Publishers.

Roth, E.A., E. Fratkin, A.Y. Eastman, and L.M. Nathan, 1999, Knowledge of AIDS among Ariaal Pastoralists of Northern Kenya. *Nomadic Peoples* (NS) 3(2): 161–175.

Roth, E.A., E. Fratkin, E. Ngugi, and B. Glickman, 2001, Female Education, Adolescent Sexuality, and the Risk of Sexually Transmitted Infection in Ariaal Rendille Culture. *Culture, Health and Sexuality* 3: 35–47.

Salih, M.A.M. Salih, and J. Baker, 1995, Pastoralist Migration to Small Towns in Africa. In *The Migration Experience in Africa,* edited by J. Baker and T.A. Aina, pp. 181–196. Uppsala: Scandinavian Institute of African Studies.

Salzman, P.C., 1980, *When Nomads Settle: Processes of Sedentarization as Adaptation and Response.* New York: Praeger.

Salzman, P.C., 1996, Introduction: Varieties of Pastoral Societies. In *The Anthropology of Tribal and Peasant Pastoral Societies,* edited by U. Fabietti and P.C. Salzman, pp. 21–37. Pavia: Ibis.

Salzman, P.C. and J.G. Galaty, 1990, *Nomads in a Changing World.* Naples: Instituto Universitario Orientale.

Sato, S., 1997, How the East African Pastoral Nomads, Especially the Rendille, Respond to the Encroaching Market Economy. *African Studies Monographs* 18(3–4): 121–135. The Center for African Area Studies, Kyoto University.

Schwartz, H.J., 1980a, An Introduction to the Livestock Ecology Programme. *IPAL Technical Report A-3.* pp. 56–61. Nairobi: UNESCO Integrated Project in Arid Lands.

Schwartz, H.J., 1980b, *Draft Final Report: On the Implementation of the UNESCO-FRG Traditional Livestock Management Program.* June 1978–July 1980. Nairobi: UNESCO Integrated Project in Arid Lands.

Scoones, I., 1994, *Living with Uncertainty: New Directions in Pastoral Development in Africa.* London: Intermediate Technology Publications.

Sellen, D.W., 1996, Nutritional status of Sub-Saharan African pastoralists: A Review of the Literature. *Nomadic Peoples* 39: 107–134.

Shell-Duncan, B., 1995, Impact of Seasonal Variation in Food Availability and Disease Stress on the Health Status of Pastoral Turkana Children: A Longitudinal Analysis of Morbidity, Immunity, and Nutritional Status. *American Journal of Human Biology* 7: 339–355.

Shell-Duncan, B. and Y. Hernlund, 2000, *Female 'Circumcision' In Africa: Culture, Controversy and Change.* Boulder: Lynne Rienner Publishers.

Shell-Duncan, B. and W.O. Obiero, 2000, Child Nutrition in the Transition from Nomadic Pastoralism to Settled Lifestyles: Individual Settled Lifestyles: Individual, Household, and Community-level Factors. *American Journal of Physical Anthropology* 113: 183–200.

Simondon, K.B., E. Bénéfice, F. Simondon, V. Delaunay, and A. Chahnazarian, 1993, Seasonal Variation in Nutritional Status of Adult and Children in Rural Senegal. In *Seasonality and Human Ecology,* edited by S.J. Ulijaszek and S.S. Strickland, pp. 166–183. Cambridge: Cambridge University Press.

Smith, K., 1999, The Farming Alternative: Changing Age and Gender Roles among Sedentarized Rendille and Ariaal. *Nomadic Peoples* (NS) 3(2): 131–146.

Sobania, N.W., 1979, *Background History of the Mt. Kulal Region of Kenya.* Nairobi: Integrated Project in Arid Lands, Technical Report No. A-2.

Sobania, N.W., 1991, Feasts, Famines and Friends: Nineteenth Century Exchange and Ethnicity in the Eastern Lake Turkana Region. In *Herders, Warriors and Traders*, edited by J.G. Galaty and P. Bonte, pp. 118–142. Boulder: Westview Press.

Spear, T., 1997, *Mountain Farmers: Moral Economies of Land and Agricultural Development in Arusha and Meru.* London: James Currey Publishers.

Swift, J., 1988. *Major Issues in Pastoral Development with Special Emphasis on Selected African Countries,* FAO, Rome.

Talle, A., 1988, *Women at a Loss: Changes in Maasai Pastoralism and Their Effects on Gender Relations.* Stockholm Studies in Social Anthropology, 19. Stockholm: Department of Social Anthropology, University of Stockholm.

Talle, A. 1999, Pastoralists at the Border: Maasai Poverty and the Development Discourse in Tanzania. In *The Poor are not us: Poverty and Pastoralism in Eastern Africa*, edited by D.M. Anderson and V. Broch-Due, pp. 106–124. Oxford: James Currey.

Waller, R., 1993, Acceptees and Aliens: Kikuyu Settlement in Maasailand. In *Being Maasai: Ethnicity and Identity in East Africa*, edited by T. Spear and R. Waller, pp. 226–257. London: James Currey.

Waters-Bayer, A., 1988, *Dairying by Settled Fulani Women in Central Nigeria: The Role of Women and Implications for Dairy Development.* Kiel: Wissenschaftsverlag Van Kiel.

The World Bank, 1984, *Towards Sustained Development in Sub-Saharan Africa: A Joint Program of Action.* Washington, D.C.: The World Bank.

The World Bank, 1993, *Indigenous Views of Land and the Environment.* World Bank Discussion Paper 188. Washington: The World Bank.

The World Bank, 1997, *Investing in Pastoralism. Sustainable Natural Resource Use in arid Africa and the Middle East.* World Bank Technical Paper No. 365, Washington: The World Bank.

Zaal, F. and T. Dietz, 1999, Of Markets, Maize, and Milk: Pastoral Commoditization in Kenya. In *The Poor are not us: Poverty and Pastoralism in Eastern Africa*, edited by D.M. Anderson and V. Broch-Due, pp. 163–198. Oxford: James Currey.

Chapter 2

The Setting

Pastoral Sedentarization in
Marsabit District, Northern Kenya

ELLIOT FRATKIN AND ERIC ABELLA ROTH

1. MARSABIT DISTRICT—POPULATION AND GEOGRAPHY

Northern Kenya has been treated by both the Kenyan government and the former British colonial administration as "another country", a distant wasteland inhabited by small and economically insignificant populations of nomadic pastoralists. The area's principle benefit was, and continues to be, viewed as a large space buffering Kenya from potentially hostile countries including Ethiopia and Somalia; it has also served as isolated location for detention and international refugee camps.

During its rule (1900–1963), Britain enforced a "Pax Britannica" in the then Northern Frontier District, maintaining peace between competing and warring pastoralist groups. Society-specific "tribal grazing areas" were created to separated and restrict pastoral movements by local Samburu, Rendille, Boran, Gabra, Sakuye, Adjuran, and Somali groups; the British also drew specific boundaries including the "Somali-Galla (Boran)" line drawn in 1934 between present day Marsabit and Wajir Districts. Ultimately, administrative units (later Districts) bounded ethnic groups into Turkana, Samburu, Marsabit, Isiolo, Wajir and Mandera Districts (Zwanenberg, 1977: 89). Marsabit District, located in north central Kenya, was shared by Boran, Gabra, Rendille, although the British attempted, with mixed success, to contain each of these groups in their own areas. Following Kenyan independence in 1963, the northern districts remained undeveloped by the government, although many famine relief organizations moved into the area following the droughts of the 197 and 1980s.

ELLIOT FRATKIN • Department of Anthropology, Smith College, Northampton, Massachusetts 01
ERIC ABELLA ROTH • Department of Arthropology, University of Victoria, Victoria, British
Canada V8W 3P5

Marsabit District is Kenya's largest, most arid, and least inhabited region, and the majority of their residents were, until quite recently, nomadic pastoralists. Its population in 1993 was 138,500 people occupying 75,078 square kilometers. The district was divided in 1995 to allow for the creation of Moyale District on the Ethiopian border, after which Marsabit District's size was 61,296 square kilometers occupied by 121,478 people (Ministry of Planning and Development, 1997). Marsabit District (before its partition) bordered Ethiopia in the north, Wajir District in the east, Isiolo District in the southeast, Samburu District in the southwest, and Lake Turkana and Turkana District in the west.

The District has a mean annul rainfall of 200 mm in the lowlands and 800 mm in the highlands, with vegetation ranging from scrub bush in the lowland deserts ranging to evergreen forests in the highlands. While the district is predominately lowlands (from 400 to 700 m altitude), it is interspersed with several mountain ranges and hills including the Ndoto Mountains (2660 m) in the west, the Hurri Hills (1260 m) in the north, and solitary Marsabit Mountain (1545 m) in the center of the District. Marsabit town, on Marsabit Mountain, is the district capital, and the mountain is home to most of the district's agriculturalists, which include Burji, Boran, and Ariaal and Rendille communities. Today an estimated 30,000 of the district's 120,000 people live on Marsabit Mountain (Adano and Witsenburg, this volume).

There are no permanent rivers in the district, although mountain run-offs provide temporary surface water in the lowlands (Milgis and Merille Rivers), and the highlands have several permanent lakes, including Lake Paradise and several water-filled craters on Marsabit Mountain. The only permanent water resources, besides Marsabit Mountain, are Lake Turkana to the west and the Uaso Nyiru River to the south in Isiolo District. The Uaso Nyiru in particular has played an important role in pastoral livelihood, and is currently utilized by Samburu, Ariaal, Rendille, Somali, and Boran herders (Schwartz et al., 1991). The District administration estimates that livestock keeping pastoralists make up 80 percent of the total population, while another 10 percent are highland farmers, 5 percent commercial traders, and 5 percent are salaried employees working with the District administration, schools, police, hospital, or with non-government organizations (Ministry of Planning and Development, 1997).

The pastoral groups live predominately in the lowlands, although there are cattle keeping Boran and Ariaal settlements on Marsabit Mountain, and Samburu settlements in the Ndoto Mountains. Between the Ndoto Mountains and Mt. Marsabit lies the broad and flat Koroli Desert, which is occupied by Rendille camel keeping settlements. To the north of the Rendille is the Chalbi Desert bordering Lake Turkana and Ethiopia, inhabited by Gabra camel pastoralists, Boran cattle pastoralists, and Dasenech agro-pastoralists on the shores of Lake Turkana. To the west (in Turkana district) live the Turkana, cattle and camel pastoralists who ~ve traditionally raided Rendille and Samburu. To the east (Wajir and Isiolo) are Somalis, ~n of different clan groupings. All of these groups have raided and counter raided each ~stock, with many raids intensifying during the extensive droughts of the 1990s.

~f Kenyan independence in 1963, Marsabit District had only two towns ~rsabit and Moyale, and several small trading posts at Laisamis, ᾽ North Horr. The district had only three primary schools, no ᾽vernment hospital. The Kenya government under Jomo ᾽ the north, allocating most resources and development ῼnsely populated urban regions around Nairobi, and to a ῼbasa. President Daniel arap Moi, who succeeded Kenyatta ῼne gestures to develop resources in the north, as he depended ῼsmaller agro-pastoral groups including Samburu to maintain

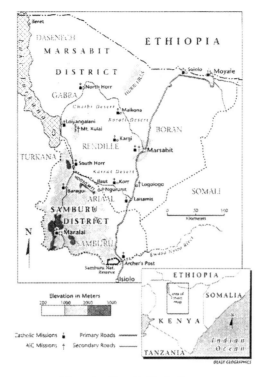

Figure 1.   Marsabit District Kenya.

his political power. By 1995, the district had 7 secondary schools, 54 primary schools, 4 hospitals, and 15 medical dispensaries. Towns along the major roads (including Merille, Laisamis, Loglogo, Maikona, North Horr, Loyangalani) grew in size, and new towns grew, particularly in the lowland areas in response to mission-sponsored famine relief efforts, including Korr and Kargi among the Rendille (see Figure 1). While Catholic and to a lesser degree Protestant churches developed primary schools and health dispensaries in these communities, international development organizations including GTZ (the German Development Corporation) and religious-based organizations, including World Vision and Food for the Hungry, contributed to infrastructure development of mechanizing wells, building dams and catchments, and laying water pipes for irrigated agriculture.

2. TRADITIONAL LIVESTOCK PASTORALISM IN MARSABIT DISTRICT

Until the onset of long-term drought beginning in the late 1960s, the majority of Marsabit's residents practiced mixed species pastoralism, i.e., they lived principally off the products of their camels, cattle, and small stock of goats and sheep. These pastoral populations include Boran cattle keepers (pop. 36,447), Gabra camel herders (pop. 30,213), Rendille camel herders (pop. 23, 585), Samburu (specifically Ariaal) mixed cattle and camel herders (5887), Sakuye camel pastoralists (1856), and some Somali families who moved in from the larger Somali areas to the East in Wajir District.[1] In addition to large

Table 1. Ethnic Populations of Kenya.

Group	Population	% of total
Agricultural/urban		
Kikuyu	4,455,865	20.78
Luhya	3,083,273	14.38
Luo	2,653,932	12.38
Kalenjin	2,458,123	11.46
Kamba	2,448,302	11.42
Kisii	1,318,409	6.15
Meru	1,087,778	5.07
Mijikenda	1,007,371	4.70
Embu	256,623	1.20
Taita	203,389	0.95
Teso	178,455	0.83
Kuria	112,236	0.52
Basuba	107,819	0.50
Mbere	101,007	0.47
Degodia	100,400	0.47
Tharaka	92,528	0.46
Pokomo	58,645	0.27
Bajun	55,187	0.26
Kenyan Asian	52,968	0.25
Kenyan Arab	33,714	0.16
Indians	29,091	0.14
Other Kenyans	28,722	0.13
Taveta	14,358	0.07
Swahili	13,920	0.06
Bulji (Burji)	5975	0.03
Pastoralists		
Maasai	377,089	1.76
Turkana	283,750	1.32
*Pokot	225,000	1.05
Samburu	106,897	0.50
Boran	60,160	0.37
Ogaden	139,597	0.65
Gurreh	80,004	0.37
Orma	45,562	0.21
Somali	45,098	0.21
Gabra	35,726	0.17
Hawiyah	27,244	0.13
Ajuran	26,916	0.13
Rendille	26,536	0.12
Njemps (LChamus)	15,872	0.07
Sakuye	10,678	0.05
Dasnachi-Shangil	418	0.00
Hunter–Gatherers		
Dorobo	24,363	0.11
Boni-Sanye	10,891	0.05
El Molo	3600	0.02
Non-Kenyan Groups	136,000	0.57
Ugandans	19,325	0.09
Not Recorded	16,716	0.08
Other Europeans	15,768	0.07
British	15,608	0.07

Table 1. (*Continued*)

Group	Population	% of total
Other Africans	14,471	0.07
Other Arabs	7881	0.04
Others	6308	0.03
Other Asians	5264	0.02
Kenyan Europeans	3184	0.01
Tribe Unknown	2411	0.01
Gosha	2081	0.01
Pakistanis	1862	0.01

*   Pokot included on census as Kalenjin.
Source: 1989 Census, Central Bureau of Statistics, Republic of Kenya, 1999.
Census figures not available.

animals, these groups kept large flocks of goats and sheep as well, used principally for trade and meat. In addition to these pastoral groups, there are several small mixed pastoral populations including the Dasenech (pop. less than 500 in Kenya) living on the northern edge of Lake Turkana who grow millet and keep cattle and small stock, the El Molo (pop. 3600), who are Samburu-speaking fishermen living on the eastern side of Lake Turkana, and Dorobo hunter-gatherers and small stock herders living mainly in the forests of the Ndoto Mountains which separate Samburu and Marsabit Districts. Their population is less than a few thousand. There also sedentary agricultural communities including Burji farmers (pop. 600) living on Marsabit Mountain, and who originally migrated from Ethiopia when the British built the Marsabit road in the 1930s. Population figures of Kenyan ethnic groups are listed in Table 1.

Although the various pastoral groups (Ariaal, Samburu, Rendille, Boran, Gabra, and Somali) have distinct cultures, languages, and customs, their social organization and live-stock production system are quite similar. Each of these groups live in semi-nomadic settle-ments, where domestic livestock are herded in home territories. [East African pastoralists, by and large, are not long-distance nomads as are Fulani and Tuareg in West Africa.] Each group depends on milk animals for daily subsistence (camels or cattle), and trade or sell animals (mainly small stock) to purchase grains, tea, sugar, and other commodities. All move their animals in mobile herding groups, and all manage their livestock in male-headed household units, utilizing their children and occasionally hiring kinsmen to help herd their animals. Furthermore, all of these societies are organized into patrilineal kinship groups, and often reside close to one another based on their kinship ties. Homesteads can vary from a few male stockowner and their domestic families (as among Samburu and Boran living in highland areas) to large communities of over fifty stockowners, as Gabra, Rendille and lowland Ariaal. Politically, these are acephalous societies lacking chiefs or kings, and are organized into what anthropologists recognize as 'segmentary descent sys-tems,' where affiliation is determined by degrees of closeness of consanguineal relation-ships. This is reflected in the well-known Somali proverb, "I and my clan against the world. I and my brother against the clan. I against my brother" (Cassenelli, 1982: 21).

The essential strategy of livestock pastoralism is to ensure adequate grazing and water for their livestock to provide a regular and food supply for the human community. Herders follow various strategies aimed at keeping their herds productive through both rainy and dry seasons, as well as deal with periodic and extensive drought. The two most important strategies for these herders are species diversity and herding mobility.

Photo 1. Ariaal warriors herding cattle in highlands.

Species diversity, keeping different types of livestock rather than specializing on one type of animal, enables a pastoralist to utilize different grazing environments as well as provide insurance against particular herd losses caused by diseases, including bovine pneumonia for cattle or trypanosomiasis in camels. Boran and Samburu raise cattle and small stock (goats and sheep), living mainly in or near highlands where cattle thrive due to greater grass and water availability. Gabra, Rendille, and Somali groups concentrate on camel and small stock production in the lowlands, where insect vectors of disease are fewer and where animals survive on a diet of browsing (leaves and bushes) rather than grazing (grasses). These ethnic/livestock divisions are not absolute, as some Gabra will own cattle and some Samburu keep camels. But most East African herders try to keep multi-species as insurance against loss to drought, disease, or raids (Fratkin et al., 1994).

Mobility is essential to pastoral production in arid lands. Because of the high seasonality in rainfall and the high evaporation rate, vegetation resources deteriorate quickly, and herds are taken to short-lived pastures following periodic rainfall. Except for camels who can graze for up to ten days without water, cattle and small stock have access to water at least every two days, and are herded near available water sources. Herding movements are consequently seasonal, with cattle alternating between highland and lowland resources and camels between home settlements and wide grazing areas away from the settlements.

Each type of stock has its own particular feeding requirements and grazing environment. Cattle are grazers (grass-eaters) which need water every two to three days, and consequently must be herded in the wetter highlands. Camels are adapted to desert conditions, preferring browse (leaves) of shrubs and trees that thrive when grasses dry out. Furthermore, camels can go without watering for ten days, offering enough time for their herders to graze them extensively in the desert lowlands between fixed water points. Small stock can thrive in the deserts, but like cattle need water every two to three days, and must be grazed near the mountain springs and wells.

Photo 2. Lewogoso girls and transport camels fetching water from Ngrunit wells.

Unlike long-distance herders like Fulani and Tuareg in West Africa (or Somali in northeastern Africa), Kenyan pastoralists remain home grazing areas, with semi-sedentary domestic settlements usually living near permanent water sources. Animals are taken from the settlements to either distant herding camp in dry seasons, and back to the settlement when there is sufficient grazing nearby. The Ariaal, like the Rendille, separate their animals into different herds, keeping milk animals, male transport camels, and small stock in or near the domestic settlements, while herding non-milking cattle in highlands and non-milking camels in distant lowlands grazing areas for long periods of time. Ariaal are not long distance nomads. Their settlements are semi-sedentary, located near permanent water sources and small urban centers along the Ndoto Mountains or Mt. Marsabit. People do not generally live closer than ten kilometers to the water holes, as they fear overgrazing the available vegetation quickly and have to graze their animals at greater distances. The chance of finding better pastures increases with the distance from the water points, and settlements with large numbers of transport camels will live farthest from towns or water sources (Fratkin, 2004).

Livestock production in northern Kenya provides three principle foods to the human population—milk, meat, and blood. Sale of livestock (usually small stock but increasingly cattle) as well as hides earn necessary money to purchase grains (maize meal) when milk production is low, and tea, sugar, tobacco. Milk is the preferred food, consumed either fresh following the morning and evening milking, or as sour (curdled) milk, usually consumed by older men. Milk can provide 75% of daily calories and 90% of their protein in the wet season, and 60% of their calories in the dry season. As milk supplies diminish in the dry season, blood is added to the milk, and more small stock are butchered to provide meat. In addition, small stock and cattle are sold to purchase maize meal, tea, and sugar. Cash income from livestock and skin sales is also used to buy cloth, rubber sandals, cooking utensils, and beads for jewelry.

Camels produce more milk than the other stock, where a lactating female can provide 5.0 liters of milk daily, providing an average of one liter of milk to each member of th: household per day. Cattle produce an average of 1.0 liters of milk each daily, but this fluc tuates directly with the quality of grazing vegetations. Typically cattle herds outnumber camels by 4:1, a feature due to their higher reproductive and survival rates (Fratkin, 2004: 83).

Ariaal pastoralists living in the lowlands have average household herds of 3 milk camels; 4 milk cows, and 12 milk goats and sheep, which yield approximately 1.5 liters of milk per person daily. However, actual milk consumption varies by both seasonal supplies and by differential consumption patterns based on age, gender, and wealth differences. A warrior in a cattle camp may drink 3–4 liters of milk mixed with blood in one sitting, while settlement children may have access only to one liter or less of milk daily. In the dry season when milk yields are reduced, households, which own only a few camels, may have no milk and depend on store-bought grains to survive. The grains, usually maize-meal sometimes wheat flour, are made as porridge, consumed with milk, sugar, and butter when available.

As the dry season progresses and milk resources are depleted, meat and maize meal are increasingly consumed as households slaughter or sell goats, sheep, and—to a lesser extent—cattle. Ariaal will also eat cattle or camels, which have died from predation or disease. The high milk, meat, and blood diet of East African herders provides more than adequate protein, exceeding the World Health Organization's recommended protein allowances of 65 g per adult male and 50 g per adult women per day. Despite high protein intake among the Ariaal, daily calorie consumption is low and there are seasonal shortages, particularly at the end of the dry season when their animals are producing very little milk. Where Americans typically consume over 2500 kcal of energy per day, Ariaal and other East African pastoralists make due on less than 1600 kcal per day, and less than 1200 kcal in the dry season, which the Ariaal call "the long hunger" (Fratkin, 2004; Sellen, 1996).

Subsistence pastoralists sell between 5–10% of their herds annually, mainly steers and male goats to local shopkeepers or urban markets. Cattle in particular bring in substantial income, $100 to $200 per animal, and are often taken down country to Isiolo town, on the main road to Nanyuki and Nairobi. Livestock marketing is the principle income generating activity for pastoralists, although it is increasingly common for a family to support one or two children through school, in the hopes they will obtain wage-earning jobs and contribute to the family.

2.1. The Ariaal and Rendille of Marsabit District

The Ariaal are a population of about 6,000 people who raise camels, cattle, goats, and sheep in both the lowlands and highlands of Marsabit District. Many Ariaal families descend from the larger Rendille people, a tightly integrated society of about 20,000 peo-ple, Cushitic-speakers distantly related to Somalis, and who subsist off camel, goats, and sheep production. But the Ariaal are also closely related to the Samburu, who are related to the Nilotic-speaking Maasai of southern Kenya and northern Tanzania. The Samburu are a population of 100,000 who subsist on cattle and small stock production in the highlands and plains to the west in Samburu District. The Samburu and Rendille have been mutually allied for generations (if not several centuries) against common enemies (Boran and Turkana) and who maintain their alliance through ties of intermarriage, intermigration, and the non-competitive economy of cattle (Samburu) and camel (Rendille) production

(Fratkin, 2004; Spencer, 1973). The Ariaal are a product of this union, and are bilingual in Samburu and Rendille (a Cushitic language of the Afro-Asiatic family), although Schlee (1989: 210) observes that Ariaal speak Rendille poorly.

Samburu and Rendille share similar cultural features including segmentary descent organization (where each community is made up of distinct and autonomous clan families) and the institution of named age-sets where whole sets of men collectively pass through the age grades of child, warrior, and elder. The Ariaal are affiliated with the Samburu clans and age-sets and are considered Samburu by the Rendille, yet because they also speak Rendille and keep camels as well as cattle the Samburu treat them as Rendille. The name 'Ariaal' is used by Rendille to distinguish those mixed groups of Samburu/Rendille who speak Samburu and who raise camels as well as cattle. (Ariaal call themselves Samburu (*Loikop*), distinguishing their clan identity from Rendille) (Fratkin, 2004: 45).

Most Ariaal live in the flat lowlands between the Ndoto Mountains and Marsabit Mountain, in large circular settlements indistinguishable from pastoral Rendille communities; other Ariaal live in the highlands and raise cattle and small stock, particularly on Marsabit Mountain. The community of Karare on the Marsabit road, about 17 km below Marsabit town, is a large Ariaal community, where families have lived for generations raising cattle, and more recently, engaging in maize agriculture.

Ariaal settlements are local descent groups belonging to the larger Samburu system, where communities are made up of relatives from the same clan. The Samburu descent group system divides the society into two halves (or moieties), the White Cattle and Black Cattle. Each moiety is made up of four clans (*l-marei*, the "ribs"), with the White Cattle clans of Lukumai, Lorokushu, Longieli; and Loimusi (the latter not represented in Ariaal), and the Black Cattle clans of Masala, Pisikishu (Turia in Ariaal), Nyaparai (LeSarge in Ariaal), and Lng'wesi (the latter clan is not found in Ariaal). The largest clans in Ariaal are Lorokushu, Lukumai, Longieli, Masala, and Turia. The main study site of Fratkin's

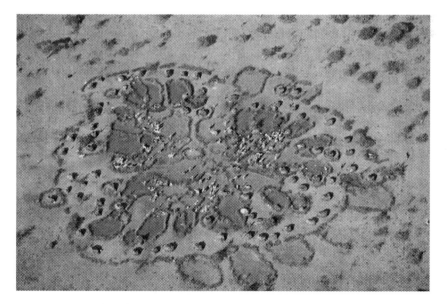

Photo 3. Pastoral Ariaal settlement of Lewogoso, with camels.

ethnographic research on Ariaal was in Lewogoso settlement, made up of patrilineal sub-clan of Lukumai section of Ariaal (and Samburu).

Ariaal share with the Samburu an age-grade organization with named age sets that cross cut clan and kinship ties, similar but independent from the Rendille age-set organization. (Boran and Gabra also categorize men by generation or age grade, but do not have named age-sets.) Among Ariaal, Samburu, and Rendille, adolescent boys are initiated (by circumcision) into a newly formed age set; they remain members of this age set for life, passing through the ladder of different age-grades together as warriors, junior elders, and senior elders. Like Rendille and Samburu, Ariaal age-sets are initiated every fourteen years, where boys who are three sets below that of their fathers (anywhere between the ages of ten and twenty-five are circumcised with other members of their clan. For the next fourteen years, the warriors are expected to herd animals in the distant camps and protect the settlements from armed attack. Two years before the initiation of the next warrior age set, these men are released from warriorhood and allowed to marry and start families of their own. Women are not formally initiated into age sets; they too pass through distinct life stages of young girls, adolescent girls, and married women (Fratkin, 2004). Women are circumcised (by clitoridectomy, also called excision) shortly before their weddings; the Ariaal see female circumcision as an essential ritual that prepares a woman for the pain of childbirth, and signals the legitimacy of the offspring (see Shell-Duncan et al., this volume). Widows may not remarry in Ariaal, nor can they usually divorce or return to their natal home, as families are reluctant to return the bride price, which is used to help brothers marry.

Ariaal, unlike the Rendille but similar to Samburu, practice polygyny. Lewogoso community between 1975–1995 averaged a 1.4 polygyny rate, where almost one half of the married men have more than one wife. A few wealthier men have three wives each. Both women and men informants in Ariaal value polygyny because of its contribution to the labor supply. Men state that multiple wives produce more children to herd animals, and women prefer having a co-wife with whom they can share household tasks.

While there are peculiarities to Ariaal society, many features are shared by other pastoralist groups in Marsabit District including the importance of patrilineal kinship, age grade organization, household autonomy, and patriarchy.

3. SEDENTARIZATION IN MARSABIT DISTRICT

Sedentarization is a recent phenomenon in Marsabit District. Unlike groups like the Orma who settled and participated in the market economy for most of the 20th century (Ensminger, 1992), northern Kenya has remained both isolated and undeveloped for much of the 20th century, This situation changed dramatically after a long series of droughts beginning in 1971, when both religious missions and international development agencies encouraged the settling of impoverished pastoralists in famine relief centers and agricultural projects. Today, about one quarter of the Rendille are settled, living in lowland towns of Korr, Kargi, and Laisamis or the highland agricultural scheme of Songa on Marsabit Mountain, A similar proportion of Ariaal are settled and live near the towns of Logologo, Karare, and Ngrunit.

3.1. Pre-Colonial History (to 1900)

Although ethnically distinct, Marsabit pastoral groups share intertwined histories of intermigration and assimilation as well as competition and warfare. Rendille, Somali,

Gabra, and Boran share distant origins in southern Ethiopia, and speak languages belonging to the Eastern Cushitic groups of Afro-Asiatic languages (Greenberg, 1955; Schlee, 1989). Boran, Gabra, and Sakuye belong to the Oromo sub-group of languages and cultures; the Rendille and Somali derive from a common 'proto-Rendille-Somali cluster' that moved into northern Kenya during the first millennium A.D. Rendille and Somali, who depended primarily on camel keeping, separated before 1500 as Somalis moved east into the Horn of Africa where they adopted Islam, while the Rendille remained in northern Kenya around Lake Turkana and did not adopt Islam (Schlee, 1989). The Samburu are part of a larger cattle-keeping Maasai migration who entered Kenya from Sudan sometime prior to 1600 A.D., with the Samburu and LChamus groups remaining in the northern Rift Valley and Maasai groups moving south into the grasslands below Mt. Kilimanjaro. A large Maasai group, the Laikipiak, occupied central Rift Valley near Lake Naivasha, but were defeated and dispersed by southern Maasai groups in the 1870s (Sobania, 1993; Spear and Waller, 1993).

Oromo speaking groups entered northeastern Kenya from southern Ethiopia during a major Borana expansion in the 16th century, differentiating into cattle-raising Boran and camel-keeping Gabra and Sakuye (Schlee, 1989). Initially the varying pastoral communities lived among each other with relatively little conflict, according to their oral traditions. This may be due to their small population sizes and ecological specializations, with the camel-keeping Rendille, Gabra and Sakuye living in the lowlands and cattle-keeping Boran, Ariaal and Samburu in the highlands. These communities were not isolated, however, and traded livestock for grain with agricultural Dasenech, Konso, and Burji, as well as intermarriage between camel keeping and cattle keeping communities (Sobania, 1991).

During the mid-19th century, many of these pastoral populations competed for grazing-lands. The Turkana expelled Samburu and Rendille from the northern plains west of Lake Turkana, while Boran fought Somalis in present day Wajir District and Samburu and Ariaal on Marsabit Mountain. To the south, rival Maasai tribes competed for the rich grazing lands around the Rift Valley Lakes Naivasha and Nakuru, culminating with the southern Maasai defeating and dispersing Laikipiak and Uasin Gishu Maasai from Kenya's central plains in the 1870s. The northern pastoral groups were only marginally involved in the Laikipiak wars, but Rendille and Ariaal fought against Laikipiak near the present town of Laisamis (Sobania, 1993; Spencer, 1973; Spear and Waller, 1993).

Following this period of warfare, pastoralists throughout East Africa faced tremendous hardships from drought, famine, and epidemics including bovine pleuro-pneumonia (1882) and rinderpest (1891) which decimated cattle herds, followed by smallpox which killed large number northern Kenyan pastoral herders in the 1890s. During this time, known in Maasai as *Emutai*, the Disaster, individual families moved in search of food, some becoming hunter-gatherers, some became livestock thieves (Waller, 1988). Simultaneously, Turkana from the west and Boran and Gabra from the north expanded into Rendille, Ariaal, and Samburu pastures (Sobania, 1988: 227; Spencer, 1973: 152–154).

3.2. Colonial Era (1900–1963)

The European powers divided up Africa following the Berlin Conference of 1884–85. In East Africa, the British had claimed by 1900 Kenya, Egypt-Sudan, Uganda, and British Somaliland; the French took Djibouti, and the Italians Italian Somalia and Eritrea, all choke points on the Red Sea and its access to the Suez Canal. The Ethiopians under Emperor Menelik II were also an imperial power, where the Shoa kingdom in the central highlands annexed the southern and eastern lowlands inhabited by Oromo and Somali populations (Marcus, 2002). The present day borders between Kenya, Somalia, and

Ethiopia were boundary lines drawn by these imperial powers, but their pastoralist populations were purposely ignored and left undeveloped.

The British occupied northern Kenya principally as a military buffer against Italy and Ethiopia, to protect their railroad line from Mombasa to Uganda (built 1895–1900) and the large highland tea estates and cattle ranches in central Kenya. The north was administered as a single entity, the Northern Frontier District (NFD), which encompassed an area as large as the rest of Kenya. But the NFD, which included present day Marsabit District, was closed to commerce or migration with the south, in part to reduce competition with European cattle ranchers in Kenya's Laikipiak District. Few roads, schools, or hospitals were built, and even Christian missions were not allowed in, for "the administration feared they might instill new desires in the local population which could not be satisfied later" (Schlee, 1989: 45). The British concentrated on preventing inter-pastoralist raiding; they established their first administrative post in Marsabit District at Marsabit Town in 1909, and in 1921 placed the entire NFD under military rule of the Kings African Rifles. They built several roads including one from Marsabit to Lake Turkana through North Horr, and a major road from Isiolo in the south to Moyale on the Ethiopian border in the 1930s. Police and administrative posts were established at Loyangalani (on Lake Turkana), Laisamis, Maikona, and North Horr in Marsabit District, and Archer's Post near the Ewaso Nyiru river near the boundary of Marsabit, Samburu, and Isiolo Districts (Falkenstein, 1996).

Boran and Gabra people had moved *en masse* from southern Ethiopia into northern Kenya at the beginning of the twentieth century to escape oppression and forced recruitment into Menelik's army. Gabra camel herders grazed their herds in the Chalbi Desert above the Rendille, Boran concentrated herded their cattle in the northern part of Marsabit District in the Hurri Hills and Marsabit Mountain, as well in Wajir District. The British established a "Galla-Somali" line separating the Boran, Gabra, Arjuran on one side and Somalis on the other (along present day Marsabit / Wajir District lines), although it was difficult to contain individual pastoralist households who were accustomed to moving to distant areas, particularly during drought (Schlee, 1989: 45–46). Earlier, in 1919, the British designated society-specific "Tribal Grazing Areas" which separated and restricted Samburu, Turkana, Gabra, Boran, Rendille, and Somali into their own defined areas, ultimately bounded as Turkana, Samburu, Marsabit, and Wajir and Mandera Districts. In 1932, the British forced the Boran to give up their wells in Wajir to avoid conflicts with the Somali, giving the Boran land as compensation (at the expense of local Samburu) in what is now Isiolo District along the Ewaso Nyiru River south of Marsabit District (Schlee, 1989: 47).

While many Boran of Isiolo District converted to Islam, the northern Kenyan Boran retained their traditional religion and *Gada* age grade organization as did other Boran in Ethiopia (Galaty, this volume; Hogg, 1986). In addition to external district boundaries, internal 'tribal' boundaries were also created to separate ethnic groups, including the Stigand Line in 1938, which separated Rendille from Gabra pasture areas. These bounded grazing areas were opposed by all pastoralist parties who needed to move to distant areas during drought, and it also defied the extensive network of social relations, both within and across ethnic boundaries. But older Rendille today acknowledge that the colonial boundary controls reduced the periodic raiding and killings over water and pasture. Nevertheless, pastoral mobility was greatly reduced by the restrictions, where the Rendille herding range was reduced from 57,600 km^2 to 8000 km^2 while their human population grew from about 8000 to 25,000 between 1960 and 1985 (Sobania, 1988).

Photo 4. Lewogoso camels near Baiyo mountain.

Although pastoralists found themselves confined to restricted areas during the colonial era, their traditional livestock economy was not disrupted. Furthermore, despite administrative restrictions, livestock marketing developed with the construction of new roads, towns, and markets. This was particularly facilitated during the 1930s by the construction of the north–south road from Moyale through Marsabit town to Isiolo, and the east–west road from Marsabit to Loyangalani through Maikona and North Horr. Merchants from down country developed shops selling imported goods including maize meal, tea and sugar, while Somali traders bought local Borana (Zebu) cattle which were desired for cross breeding by European ranchers. As early as 1923, Gabra and Rendille were selling livestock in Isiolo, and Boran were selling milk and meat to local traders and administrators on Marsabit Mountain.

During colonial rule, the British consciously discouraged pastoralists from settling, although Boran and Ariaal were allowed to graze cattle on Marsabit Mountain. The British encouraged Burji and Konso farmers from Ethiopia to settle on Marsabit Mountain in the 1930s to provide grains for road construction and local administrators. (See Adano and Witsenberg, Galaty, this volume). The majority of the districts pastoralists were left relatively undisturbed, as long as they paid their taxes, stayed in their designated grazing areas, and their warriors did not raid each other's livestock. Some warriors from these groups found employment in the colonial police and army, particularly during World War II and as home guards during the anti-colonial Mau Mau rebellion of the 1950s (Spencer, 1973: 163).

3.3. Shifta Period (1962–1969)

Following Kenyan independence in 1963, a six-year period of violence, livestock raiding, and civil war occurred in the north during the *shifta* ("bandit") conflict when

Somali and other Muslim populations attempted to secede from Kenya and join the Somali Republic. In 1962, the British held a series of meetings in the NFD which was attended by 40,000 Muslim Somalis, Boran, and Sakuye, who expressed their desire to join Somalia rather than remain in Kenya under a Christian and Bantu-speaking government (Kenya, 1962). However, in March 1963, without consulting the Somalia government as agreed, the British announced their decision to make the Northern Frontier District a seventh province in Kenya. Political opposition was swift. Somalis and their Muslim Boran and Sakuye allies boycotted the Kenyan national elections and called for secession. They began an armed insurrection which included the mining of roads and attacking government officials and missionaries in Marsabit, Wajir, and Garissa Districts; they also raided livestock from non-Muslim Boran, Rendille, Ariaal, Samburu, and Gabra pastoralists who had remained loyal to the Kenya government. The secession activists became known as *shifta* from the Amharic Ethiopian word for "bandit." The Kenya government responded by forcing Somali, Sakuye, and Waso Boran into enclosed "strategic villages" (or *daba*). Camel herds were shot as "supporting the enemy," and residents found a mile outside the villages were considered *shifta* and arrested or shot. Waso Boran concentrated around Isiolo town were particularly brutalized and left destitute as the government made large confiscations of their animals whenever a lorry was mined on the roads. (Hogg, 1986). The *shifta* period created a climate of physical insecurity, particularly for Rendille and Gabra who moved their manyattas closer to police posts and towns.

3.4. Drought, Famine, and Sedentarization 1970–2002

During the 1970s, the situation in Marsabit District irreversibly changed with the onset of protracted drought and the arrival of religious missions distributing famine relief foods. In short time, destitute pastoralists began to settle around the missions and migrate to newly created farm projects on Marsabit Mountain. Drought is recorded in Marsabit District for eight years between 1900–1970 (1919–22, 1928–29, 1934, 1945, 1949, 1960); the same number of drought years occurred in the following thirty years (1971, 1975–76, 1980, 1983–84, 1992, 1996 (O'Leary, 1990; Ministry of Planning and Development, 1997). Northern Kenya did not suffer the severe famines that occurred in Ethiopia during the 1970s and 1980s, although many animals were lost during drought periods. While the majority of the district's residents continue to live as livestock pastoralists, many have reduced their mobility and moved closer to lowland towns, in part to have access to famine relief foods, in part to gain access to social services, particularly health care and education, and in part to seek safety from increased inter-ethnic raiding, which has become more deadly due to the flow of small arms into the region from civil wars in Ethiopia, Somali, and Sudan.

Many Marsabit pastoralists began to settle in large numbers following the droughts of the 1970s, attracted initially to famine-relief centers in the lowlands and agricultural schemes on Marsabit Mountain established principally by religious organizations. The Marsabit Catholic Diocese in the lowland towns of North Horr and Maikona for the Gabra, and Korr and Kargi for the Rendille, or agricultural schemes on Marsabit Mountain developed principally by members of the African Inland Church. Foreign church missions were kept out of the NFD during much of the colonial era, but were invited in during the 1950s and 1960s as the colonial and newly independent governments began active development of pastoral regions. These included improvements in education, health care, water development, and veterinary measures, in part to draw pastoralists into the commercial economy

Photo 5.   Korr Catholic church, 1994.

(for a Maasai example, see Waller and Homewood, 1997: 74). In 1953, the Consolata Order of the Roman Catholic Church built a mission with a church, dispensary, and primary school in Marsabit town, and by 1968, had established a diocese with missions among Rendille and Ariaal at Laisamis and Archer's Post; among the Samburu at Baragoi, South Horr, and Maralal; among the Gabra at North Horr and Maikona; and among the Elmolo on Lake Turkana at Loyangalanai. Similarly, the protestant African Inland Church established missions (with churches, clinics, and primary schools) at Loglogo, Marsabit town, Ngrunit in the Ndoto Mountains, Illeret on Lake Turkana, and a hospital at Gatab on Mt. Kulal.

Following the drought of 1971, the Catholic church assumed responsibility in Marsabit District for distributing relief grains of corn, rice, and soybean flour, which were donated by international relief agencies including USAID, UNICEF, and CARE. Concentrating on the low-lying pastoral areas, the Catholic Church reached Gabra through Maikona and North Horr and Rendille through Laisamis. The Laisamis Catholic mission also began mobile food distribution to Rendille pastoralists at two wells in the Kaisut Desert, at Korr and Kargi, which soon attracted an estimated population of 6000 Rendille, primarily women, older men, and children too small to herd livestock in distant camps (Fratkin, 1991).

Both the Catholic and Protestant missions delivered humanitarian assistance, but their methods and impact differed. The Catholic Church, made up largely of expatriate clergy from Italy and India, encouraged the creation of permanent sedentary communities at Korr, Kargi, Laisamis, Maikona, North Horr, and Loyangalani to facilitate famine relief, education, and medical care, as well as religious conversion. Missionaries from the smaller Africa Inland Church (AIC), drawn predominately from the United States and Canada, were also interested in conversion, which they encouraged through church services, Sunday schools, and bible study. But the AIC also focused on water development, working with religious NGOs including World Vision and Food for the Hungry to mechanize wells, lay water pipes, and build water catchments at Arsim and Ngrunit in the Ndoto Mountains, Loglogo and Karare village on Marsabit Mountain, and in the Hurri Hills. While these projects resulted in some settling of poorer pastoralists, in the main they left the local pastoral communities intact, and indeed improved water access for their livestock.

In addition to development sponsored work through church missions, Rendille, Ariaal, and Gabra communities became targets of a large multilateral project, UNESCO's Integrated Project in Arid Lands (IPAL), which emerged following the Conference on Desertification in 1977 held in Nairobi by the newly created United Nations Environment Program's (UNEP) and UNESCO's Man and the Biosphere Programme. "Desertification" embraced the view that deserts were expanding in part due to pastoralists' mismanagement of rangelands. The IPAL project intended to combine basic research in land use with practical policies aimed at reducing environmental degradation. Its first proposed field station (out of several world wide) was Marsabit District, and IPAL chose Rendille, Ariaal, and Gabra as subjects to demonstrate human environment interaction (IPAL, 1984).

Because IPAL viewed pastoral practices as responsible for overgrazing in areas including Korr and Kargi towns (when in fact people settled there in search of famine foods), IPAL implemented projects aimed at reducing herd size by encouraging more livestock marketing as well as improving livestock production by building roads, water catchments, and improving veterinary care. Much of these developments was carried out in Ngrunit and Korr, and contributed to some settlement by pastoralists, but not to the extent of the mission towns. Despite IPAL's efforts, Rendille did not increase their livestock off take (marketing or butchering), and continued to sell animals only during dry seasons to purchase grains when milk supplies ran low (Fratkin, 1991; Little, 1994). By 1985, the IPAL project had disbanded, and most international development efforts (including GTZ had sponsored much of the IPAL work) concentrated on improving agricultural crop production on Marsabit Mountain rather than pastoralism as the key to improving the region's economy. This was due to the fact the large numbers of pastoralists from Rendille, Gabra, and Boran communities were migrating to Marsabit Mountain.

4. THE IMPACT OF SEDENTARIZATION ON SOCIAL LIFE

4.1. Highland Farming Communities

In 1973, the AIC and other groups encouraged impoverished Rendille pastoralists living around missions at Loglogo and Laisamis to settle on new agricultural schemes on Marsabit Mountain, in order to learn and practice maize and vegetable agriculture as an alternative to pastoralism. Initiated by AIC missionary Herbert Anderson, a joint effort was coordinated between the government, the National Christian Churches of Kenya (NCCK), African Inland Church (AIC), CARE, and the Catholic Mission Marsabit, to develop agricultural settlements at Naskikawe and Kitiruni (near the Ariaal community of Karare) and Songa (in the Marsabit Forest Reserve 17 km below Marsabit town). Immigrants were given small plots of land (2–5 hectares each) and building materials for houses, the use of communally owned oxen for clearing and plowing, and demonstrations in agricultural techniques. In 1976–77, agricultural communities for Boran were established at Manyatta Jillo, Sagante, and Badessa on the northern side of the mountain (see Adano and Witsenberg, and Smith, this volume). In time, new immigrants joined these communities from both lowland areas as well as Boran and Ariaal cattle keeping communities on Marsabit Mountain. Today, these agricultural communities have 1500–2500 residents each, many with their own primary schools and clinics. Although distinct (Nasikawke and Kitiruni are rain-fed, while Songa has drip irrigation), these farms resemble those of other Kenyan agriculturalists, living in permanent houses with fields growing with maize, kale (*sukuma wiki*), peppers, squash, and fruit trees. Some households also keep a few milk animals.

Life for these formerly pastoralist farmers has changed dramatically on Marsabit Mountain. Anthropologist Kevin Smith (1998) reports changes in male authority, particularly a decline in the collective authority of male elders, and woman's greater autonomy to grow and sell crops. Geographers Wario Roba Adano and Karen Witsenberg (this volume) report that residents of Marsabit's agricultural schemes are highly satisfied with life as farmers and few wished to return to their former pastoral lives. Farms are seen as advantageous because they provided steady food and were less risky than dependence on animals which can be lost to drought, diseases, and war. Furthermore, people respond that settled farm life gives opportunities for the poor that were previously not available, although they also note that farm work is far more onerous, and that they missed the milk and meat of pastoral life. Fujita et al. (Chapter 11, this volume), note, however, that women in Songa farming community showed greater malnutrition than their pastoral counterparts, a fact they attribute to both declines in milk and greater work expenditures of farming versus pastoral women.

The success has attracted a continuing stream of Rendille and Boran immigrants, threatening a water and fuel wood supply of finite quantity. Conflicts over water and

Photo 6. Woman of Songa.

farmland have intensified between the Boran and Rendille at Songa, where since 1992, a dozen people, mainly women and children, have been killed in ambushes. While elders from both communities, as well as government officials and NGOs have worked at mediating between the two groups, competition and conflict remains a continuing threat on Marsabit Mountain.

4.2. Lowland Towns

Where the agricultural settlements on Marsabit Mountain have attracted destitute pastoralists, many lowland pastoralists who still keep livestock, particularly Rendille, have moved closer to towns and roads for security and social services including health care, education, and periodic famine relief. Similarly Gabra have moved towards Maikona and North Horr, Ariaal to Laisamis and Loglogo, and a large number of Rendille have settled near the mission towns of Korr and Kargi in the Kaisut Desert. An estimated 6,000 Rendille live within 20 km of these two towns, and although they still keep significant herds of camels, small stock, and some cattle, these cannot be maintained in the arid areas of Korr and Kargi, and are herded for most of the year in distant camps managed by young men. Consequently, the diet of the sedentary communities, made up of married women, elderly men, and small children, has changed from predominately milk maize meal (*posho*) as the staple food.

Towns including Korr, Kargi, Laisamis, Maikona, and North Horr are located in wind-swept lowland desert areas, and do not have enough vegetation to support animal herds. While these towns do have churches, schools, and dispensaries, there are few jobs available. Most shops are run by Somalis, Ethiopians, or other foreigners to the District. Alcoholism is a growing problem for both men and women, and where illegal beer brewing is associated with prostitution and the increasing risk of HIV/AIDS, according to interviews with the Marsabit District health officer.

The effects of the church sponsored famine-relief projects and the towns of Korr and Kargi have been large. The bilateral donor organization GTZ noted in a 1994 workshop that the pastoralists had become too used to charity and were too ready to accept aid without seeking new forms of income. While there may be some truth to this, Boran, Gabra, Rendille, Ariaal, and Somali find themselves in an increasingly restricted herding environment. Furthermore, pastoralists have sought new ways to continue their pastoral economy, including digging new water wells near the towns so they can keep animals in residence, at least during wet season when there is sufficient grazing for their livestock.

About one quarter of Rendille, Ariaal, Boran, and, to a lesser degree Gabra, have now permanently settled. Some have found security as farmers, or wage employees working in shops, government services, or various non-government organizations. Others who are less fortunate search for odd jobs including cleaning, selling charcoal, or herding animals to make ends meet. Many town residents and farmers have converted to Christianity; most continue to adhere to their traditional customs of age-grades, marriage rules, and ritual life. Smith (this volume) describes a diminishment in the importance of male age-sets in the farming communities, where young men are marrying earlier and where male elders have lost former authority over the warrior age-sets. Shell-Duncan et al. also describe changing patterns in female circumcision, where although the custom continues, it is becoming medicalized (despite prohibitions by the Kenyan government against female circumcision). Importantly, opportunities for women are increasing with settled life. Women have increased their participation in the market economy, were farming women sell produce and pastoral women are selling milk in town markets, particularly in Marsabit. Poor widows, who would have had to rely on the charity of their relatives in the pastoral community, have

Photo 7. Korr town in Marsabit Lowlands.

found some new outlets working for shopkeepers or making and selling charcoal for town dwellers. Some women, however, have turned to beer-brewing or the selling of *mira'a* (or *khat* (*Catha edulis*, a widely consumed stimulant) to earn income, and some have become prostitutes, particularly in the towns.

Despite these major changes, many pastoralists are not opposed to these new influences, but see the towns as one more resource to utilize, an essential alternative for poor households who have few animals, or an important center to gain employment, sell livestock, seek health care, and obtain education for their children.

Patrick Ngoley, a resident of Korr town, remarked,

> "Before we were nomads, we would move with our animals. Now we stay in one place, at Korr or at Kargi. This is *mandeleo* (development). Here we have shops, schools, hospitals. But we still keep our animals in *fora*. When we need money for food we tell our warriors to sell stock at Laisamis, Isiolo, or Merille. They sell them and send us home money. But now everybody is paying for things they used to get for free—meat, transportation, *posho* (meal), milk. People must spend money until rain comes (and animals with milk can return). People must look for work in town making buildings, cleaning houses, even digging urinals. If there is no work, people must sell an animal. If a person is too poor and has no animals or money, he must beg from others." [Fratkin, 2004: 126–127]

5. COMMUNITIES IN THIS STUDY

5.1 Lewogoso Lukumai—A Nomadic Pastoral Community

Lewogoso is a mobile camel-, cattle-, and small-stock-keeping Ariaal settlement of approximately 250 people practicing mixed-species husbandry. The live in several large circular settlements, located near one another in the area along the Milgis River,

Photo 8. Lewogoso settlement, 1996.

between the towns of Laisamis and Ngrunit. Lewogoso includes male stockowners and their families, predominately members of the same kinship group—Lewogoso ('the long necks') clan of the larger Lukumai section (one of eight Samburu sections, and members of the White Cattle moiety)." This community was originally studied by Fratkin in 1974–76, by 1985 it had segmented into four settlements, some concentrating on camel ownership, and others more on cattle and small stock (Fratkin, 1991, 2004). While some members of Lewogoso have left the pastoral community, the majority have remained. Lewogoso forms a control group to which many of the comparisons with the sedentary towns and communities in this study were made (including Fratkin, Roth, Nathan, Shell-Duncan, Fujita).

5.2 Korr—A Lowland Mission Town

Korr is a town in the arid lowlands of the Kaisut Desert below Marsabit Mountain, near the Rendille water holes at Halisuruwa. Korr was developed initially by the Marsabit Catholic Diocese to feed destitute Rendille during the famine of the 1970s, today Korr has a sedentary population of about 2500 people, with semi-nomadic Rendille settlements of perhaps 2000 people living nearby. Korr has poor marketing facilities, although the town has a small stock and meat market. The town has a primary school, medical clinic (maintained by the Catholic Mission) and large Catholic church, and is home to several NGOs, including a Protestant adult education school (Roth, 1991, 1996).

5.3 Karare—A Highland Agro-Pastoral Community

Karare is a settled highland community on Marsabit Mountain 17 km from Marsabit town, the district capital. Its 2000 residents are primarily Ariaal (Samburu/Rendille mix) who both keep cattle herds and raise dryland maize. Karare has access to good marketing

Photo 9.   Karare town on Marsabit Mountain.

facilities as well as a large urban population in Marsabit town and is located on the major truck road from Nairobi to Addis Ababa. Karare women sell milk on a regular basis to Marsabit townspeople (Fratkin and Smith, 1995; Roth, 1996). Karare has existed as an Ariaal settlement since at least 1897, when members of the Leruk family greeted Lord Delamere on his journey south from Ethiopia. It remained a cattle keeping settlement until the 1970s, when a coalition of churches the National Council of Churches of Kenya (NCCK), the African Inland Church (AIC) and the Catholic Diocese of Marsabit developed a dryland agricultural scheme near Karare called Nasikakwe. Today, about one half of Karare's residents live in Nasikakwe as farmers, and the other half keep livestock.

5.4 Songa—An Agricultural Community on Mt. Marsabit

Songa is a sedentary highland agricultural community on Mt. Marsabit of 2500 people, founded by American Protestant missionaries from the African Inland Church in 1973 in a forest on Marsabit Mountain for destitute Rendille. Songa is located in the thick forest region south of Marsabit Town; many of its farms have drip-irrigation agriculture, and raise a variety of crops including maize, kale, fruit trees, tobacco, tomatoes, and peppers. Songa's population grows vegetables for sale in Marsabit town (Smith, this volume).

5.5 Ngrunit—A Settled Pastoral Community

Ngrunit is a sedentary agro-pastoral community of approximately 1200 people located in a forested valley in the Ndoto Mountains made up of Rendille, Ariaal, Samburu, and Dorobo peoples. This community has a church, school, and small dispensary but is isolated and not well integrated into marketing activities. Its inhabitants raise vegetables from their gardens and market livestock.

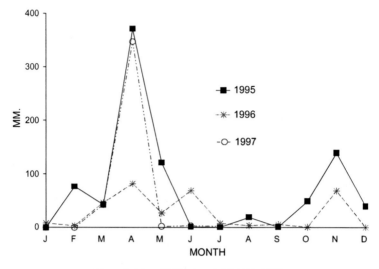

Figure 2. Marsabit District rainfall 1994–1997.

5.6 Other Communities in This Study

John McPeak and Peter Little's chapter compares several communities in their multi-ethnic comparison of Samburu, LChamus, Ariaal, Rendille, Gabra, and Boran. **Dirib Gumbo** is a Boran settlement approximately 10 kilometers from Marsabit town; **Ngambo** is an Il Chamus settlement approximately 10 kilometers east of Marigat town. Marigat town is located 100 kilometers north of Nakuru, and is the major market center used by Ngambo residents. **Sugata Marmar** is a Samburu settlement on the Laikipia—Samburu District border, approximately 50 kilometers south of Maralal on the Maralal—Rumuruti road. Significant populations of impoverished Turkana and Pokot are resident in this location as well. Sugata Marmar has a large weekly livestock market offering households the opportunity for alternative income sources and a place to sell animals. **Logologo** is an Ariaal settlement approximately 40 kilometers south of Marsabit town on the main Isiolo—Marsabit road. Logologo residents utilize markets in both Marsabit town and in Logologo town. Rain-fed agriculture is possible in the higher areas of this location, and a very small amount of small-scale irrigation is practiced in town. Most households in Logologo settled there in the 1970s following a series of poor rainfall years and herd losses. **Kargi** is a Rendille settlement approximately 75 kilometers to the west of Marsabit town in a flat, arid basin. Kargi residents mostly conduct market activity in Kargi town, although they make occasional use of Marsabit markets. Rendille in the Kargi area keep small herds in the area around town and rely on young men to stay with the remainder of the herd in highly mobile satellite camps. **North Horr** is a Gabra settlement approximately 200 kilometers west of Marsabit town on the northern edge of the Chalbi desert. Similar to Kargi, most market activity takes place in North Horr town, although residents do make occasional marketing trips to Marsabit town.

Adano and Witsenburg's chapter, as well as Smith (this volume) discuss **Marsabit Town**, which is the district capital. Marsabit's population grew from 4000 in 1974 to over 30,000 by 2002. It principle residents include Burji farmers, Boran, Samburu (Ariaal), and Somali, with some Indian and Ethiopian, and Kikuyu shop keepers. It is also the location of the administration, employing several hundred people, the majority from "down country" including Kikuyu, Luo, Kalenjin, Kamba, and other groups.

6. CONCLUSION

Marsabit District is the setting for the majority of research discussed in this volume. Before 1970, the district was made up of almost exclusively nomadic livestock herding populations, about one quarter of whom settled near the growing towns and famine-relief centers created in large part by missionary activity. Both urban towns and agricultural communities have grown in the past half-century, and contribute to a dynamic interaction between herders, farmers, traders, and townspeople. The volume As Pastoralists Settle proceeds with focused descriptions and analysis of the social, health, and economic processes that occur with pastoral sedentarization.

NOTES

1. Population figures are based on 1989 Kenyan Census; figures from 1999 census have not been made available. See Table 1. Descriptions of these pastoral societies can be found for Ariaal (Falkenstein, 1995; Fratkin, 1991, 2004; Spencer, 1973), Boran (Baxter, 1978, 1979; Dahl, 1979; Hjort, 1979), Gabra (Tablino, 1999; Wood, 1999), Rendille (Sato, 1997; Schlee, 1989; Spencer, 1973), Samburu (Spencer, 1965, 1973), Laikipiak (Sobania, 1993,), Somali (Cassanelli, 1982; Lewis, 1961, 1994; Little 2003; Merryman 1987), and Turkana (Gulliver 1955; Lamphear 1992). For comprehensive histories of pastoral interaction in northern Kenya, see Schlee (1989) and Sobania (1988, 1991, 1993).

REFERENCES

Baxter, P.T.W., 1978, Boran age-sets and generation-sets: Gada, a puzzle or a maze? In *Age, Generation and Time*, edited by P.T.W. Baxter and U. Almagor, pp. 151–181. New York: St. Martin's Press.

Baxter, P.T.W., 1979, Boran age-sets and warfare. In *Warfare Among East African Herders*, edited by K. Fukui and D. Turton. Senri Ethnological Studies 3, pp. 69–95. Osaka: National Museum of Ethnology.

Cassanelli, L.V. 1982, The Shaping of Somali Society: Reconstructing the History of a Pastoral People. Philadelphia: University of Pennsylvania Press.

Dahl, G., 1979, *Suffering Grass: Subsistence and Society of Waso Borana*. Stockholm Studies in Social Anthropology. Stockholm: University of Stockholm.

Ensminger, J., 1992, *Making a Market: The Institutional Transformation of an African Society*. New York: Cambridge University Press.

Falkenstein, M., 1995, Concepts of Ethnicity and Inter-Ethnic Migration among the Ariaal of Kenya. *Zeitschrift für Ethnologie* 120 (2): 201–225.

Fratkin, E., 1991, *Surviving Drought and Development: Ariaal Pastoralists of Northern Kenya*. Boulder: Westview Press.

Fratkin, E., 2004, *Ariaal Pastoralists of Northern Kenya: Studying Pastoralism, Drought, and Development in Africa's Arid Lands*, Second edition. Needham Heights MA: Allyn and Bacon.

Fratkin, E., E.A. Roth, and K.A. Galvin, 1994, *African Pastoralist Systems: An Integrated Approach*. Boulder; Lynne Rienner Publishers.

Fratkin, E. and K. Smith, 1995, Women's Changing Economic Roles with Pastoral Sedentarization: Varying Strategies in Four Rendille Communities. *Human Ecology* 23 (4): 433–454

Greenberg, J.H., 1955, *Studies in African Linguistic Classification*. New Haven: Compass Press.

Gulliver, P.H. 1955, *The Family Herds*. London: Routledge and Kegan Paul Ltd.

Hjort, A., 1979, *Savanna Town: Rural Ties and Urban Opportunities in Northern Kenya*. Stockholm Studies in Social Anthropology Vol. 7. Stockholm: University of Stockholm.

Hogg, R., 1986, The New Pastoralism: Poverty and Dependency in Northern Kenya. *Africa* 56 (3): 319–333.

IPAL, 1984, *Integrated Resource Assessment and Management Plan for Western Marsabit District, Northern Kenya*. Integrated Project in Arid Lands Technical Report No. A-6. Nairobi: UNESCO.

Kenya, 1962, *Report of the Northern Frontier District Commission*. London: Her Majesty's Stationary Office.

Kenya, Republic of, 1989. National Census, Central Bureau of Statistics, Nairobi: Government Printing Office.

Lamphear, J., 1992, *The Scattering Time: Turkana Responses to Colonial Rule*. Oxford: Clarendon Press.

Lewis, I.M., 1961, *A Pastoral Democracy, A Study of Pastoralism and Politics Among the Northern Somali of the Horn of Africa*. London: Oxford University Press.

Lewis, I.M., 1994, *Blood and Bone: The Call of Kinship in Somali*. Society. Lawrenceville NJ: The Red Sea Press.

Little, P.D., 1994, The social context of land degradation ("Desertification") in dry regions. In *Population and Environment: Rethinking the Debate*, edited by L. Arizpe, M. P. Stone, and D. C. Major, pp. 209–251. Boulder: Westview Press.

Little, P.D., 2003, *Somalia: Economy Without State*. Bloomington: Indiana University Press.

Marcus, H.G., 2002, A History of Ethiopia, Updated edition. Berkeley: University of California Press.

Ministry of Planning and National Development, Kenya, 1997. *Marsabit District Development Plan, 1997–2001*. Nairobi: Republic Of Kenya.

Merryman, J.L., 1987, The Economic Impact of War and Drought on the Kenya Somali. *Research in Economic Anthropology* 8: 249–275.

O'Leary, M.F., 1990, Drought and change amongst Northern Kenya nomadic pastoralists: The case of Rendille and Gabra. In *From Water to World-Making: African Models in Arid Lands*, edited by G. Palsson, pp. 151–174. Uppsala: The Scandinavian Institute of African Studies.

Roth, E.A., 1991, Education, Tradition, and Household Labor among Rendille Pastoralists of N. Kenya. *Human Organization* 50: 136–141.

Roth, E.A., 1996, Traditional Pastoral Strategies in a Modern World: An Example from Northern Kenya. *Human Organization* 55: 219–224.

Sato, Shun, 1997, How the East African Pastoral Nomads, Especially the Rendille, Respond to the Encroaching Market Economy. *African Studies Monographs,* Vol. 18, No. 3–4, pp. 121–135. The Center for African Area Studies, Kyoto University.

Schlee, G., 1989, *Identities on the Move*. Manchester University Press.

Schwartz, H.J., S. Shaman, and D. Walter, 1991, *Range Management Handbook of Kenya, Volume II (1): Marsabit District*. Nairobi: Republic of Kenya, Ministry of Livestock Development.

Sellen D.W., 1996, Nutritional Status of Sub-Saharan African Pastoralists: A Review of the Literature. *Nomadic Peoples* 39: 107–134

Sobania, N.W., 1988, Pastoralist migration and colonial policy: A case study from northern Kenya. In *The Ecology of Survival: Case Studies from Northeastern African History*, edited by D. Johnson and D. Anderson, pp. 219–239. London: Crook Greene, and Westview Press.

Sobania, N.W., 1991, Feasts, famines and friends: nineteenth century exchange and ethnicity in the eastern lake Turkana Region. In *Herders, Warriors and Traders*, edited by J.G. Galaty and P. Bonte, pp. 118–142. Boulder Colorado: Westview Press.

Sobania, N.W., 1993, Defeat and Dispersal: The Laikipiak and their neighbors in the 19th century. In *Being Maasai: Ethnicity and Identity in East Africa*, edited by T. Spear and R. Waller, pp.105–119. London: James Currey.

Smith, K., 1998, Farming, Marketing and Changes in the Authority of Elders among Pastoral Rendille and Ariaal. *Journal of Cross Cultural Gerontology* 15: 1–24.

Spear, T. and R., Waller, 1993, *Being Maasai: Ethnicity and Identity in East Africa*. London: James Currey.

Spencer, P., 1965, *The Samburu: A Study of Gerontocracy in a Nomadic Tribe*. Berkeley: University of California Press.

Spencer, P., 1973, *Nomads in Alliance*. London: Oxford University Press.

Tablino, P., 1999, *The Gabra: Camel Nomads of Northern Kenya*. Nairobi: Pauline's Publications Africa.

Waller, R., 1988, Emutai: Crisis and Response in Maasailand 1883–1902. *The Ecology of Survival*, edited by D. Johnson and D. Anderson, pp. 73–112. London: Lester Crook Academic Publishing.

Waller, R. and K. Homewood, 1997, Elders and Experts: Contesting Veterinary Knowledge in a Pastoral Community. In *Western Medicine as Contested Knowledge*, edited by A. Cunningham and B. Andrews, pp. 69–93. Manchester: Manchester University Press.

Wood, J.C., 1999, *When Men Are Women: Manhood Among the Gabra Nomads of East Africa*. Madison WI: University of Wisconsin Press.

Zwannenberg, R.M.A. van, 1977, *An Economic History of Kenya and Uganda 1800–1970*, first paperback edition. London: MacMillan Press Ltd.

Chapter 3

Time, Terror, and Pastoral Inertia

Sedentarization and Conflict in Northern Kenya

JOHN G. GALATY

1. INTRODUCTION

Until the Twentieth Century residents east of Lake Turkana in what is now known as Marsabit District were primarily nomadic pastoralists, with several communities along the lake shore also pursuing fishing or rain-fed and flood-plain cultivation. Pastoralism still predominates but during the last century sedentarization has increased, with some herders settling on the slopes of Marsabit Mountain or around trading centers established near wells in the district. The question of whether sedentarization is a desirable or regrettable option for pastoralists is linked to a prior query, of whether intensive pastoralism and semi-nomadism offers its practitioners a rewarding or a marginal existence. Pondering policy for the arid lands, administrators and development specialists ask whether planned settlements can offer improvement in the quality of herders' lives, or whether retreat from a livestock-based economy leads to decline in household resilience, the quality of nutrition and health, and self-reliance, often due to the dependence of sedentarists on external relief.

Six hundred years ago, Ibn Khaldun described the paradox of nomadic life, its superior character, morality, fortitude, and social cohesiveness undermined by the inexorable change from nomadic to sedentary life, not the opposite. It was Bedouin who sought the 'luxuries' of urban life, not sedentarists who sought the spiritually uplifting but demanding life of pastoralism. In his study of the nomads of South Persia, Barth (1961, 1986)

JOHN G. GALATY • Department of Anthropology, McGill University, Montreal, PQ, Canada.

explained that herders tended to settle both from the top and the bottom of pastoral society, responding either to conditions of prosperity or the realities of impoverishment. Although insightful in depicting many pastoral experiences, this account is somewhat circular, for impoverished pastoralists must settle or die and rich pastoralists must invest outside the herding economy or see their wealth dispersed (Waller, 1999: 24). But decisions to sedentarize are rarely taken just once (and for all), and many never definitively choose between nomadic or settle alternatives. The Somali experience of restocking while in famine and refugee camps to make possible the resumption of pastoral life is a cautionary tale against reading the nomadic experience of settlement as historically unidirectional (Hogg, 1985). The settled, maintaining ties to the livestock economy, are often not ex-pastoralists but pastoralists-by-other-means. Over the last century, pastoral regions have seen the growth of trading centers, with shops, clinics, churches or mosques, and schools, the creation of regional markets for grain and livestock, the establishment of missions, and the institutionalization of food relief provided by government and international donors. No longer does the critical triangle of human-livestock-environmental relations unilaterally determine patterns of mobility and husbandry practiced in a pastoral community: herders move not just to access pastures but schools, not just water but shops, or development projects or clinics. What is now at stake is not nomadism in a nomadic world but continuing pastoralism in a world of sedentary sites and institutions.

Sedentarization in Marsabit District must be seen in historical context, notably the establishment of administrative structures of governance, including international frontiers and internal boundaries for grazing, and the incidence of conflict and violence to which settlement and movement patterns accommodate themselves. The administrative demarcation over time of spatial limits on pastoral land use and the terrorization of local communities by periodic insecurity, stimulated by local and transnational political processes, has increased pastoral inertia, slowing the centrifugal use of grazing territory by herders and resulting in increased stasis of herds and households and in demonstrable loss of the economic, environmental, and social benefits of efficient resource use. Development specialists often see spontaneous settlement as signaling the non-viability of nomadic pastoralism or a positive embrace of the 'modern'. But such an analysis translates local responses into the technical terms and universal time of cultural ecology and development planning, while settlement often represents a contingent response in historical time to reversible conditions of the environment, regional politics, international relations, and insecurity. In what follows, this paper first examines the roots of sedentarization in processes of colonial boundary-making and administration, then reviews recent instances of conflict within Marsabit District that have affected settlement by herding communities, and concludes by comparing different settlement patterns that have emerged out of conditions of insecurity and diminished range use.

2. THE COLONIAL BACKDROP TO SEDENTARIZATION: MIGRATION, CONFLICT, BOUNDARIES

2.1. Cultural Tectonics and Ethnic Alliances

During the colonial period, the British sought to contain the Samburu within their own district, ignoring the deep ties they latter enjoyed with Rendille and Ariaal and their historical links to the Marsabit region. Enforcing the boundary between Marsabit and Samburu districts proved difficult, however, not least because of Ariaal bilingualism (in the Rendille and Samburu languages), which rendered them indistinguishable from Samburu for many administrators. A central theme of Marsabit history would be continuing conflict between Rendille and Gabra to the west of Marsabit mountain (despite close ties between them), and between Ariaal-Rendille and Borana on Marsabit mountain. But these axes of antagonism occurred within a longer history of alliance between the Oromo-speaking Gabra and Borana, and between Rendille (including Ariaal) and Samburu. Once the Kenya/Ethiopia border was in place, British policy was ambivalent towards frontier-groups. The Gabra and Borana were jurally sub-divided into Kenyans and Ethiopians but Dassanach were seen definitively as Ethiopians, although their homeland straddled the Omo Delta (in Ethiopia) and the north-eastern shore of Lake Turkana (in Kenya). As a result, much effort was expended in trying to prevent them (futilely) from inhabiting Kenyan territory. Recurrent motifs in the administrative history of Marsabit were the incursion of Dassanach into Kenyan territory, to raid (often in the company of generic 'Ethiopians') or to reside along the eastern shores of the lake for grazing, and conflict between Dassanach (sometimes with Turkana and Hamar Koke) and both Gabra (often together with their Borana allies) and Rendille (together with their Ariaal and Samburu allies) over grazing in the northwest of the district (Salvadori, 2000).[1]

Several points follow from the history of British/Abyssinian border-making in northern Marsabit. British efforts at securing the frontier by moving 'their' Gabra and Borana away from the border zone would reconfigure the pattern of land occupation in the district, Gabra primarily occupying the north and northwest, Borana the north and northeast of Marsabit mountain. Southward movement of Gabra and Borana was at the expense of Samburu, Rendille and Ariaal, who in a ripple effect were pushed farther southward. The British policy of excluding the Dassanach from the colony effectively destabilized the northwest corner of the district, sometimes opened up for use by Gabra and Rendille, other times closed to them due to the insecurity created by incursions by the Dassanach, acting on their historical rights to lake shore grazing. The frontier separating 'British' (and later Kenyan) from 'Ethiopian' subjects would engender ethnic fragmentation and establish a set of complex, transborder relations that would come into play during conflicts across the frontier, affecting patterns of sedentarization.

Schlee (1989: 32–33) describes the existence of a proto-Rendille-Somali culture (PRS) across northern Kenya some 500 years ago, which was split into Rendille and Somali wings when the Oromo expanded southward in the Sixteenth Century.[2] Several Somaloid-speaking groups fell under Borana influence, adopting the Oromo languages or becoming bilingual in Somali and Borana/Oromo: camel-keeping Gabra adopted Oromo, while Ajuran, Sakuye, and Garre evolved high degrees of bilingualism to the east of Marsabit. One element in the continuing cultural autonomy of the Rendille, who retained a Somaloid language and culture despite the regional prominence of the Borana, was their establishment of close social ties with Maa-speaking Samburu (and earlier most likely with the Maa-speaking Laikipiak) to their south. Some Rendille intermarried with Samburu, evolving into the distinctive mixed group of Ariaal. Notwithstanding friction between Gabra and Rendille, the two have experienced centuries of close interaction due to their common attachment to the camel culture, with several streams of Rendille becoming assimilated into Gabra clan structure (Schlee, 1989: 158).

The cultural 'tectonics' described above can be summarized as follows. The proto-Rendille-Somali stratum occupying the northern Kenyan/southern Ethiopian region was physically split by the expanding Oromo and culturally overlain by Oromo language and culture. Where the Borana and Somali influences met, bilingual groups evolved, the Somali cultural plate sliding under the Oromo. Similarly, when the expanding Maa-speaking peoples (represented by Laikipiak and Samburu) encountered the Rendille, the interstitial bilingual, bicultural Ariaal evolved, representing a Rendille overlaid by a Samburu cultural stratum. In mid nineteenth century, northern Kenya was the site of conflict both between Laikipiak and Borana and between Laikipiak and Rendille, with Rendille and Samburu assuming regional prominence after the defeat of the Laikipiak (Sobania, 1993). Thus early colonial migration of communities caught between two Imperial forces (whose negotiations defined the Kenyan/Ethiopian border) took place in front of a backdrop of previous historical interactions between the same actors. The border partially inhibited and partially

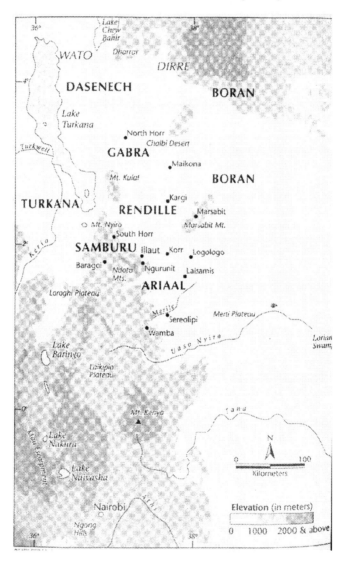

Figure 1. Pastoralist groups of northern Kenya.

facilitated the movement of the Gabra and Borana into Marsabit, which in the early colonial period represented a resurgence of the historical Oromo expansion southward. The Rendille-Samburu alliance retreated, first from Wato in southern Ethiopia (east of the Dassanach), then from the Chalbi area northwest of Marsabit, where they left behind an array of place names, as noted in Figure 1.

When two societies, languages, or cultures meet, it is not always obvious who will assimilate whom, who will prove the assimilator, who the assimilee. If the northern Samburu are undergoing 'cushitization' due to their intimate contact with the Rendille, their language and practice of cattle-keeping is increasingly being adopted by Ariaal Rendille undergoing 'nilotization'. Similarly, Gabra are assimilating Rendille, but are also becoming 'Rendille-ized'. Nonetheless, two cultural plates continue to slide over the core Rendille communities of Marsabit, one being Oromo, the other Samburu. The progressive sedentarization of Marsabit pastoralists, especially Ariaal Rendille, should be interpreted in light of the colonial demarcation of the Kenya/Ethiopia border and the establishment of district boundaries in Kenya, through which the colonial government sought to separate pastoral groups by confining them to their own locations. The British sought to define districts according to the distribution of pastoral groups, but in the end unwittingly defined groups according to districts, a process that Marsabit clearly illustrates. While the establishment of the northern border of Kenya bears most directly on the Gabra, Borana, and Dassanach, divided between Kenya and Ethiopia, it exercised a domino affect on the Rendille, Ariaal, and Samburu, who were progressively pushed southward to accommodate the latter groups in the northern part of the district. Two frontiers of colonial boundary-making have influenced the pattern of conflict and settlement in Marsabit, involving Oromo-speakers and Dassanach.

2.2. Oromo-Speakers along the Northern Border

The border between the Abyssinian Empire and British East Africa was the last to be demarcated in Eastern Africa, and its gradual establishment influenced the distribution of groups within the borderlands of Marsabit District. Having an undefined border to their south allowed "trading and raiding" to take place by the Abyssinians who governed the empire's southern marches. In 1903, the Commissioner of the East African Protectorate urged that a frontier be marked since the Abyssinians were "flowing southward" in an "aggressive advance" (Barber, 1968: 46–49). An approximate border agreement was reached in 1907, and British posts were established in Moyale and Marsabit in 1909 and on the eastern shores of Lake Turkana (known as Lake Rudolph until Independence) in 1911 (matched by Ethiopian posts), but the effect of these steps was to initiate uncertainty as to which country pastoralists who had long used both sides of the border 'really' belonged to and to stimulate transborder flight, depending on the policies being pursued on one or the other side. In 1914–15, when the Abyssinian empire sought to assign Oromo groups to various Imperial soldiers for involuntary service, a majority of the Gabra and many Borana fled to the British side, who, to protect them from further attack from the Abyssinian side, accepted them as 'refugees' and moved them away from the frontier into the northern grazing zones they still occupy (Sobania, 1979: 97–99).

The Gabra were ordered to "keep away from the frontier" and told to move south to water-holes between Marsabit and North Horr, such as at Maikona, space provided by the evacuation of Samburu and Rendille from the west and northwest of the district (Sobania, 1979: 78). At the end of 1918, Borana from Ethiopia began to cross the frontier in large numbers to escape persecution by "Tigre brigands". When they entered Kenya for the same reason in 1920, they were moved southward in order to "keep the Frontier clear of natives", an action preceding a joint "Anglo-Ethiopian effort against the Tigre". At this point, district records clearly distinguish between the "British Boran" and the "Abyssinian Boran", and in this way an 'ethnic' reality was created (Sobania, 1979: 98–99).

In 1922, it was stated that Britain would not "force" the Gabra to return to Ethiopia, but would allow them to do so "of their own accord", on the understanding that "if the Gabra return to Ethiopia the British will have no part in checking their movements" (Sobania, 1979: 81). Annual records note that the "general migration of Borana and Gabra to British territory did not change in the past year" (at 4–500 people and 4–5000 cattle crossing the border), but proposed that "the government stop the migration of Borana with stock but let in those without stock if they have been ill-treated and are not criminals, i.e., they are genuine refugees" (Sobania, 1979: 82):

> This migration abates not at all and is as relentless as the oncoming tide . . . Marsabit (District) is undoubtedly overstocked and still people filter in . . . The Gabra come like locusts and devour all before them. They eat up the game and gently push the rightful occupants out . . . It is the same with game as with the forest. It is the same with Rendille and Samburu grazing (Sobania, 1979: 84).

Defensive on defense of 'their' people and their resources, in 1925, the British anxiously reported that "The British Boran are watering in Ethiopia".

The Ethiopians resented the escape of some Borana to Kenya and the presumption that they could return at will, holding that "any refugee Boran who had moved to British territory since 1912 would be arrested on his return to Ethiopia and his stock confiscated". For both parties, continuing transboundary movements seemed to reopen the boundary question, so the British decided that all Borana who had crossed the border since 1912 would be "sent to Marsabit *as a temporary measure*" (Sobania, 1979: 101: my italics). They are, of course, still living on Marsabit Mountain, fighting for land with Ariaal and Rendille.

The southward flow continued. In 1933, British observed that "it would be to our advantage if (the Boran) were all to go into Ethiopian territory" and, in 1934, that attempts "to return Boran . . . to the Ethiopian authorities" were met with passive resistance and continuing "gradual infiltration (was) impossible to stop". Borana and Gabra were seen to "remain in a fluid state with their kin over the border" in 1935. On the Ethiopian side of the frontier, they were registered and taxed, leading many to "plead atrocities by Ethiopian tax collectors" as justification for reverse "infiltration" back to Kenya (Sobania, 1979: 122).

The establishment of a border between Kenya and Ethiopia initiated the emergence of distinct categories of Gabra and Borana, identified with one or the other country. Where each 'belonged' became an issue of policy and of policy reversal. In the end, the border served from the British side less to keep out 'Ethiopians' than to invite them in, influencing patterns of occupation and alliance in Marsabit, with effects felt to the present day.

2.3. A Deadly Shore: Dassanach Conflicts in Northwestern Marsabit

To what extent has violence between communities been heightened by the creation and enforcement of boundaries, and to what extent have boundaries become material factors, whether causes or foils, in the unfolding of conflict? These questions can be addressed with respect to the history of interaction between the Dassanach and the Gabra at the border near Lake Turkana in northwestern Marsabit District, as recounted in the colonial records.

In 1915–16, the Gabra were attacked by the Dassanach and Ethiopians at Korangogu and Dukana; these Gabra were from Ethiopia but had come into British territory six months before the raid (Sobania, 1979: 78). In 1925, a raid was reported on the Gabra at Moite in which 29 Gabra were killed and 4–5000 camels were taken. The raid was carried out by 40 Ethiopians and 300 Dassanach, but it is reported that they were defeated by the King's African Rifles, who inflicted 14 killed and 20 wounded, 117 camels being recovered. Presaging the present, in 1927, administrative reports state that "The Gabra tend to run when raided by the Dassanetch but they used guns to raid unsuspecting Rendille". Due to drought, the Gabra were allowed to graze up to Lake Turkana (known as Lake Rudolf until Independence) in the Moite area, but it was observed that "they need guards to protect them from raids along the Lake shore" (Sobania, 1979: 85).

In the pre-colonial period, the Dassanach stretched from the Omo Delta in the far south of Ethiopia along the northeastern shore of Lake Rudolph (now Lake Turkana), to Kokoi and Koobi Fora. Several questions preoccupied British administrators in the area. Should Dassanach be allowed to occupy regions in British territory that had once been theirs? If not, what should be done with the area in the far northwest of Marsabit District, south of the Banya Lugga frontier through Ileret to Koobi Fora? In any case, how were the Dassanach to be deterred from raiding communities under British administration?

In 1929, it was reported that:

> The far northwest is 'left severely alone and must remain so' until there are roads and staff . . . The 'tribes' are not allowed to graze here (the lake shore) and if they do so it is at their own risk. Generally the Gabra are in the area but also the Rendille at times. The Dassanetch are often in the northwest of the district 'but as we don't go up there it doesn't matter much' . . . The problem is that practically all the Gabra have blood or religious ties to Ethiopia and the British authorities say they have no hold over them (Sobania, 1979: 86).

In 1936, there were no Dassanach south of Banya Lugga, so Kenyan groups were allowed into these otherwise prohibited areas (Sobania, 197: 122), and in 1937, it was reported that the British would not disturb the Dassanach if they didn't come further south than Banya Lugga. After the Italian occupation of Ethiopia, the British allowed Gabra to occupy the area around Koobi Fora, but also allowed some Dassanach who asked to live in British territory to come south to Kokoi. In 1944, after the British had occupied Ethiopia, displacing the Italians in a short-lived campaign of World War II, the Dassanach were told that "they could use their customary grazing as far south as Kokoi", which brought them close to the Gabra. Dassanach grazed near the Gabra for several years, but in 1947 "raided the Gabra (Algara section) at Uruptirsa and killed 17" (Sobania, 1979: 123).

In 1948, some Dassanach were allowed to remain in British territory, though the policy towards these "British Gellebba" (sic) was that they were not to be considered British subjects, they would not be taxed, and their presence in Kenya would only be "tolerated" (Sobania, 1979: 12). This group was valued for facilitating contact with the Ethiopian Dassanach, and considered more trustworthy for theoretically refusing to raid with the latter. But violence arose in 1952. A Dassanach raid on the Rendille that year left 75 dead, mostly women and children, and three other attacks later occurred: a Kenyan police patrol was attacked by 200–300 armed Dassanach, precipitating a seven hour battle; on another occasion 26 were killed; and in a third a Dassanach raid took place on a Gabra village.

'British' Dassanach occupying the region near the Kenyan border post at Banya, previously specialists in fishing, consolidated themselves by becoming stock owners. In 1955, Ethiopian Dassanach carried out raids on the Gabra, and when Kenyan police responded they were attacked by 140–200 Dassanach, leading to a 24 hour battle in which three Dassanach were killed (Sobania, 1979: 124). The Dassanach were said to possess 1000 rifles and to be "scarcely administered and heavily armed". Although "the Ethiopians made a new attempt to administer the Dassanetch" by creating a combined province, the Dassanach continued to occupy the area of Ileret under Kenya control. Echoing sentiments expressed a half century before, in 1961, on the eve of Kenyan Self-Rule, administrators commented that "It is now doubtful if we can never get rid of these people but they must be strictly confined to the area allotted to the north of Ileret and encouraged to return to Ethiopia" (Sobania, 1979: 126).

Sections of Dassanach had always occupied the northeastern shore of Lake Turkana, as herding and fishing outposts of this community of river delta cultivators. The border left most of the lake in British-Kenyan hands, but on geopolitical grounds allocated the headwaters and the mouth of the Omo River to Ethiopia. Thus was drafted a simple scenario for Twentieth Century colonial politics in the area: Kenyan authorities demanding that the 'Ethiopian' Dassanach "return" north of the border at Banya Lugga, the Dassanach in turn further consolidating their presence along the northeastern lake shore, where they fished and exploited wet season pastures. Clearly, Dassanach had historical claim to the region, but also identified with and often depended on their brethren in Ethiopia. Though using resources seasonally, Dassanach established settlements of greater permanence than did Gabra and Rendille herders, who sought only to use lakeside grazing in the late-dry season, before rains came to the Chalbi desert.

But after certain Gabra communities were defined as legitimate residents of Kenya, their seasonal conflicts with Dassanach south of Ileret came to be defined in national terms. Dassanach have not hesitated to use the settlements at the northern tip of Lake Turkana in Ethiopia as a haven after raiding Gabra and Rendille pastoralists further south, and Gabra have been opportunistic in raiding Dassanach homesteads along the lake. So a frontier delimited in order to strike a political balance between two imperial powers and to establish an ecological division between river delta and lake resources came to serve as a pivot around which conflict between two frontier communities (Dassanach and Gabra) who straddled the border would turn. Conflicts in the far northwestern corner of the district to some extent limited the extension of Gabra grazing, pushing them further against Rendille and their grazing territories to the south.

3. SITES OF CONFLICT

Some members of Marsabit communities seek to settle when they lose the capacity to continue pastoralism, and some do so to preserve the pastoral option, by guarding their animals. In both cases, the social and spatial configuration of groups and the pattern of conflict between them provides a cause for and defines the pattern of sedentarization. The colonial history of Marsabit District, just reviewed, involved the emergence of three distinct nuclei of power and culture: of Samburu-Rendille, Gabra-Borana, and Dassanach.

At the risk of overgeneralizing and essentializing northern Kenyans, Rendille and Ariaal tend to have one ally, Samburu, one distant Somali cousin, and four antagonists: Gabra and Boran, Dassanach, and Turkana. Of these, Gabra and Boran, with their linguistic and cultural affinity, have been fairly consistent allies (setting aside recent friction), Turkana and Dassanach, despite enmity, are occasional collaborators. Apart from the Rendille, who speak a Somaloid language, all are frequent combatants with Somali-speaking groups. Let us briefly note three major conflicts that have occurred in Marsabit over the last ten years that bear on Rendille and Ariaal strategies of continuing mobility or sedentarization as patterns of land use.

3.1. Perennial Strife over Lake-Shore Grazing

The long history of conflict between Gabra and Dassanach continued during the last decade. In 1996, the Gabra—newly armed with guns procured from remnants of the Ethiopian army that had fled as the Tigrean People's Liberation Front (TPLF) assumed power in 1992—had come for grazing to pastures east of the shores of Lake Turkana. Reportedly, they stole some sheep and goats from neighboring Dassanach, who being deeply aggravated warned the Gabra to return them within a week. Turkana settlements that lay between the two were warned about an imminent attack and quietly moved away, but the Gabra did not respond. Dassanach insist that the Kenyan police and administration had also been informed that an attack would occur if they did not act to return the stolen animals. On the seventh day, at daybreak, the Dassanach attacked Gabra herding camps in the Kokoi area, seizing a great number of cattle and killing twenty Gabra. When Kenyan police pursued them, the raiders hid the animals, lay in ambush, and killed 19 of their pursuers, stealing their uniforms and weapons.

The episode became an international incident. After many months, the Ethiopian government managed to return the police uniforms but it was many years before its promise to return the animals and compensate the Gabra for their loses were even in part fulfilled. Gabra suspect that Ethiopian police and officials were forewarned but chose to collaborate with rather than deter the raiders, and undoubtedly the very fact of the raid points out the relatively weak administrative presence both governments have established over this frontier region. Neither effectively enforces the border in this remote region, and the collaboration between Kenyan and Ethiopian Dassanach in carrying out the raid demonstrated the continuity of ties that link the two sides of the frontier. Nonetheless, the greatest fear of those Dassanach who fled from Kenya to Ethiopia after the raid was that they would not be allowed to return to their 'homeland', Kenya. About five years later, after negotiations, they were in fact allowed to return and without prosecution.

While the conflict involved a certain play over variables of the contemporary world (borders, national security forces, 'citizenship', markets), the conflict had the stamp of a long heritage of two groups fighting over resources. Gabra and Dassanach are said to primarily fight in the rainy season: when it rains, labor demands drop and men have the time and luxury to raid. But differences in the arrival of rainfall in the Omo Delta, where Dassanach live, and in the Chalbi and Koroli deserts, inhabited by Gabra, influences the timing of raiding. A Gabra informant told me that, "for the Shankilla" (pejorative term for Dassanach), "the rains come sooner, freeing them to raid groups outside of the delta, and with rain they move out of the delta with their herds southeastward into Kenya towards new pastures". When rains come to Gabra territory, somewhat later, Gabra shift away from the lake back to their Chalbi lands, which are relatively inaccessible to the Dassanach. So Gabra are especially vulnerable to Dassanach raiding during a critical period, after rains have come to the northern Lake Turkana and Omo Delta region but the dry season is still in full force in the Chalbi area east of the lake, when they seek out pastures along the northeast side of the lake. On relations between the two, I was told, that "the Shankilla and Gabra have always fought . . . They just can't remain without fighting. They can stay for a year or so, but then they will fight."

The effect of ongoing conflict is to make it difficult for Gabra to exploit lakeshore grazing, further constraining pasture use elsewhere in Marsabit. The camel-herding Gabra have usually depended on the Borana for backing, but Borana had recently expressed annoyance with the Gabra refusal to provide refuge for members of the Oromo Liberation Front (OLF) who had been operating over the border during its strife with the new Ethiopian government. Their recent acquisition of arms by the Gabra, sold locally by Ethiopian soldiers fleeing southward in 1992, has meant an all too easy escalation of conventionally limited clashes into mortal conflicts, as Gabra have been pitted against their prime antagonists, Dassanach and Rendille.

3.2. Rendille and Gabra Strife between Camel-Keeping Counterparts

When the Rendille were attacked by the Borana, they often retaliated against the nearer Oromo community, the Gabra. One account emphasizes cycles of retribution between the two.

> ". . . in 1918, they again made a peace. The colonials controlled the area. The first D.C. was Bwana Bonres [Barnes?] And they respected the colonial official and became peaceful. In 1926 they [the Gabra] attacked again, and Rendille children were killed, and they fought up to 1968. Then they made peace in 1968. In 1991, two Rendille warriors killed a Gabra boy near Kalacha, when raiding their animals. To revenge that one, in 1992 they [the Gabra] killed many people; only 29 km from Marsabit, Gabra killed nine Rendille, at Okwisho [near Kargi] . . . But now, they [the Rendille and Gabra] are good friends; this was caused by politicians".[3]

One of the outcomes of conflict between the camel-herding Gabra and Rendille is that their previously coordinated and often mutual use of pastures and wells to the west and northwest of Marsabit mountain, with the Rendille moving between Kargi and Korr, has been curtailed. Schlee (1991: 131) suggests that drawing grazing boundaries between Rendille and Gabra led to an "estrangement" between the two groups, contributing to the violent clashes that occurred in the 1960s and 1970s.[4] He observed for the period of the late 1980s, prior to the clashes of the 1990s, that "Rendille/Gabra relationships are peaceful again and characterized by interlocked settlement and intermarriage" (Schlee, 1991: 131). Poorer Rendille have increasingly settled in Korr over the last two decades, where they receive famine relief and assistance in restocking through the missions. The area is intrinsically good for camels and better than Marsabit for cattle, due to the absence of ticks, so represents an area where all species of domestic stock can be maintained. But those with herds tend not to move far from town, even in the dry season, because they fear raids from other communities. Even when animals are moved away to better grazing, most people stay in Korr, so the community has grown.

By June 1992, some Rendille and related Ariaal had moved southward, shifting away from Gabra who were viewed as threatening due to their newly gained advantage in firearms. But many Rendille, who mirror Gabra in their camel-herding pursuits, remained in the arid regions between Marsabit and Lake Turkana, where camels are nourished on browse and deep wells are found. Three Gabra were killed in 1992, and when Gabra raided Rendille living in Kargi twice in 1994, more were killed. The camels lost in the two raids, that affected some 17 families, were said to number around 5000. Based on estimates that Rendille owned approximately 24,000 camels, the camels lost would represent about 20% of the Rendille total, dramatically affecting the basis of their subsistence and social autonomy.[5] The Borana are seen by Rendille to collaborate with Gabra in their conflicts, and those coming from Ethiopia are said to gain information from Gabra, on the basis of which they carry out raids (often without Gabra participating), before returning to Ethiopia. Even if not directly implicated in raids on Rendille, if Gabra are suspected of collusion they may be objects of retaliation. For this reason, in recent years some Gabra have refused the Borana permission to pass through their territory on raids.

Both camel-herders, the Rendille and Gabra share similar patterns of land use, household economy and social structure; this makes them ideal allies and almost inevitable enemies, and here is the paradox of their long relationship. Mirroring one another, the tension between them contrasts with the amiable complementarity each enjoys with a cattle-herding society, Rendille with the Samburu and Ariaal, Gabra with the Borana. But having occupied the same arid, volcanic strewn desert region, the Gabra and Rendille have become, in Schlee's (1991: 131) term, "interlocked". The "interlocked" quality of Rendille-Gabra relations, through a history of shared settlements and intermarriage, has numerous implications. Their geographic and social intimacy, sharing of values in livestock, makes each a potential target of the other's raids, but, despite the potential for volatility between them, the two groups have the inherent capacity to subdue conflict and forge peaceful relations. Their historical relations are depicted as times of harmony punctuated by outbreaks of stock-theft and violence. The detailed knowledge each holds about the practices and locations of the other, due to their social inter-penetration, makes each vulnerable to the other's raids. Spying is an indispensable aspect of a successful raid, and sisters given as wives are inevitable sources of intelligence. But like many good spies, they transmit intelligence in both directions, often informing potential victims about prospective raids. Those who suffer attacks quickly learn the background details of raids, including the disposition of the spoils, which makes retaliation all the more likely. The balance between Gabra and Rendille was upset in 1992 when Gabra gained an edge in firearms, leading to the southward retreat of the Rendille, the creation of larger settlements, and acceleration of the process of sedentarization.

3.3. Rendille and Borana: Borders and Land Conflicts

What distinguishes recent conflict from that experienced during an enduring history of sporadic, low-level raiding? Firstly, the international border makes a difference. It creates an almost natural haven for stolen goods, as animals are often (but not always) smuggled to where they can be hidden and sold on the market. Due to the economic entropy of divergent national policies, borders create almost inevitable differences in prices available for stolen livestock. If radically different economic systems are being followed (the case between Kenya and Ethiopia until 1992, when the socialist Dergue government fell in Ethiopia), these price differences may represent a substantial incentive to theft and smuggling. Secondly, there are differences in access to firearms different groups enjoy. Throughout the twentieth century, many Ethiopians possessed firearms when most Kenyans did not, so due to their trans-boundary associations Kenyan Gabra and Boran have had more guns than their neighbors. But when Gabra were recipients of increased arms flowing into the frontier market in 1992, with the flight southward of the routed Ethiopian army, the technological balance between Gabra and Rendille was overturned, to the detriment of the latter. The recent investment of Rendille in firearms, notwithstanding little-training in their use, may reestablish a balance of terror with the Gabra (not to mention the Turkana), even at a regrettably higher level of mutual threat.

Ariaal Rendille encountered Borana on Marsabit Mountain, where the latter settled seeking land to cultivate. Boran came in large numbers to Songa, on the southeast of the mountain. Faced with land adjudication and the prospect of definitively losing land to the Borana, the Ariaal attacked them and drove hundreds out of the region (Salvadori, 2000). While the Borana seem to be the aggrieved party, the Ariaal cite attacks on Rendille children by the Boran and observe that it is up to them to protect the lands which have been historically theirs. If we compare populations figures for Marsabit District as a whole we see that in 1961 15,739 Rendille outnumbered 12,123 Gabra in the 'Reserve', Borana numbering only 2893, while in the Mountain Township, the Borana represented the largest single group, at 884. (Since Borana migration was then controlled by the administration, it is likely that some Borana declared themselves to be Gabra.) Only 18 years later, in 1979, the census indicated that the Rendille had increased to 19,856 but had been surpassed by the Gabra at 23,410. Both, however, were dominated by the Boran, whose district population had grown during the years since Independence to 30,444. It is in this context that the Rendille and Ariaal initiated conflict with the Borana, to maintain their hold on land long prized for its dry season grazing but more recently valued as fertile sites for settlement, as they became farmers. Strife between the two groups has heightened over the past five years, in part due to the onset of adjudication and titling of the land in question, and in part due to ethnic competition over electoral politics.

Photo 1. Ariaal cattle on Marsabit Mountain

4. RELATIVE SEDENTARIZATION: AN OUTCOME OF
CONFLICT AND PASTORAL INERTIA

Sedentarization has increased in Marsabit due less to the desire of pastoralists to gain the fruits of modernity or to the constraints of herding in an arid environment, than to increasing political conflict that has made extensive grazing less secure. Conflict has been engendered in part by the creation of borders, in part by the spill-over into northern Kenya of political strife associated with the Ethiopian and Somali civil wars, in part by the dispersion of modern weaponry in the region, and in part by the threat to land holding posed by the process of registration and titling. The effect has been to encourage Ariaal agricultural settlements on Marsabit mountain, most notably at Karare, Kituruni and Songa, the establishment of large, consolidated pastoral settlements around sites such as Korr, where social services, markets, and famine relief can be received, and a diminishing of nomadic inertia experienced by pastoral households that move less often, less far, and less autonomously. But sedentary sites on the mountain and on the plains do not seem to represent the 'shuffed off' herders described in historical studies, but rather active members of nomadic households who establish secondary sites of residence as strategic ventures in order to access resources—whether farmland, security, or relief—that are unavailable if the entire group is in perpetual motion.

One measure of the degree of pastoral inertia can be seen in the size of pastoral homesteads and settlements, the concentration of settlements near trading centers, towns, or wells, and the frequency of household movements. Gabra have main camps, composed of an unfenced crescent arrangement of houses, with livestock pens directly in front of the houses where they can be observed, and circular satellite 'fora' camps, composed of fenced enclosures for animals. O'Leary (1985: 166) observes that the only fenced camps he observed among Gabra were very large camps built at the settlement of Dukana "for the purpose of defense against raiders". Camps tended to have 4–6 houses each. In four surveys carried out between January and August 1982, the total number of Gabra camps rose from 163 to 229, houses from 916 to 1366, the number of houses in camps from an average of 5.6 to 6 (median from 4–5 houses per camp), the proportion of camps having greater than 10 houses decreasing from 14.7 to 13%.

Based on study quadrants examined, the Gabra occupied 56.7% of the range in January, only 40.7% in March, and up to 45 and 50% in May and August. Thus Gabra camps were maximally dispersed in January (at the end of the dry season) and most congregated in March. In that year, January was at the end of a very long dry spell, so Gabra had moved to wherever grass could be found. Few were located near the permanent Chalbi springs, where the vegetation cover had disappeared, or in the Hurri Hills, which lacked permanent water, but with the February rains many left their dry season sites and rushed to areas that had received new pasture. Interestingly, many who lived at Balessa Ildere decided not to make the long move to areas where rain had fallen, since it was risky with to move animals long distances in dry conditions. Thus they stayed in an areas of "high security risk" because of clashes that had taken place there in December, 1981, between Rendille and Gabra (O'Leary, 1985: 174). It is recognized that areas of rich pasture where fora stock are grazed by young men in the rainy season, at a distance from the main camps, are often at greater risk of being raided (O'Leary, 1985: 186).

Rendille also have large, main camps and temporary satellite or 'fora' camps, the former composed of a set of houses surrounded by a perimeter barrier, both having fenced enclosures for livestock. Rendille camps tend, on average, to be four to five times larger than Gabra camps. In four surveys carried out over a one-year period from May 1982 to May 1983, the number of northern Rendille camps observed rose from 105 to 153 while the number of houses overall increased from 2456 to 2704. But the mean size of camps decreased from 23.4 to 17.7 houses, the median size dropping from 20 to 13 houses per camp. During that period, then, people tended to break into a larger number of camps, and the number of very large camps diminished: the percentage of camps having greater than 30 houses dropped from 26% to 15% over that period.

Rendille tend to have large camps, and tend also to reside together in local concentrations on the range, making the impact of sedentarization easily observed (O'Leary, 1985:41). For instance, in May 1983, 47.6% of houses were located in or close to the main trading centers of Kargi, Korr, Logologo, and Laisamis, with a large concentration of rural 'fora' camps being found on the higher reaches of the northern and north-western slopes of Marsabit mountain. Other concentrations occurred in smaller sites, near Ngurunit, Ilaut, and Laisamis. This reveals that Rendille congregate around trading centers, for added security and to gain access to provisions of famine relief distributed there by local missions and government.

The process of sedentarization does not always involve exit from the pastoral economy, but sometimes the adoption of a strategy of decreasing inertia in pastoral movement, paradoxically making the livestock economy somewhat more secure through partial settlement. Rendille and Ariaal sedentarization thus can be seen as a movement along a continuum: larger settlements are formed and individuals gravitate to trading centers, moving less often, and seeking strength in numbers. One factor in the choice of where to situate major pastoral camps is the provision of relief, which brings Rendille and Ariaal to some of the larger trading centers and towns and into the orbit of missions, schools, clinics, and other social services, interpreted by many (sometimes wrongly) as the reason for rather than the outcome of settling. 'Relative sedentarization' also occurs as Rendille respond to

Photo 2. Ariaal Rendille elders, armed, in Karare, 1999

insecurity, moving from more vulnerable northern sites and joining larger settlements to the south, further from Gabra. After the last series of Gabra raids in 1992, Rendille did not immediately retaliate, in fact responded with resigned passivity, likely another sign of pastoral inertia.

It is questionable whether sedentarization in such an arid region is adaptive in the face of the exigencies of pastoral land use, which requires continuing mobility, or of pastoral subsistence systems, in which milk and meat represent critical sources of high quality protein to assure good health. Areas of semi-permanent settlement, though seeming to promise certain amenities of modernity, often offer little employment, no sanitation, inferior pastures, fouled water sources, and lower levels of nutrition, despite famine relief. Border-making, colonial and post-colonial movement of peoples, and increasing violence have stimulated increased settlement, but it is far from clear that this represents a step forward to the life that outside observers may associate with the calm peace and prosperity of sedentary life.

ACKNOWLEDGEMENTS. Research in Marsabit District, Kenya, was pursued in June 1997, June 2001, and February 2002, supported by several projects. The project on "environmental knowledge and perception in Kajiado and Marsabit districts" was supported by the Social Sciences and Humanities Research Council of Canada (SSHRC) and the Québec Fonds pour la Formation de Chercheurs et l'Aide à la Recherche (FCAR), through affiliation with the National Museums of Kenya (research authorization #: OP/13/001/25c 261/8) and in cooperation with the Arid Lands and Resource Management Network in Eastern Africa (ALARM), itself supported by the International Development Research Centre of Canada. The project on "health and nutritional effects of drought among Rendille people of Marsabit District, Kenya", E. Fratkin principal investigator, was carried out with support of the National Science Foundation, through affiliation with the Institute for African Studies at the University of Nairobi (research permit # P/13/001/C 1594). I appreciate the help of Kevin Smith in pursuing research in Marsabi, in areas where he had pursued his doctoral research, and the assistance of Larian Aliyaro and the late Kawab Bulyar of Korr, Daniel Lemoille of Songa, and Abdirizak Kochale of Logologo in gathering information reported here.

NOTES

1. There is considerable variation in the ethno-orthography of the region, especially for the group considered here: Sobania's (1979) 'Dasenech', Almagor's (1978) 'Dassanetch', Carr's (1977) 'Dassanech', and Schlee's (1979) 'Dasanech'. Using the form adopted by members of the group for Bible translation, the 'Dassanach', who occupy the northern shore of Lake Turkana and the Omo delta, are also referred to in oral usage and the existing literature as 'Reshiat', 'Gelubba', 'Marille' and 'Shankilla', surely one of the continent's most varied set of ethnonyms. British records often refer to them as Gelubba, the Ethiopians as 'Shankilla', the latter connoting 'barbarian'.
2. It was the Oromo-speaking Warra Daaya who carried out the first expansion southward into what is now Kenya, where they came to inhabit the country from Marsabit to beyond the Juba (Schlee 1989:37). Their descendants are the Orma of the Tana River. The Boran expansion came later, and would exercise a lasting influence on the northern region. It was with the expansion of the Somali-speaking Darood and Degodia in the 19th c., southward to the Tana and westward towards Isiolo and Marsabit, that Borana expansion was halted.
3. Interview with the late Kawab Bulyar, Ariaal Rendille elder and project assistant. Korr. 11 June, 1997.
4. For discussion of the role of administrative boundaries and conflict of interest in land in engendering enmity between Maasai sections, see Galaty (1994).

5.   Female camels are worth about Kshs 12,000, which is approximately US$240, a big male or a heifer
      Shs. 8,000, about US$160, at an exchange rate then of Kshs 50 = US$1; estimates of livestock numbers
      were attributed by Larian Aliyaro to Jürgen Schwartz.

REFERENCES

Almagor, Uri, 1978, *Pastoral Partners*. Manchester Univ. Press.

Barber, James, 1968, *Imperial Frontier*, Nairobi: East African Publishing House.

Barth, F., 1961/1986, *Nomads of South Persia*. Prospect Heights: Waveland Press.

Galaty, J.G., 1994, Rangeland Tenure and Pastoralism in Africa. In *African Pastoralist Systems*, edited by
      E. Fratkin, K. Galvin, and E.A. Roth, pp. 185–204. Boulder: Lynne Rienner Publishers.

Hogg, R., 1985, Restocking Pastoralists in Kenya: a Strategy for Relief and Rehabilitation. London: ODI, Pastoral
      Development Network Paper 19C.

O'Leary, Michael, 1985, *The Economics of Pastoralism in Northern Kenya: the Rendille and the Gabra*, IPAL
      Technical Report # F-3, Nairobi: UNESCO, Integrated Project in Arid Lands.

Salvadori, Cynthia, 2000, *The forgotten people revisited: Human rights abuses in Marsabit and Moyale Districts*.
      Nairobi: Kenya Human Rights Commission.

Schlee, Günther, 1989, *Identities on the Move: Clanship and Pastoralism in Northern Kenya*. Manchester:
      Manchester Univ. Press.

Schlee, Günther, 1991, "Traditional Pastoralists: Land Use Strategies", *Range Management Handbook of Kenya
      Volume II*, edited by H.J. Schwartz, S. Shaabani, and D. Walther, pp. 130–164. Nairobi: Ministry of
      Livestock Development.

Sobania, Neal, 1979, *Background History of the Mt. Kulal Region of Kenya*. IPAL Technical Report Number A-2,
      UNEP, Nairobi: MAB Integrated Project in Arid Lands.

Sobania, N., 1993, Defeat and Dispersal: The Laikipiak and their Neighbours at the End of the Nineteenth
      Century, edited by T. Spear and R. Waller, *Being Maasai: Ethnicity and Identity in East Africa*, pp. 105–119,
      London: James Currey, and Athens: Ohio Univ. Press.

Chapter 4

Ecological and Economic Consequences of Reduced Mobility in Pastoral Livestock Production Systems

H. JÜRGEN SCHWARTZ

1. INTRODUCTION

Nomadic livestock production with domestic ruminants, dromedaries and donkeys used to be the dominant economic activity in the dry lowlands of the Old World Dry Belt, extending from Mauritania to North-West India. Common to these areas are the low and erratic rainfall, the high risk of recurring droughts, the high rate of actual and potential evapo-transpiration and the general scarcity of permanent surface water. These result in low and seasonal biomass production of both the herbaceous and the woody vegetation, which in turn cause large seasonal changes of forage availability and of forage quality. Despite significant differences in the political and economic conditions in the countries of that large eco-region it is largely the similar ecological potential which determines the status, development and performance of livestock production in the dry lowlands. Migratory pastoral production systems, which have evolved under these marginal natural conditions, share a number of unique characteristics that are aimed at minimizing production shortfalls caused by large variations in forage productivity. One of the most prominent risk minimising system attributes relates to the mobility of pastoral livestock herds and households, allowing full exploitation of forage resources that are unequally distributed in space and time (Coughenour et al., 1985; Ellis & Swift, 1988). Orientation towards subsistence production, reliance upon a combination of different animal species, considerable sharing of

H. JÜRGEN SCHWARTZ • Humboldt University of Berlin, Berlin D-10115, Germany.

resources and products within small groups, and an emphasis on milk rather than meat production are additional features contributing to risk avoidance in pastoral economies (Swift, 1982; Western, 1982).

Although abiotic and biotic conditions have not changed much in the past decades traditional, subsistence oriented, migratory pastoralism has virtually disappeared as a land use system throughout the Old World Dry Belt. In Saudi Arabia it was transformed into high input and mechanized long distance grazing systems (Ahmad, 2001), in Tunisia the Southern rangelands are largely depopulated because many pastoralists have opted for livelihood opportunities in other sectors of the economy, Somalia has experienced spontaneous and often violent privatization of formerly communal rangelands for sedentary forms of livestock production (Schwartz, 1993). In many other countries subsistence orientation has given way to market integration with an accompanying shift to specialization, i.e., meat production with sheep and goats as in Algeria or milk production with camels in the vicinity to urban markets as in Mauritania and Syria.

In Kenya pastoral production systems have remained remarkably stable for a long time. However, fundamental demographic, economic, and political changes have disrupted the delicate balances between human populations, livestock numbers, and rangeland resources in these socio-natural systems. These changes originated partly in programmes of the colonial administration aimed at raising productivity in the pastoral sector, which, subsequently, were taken over almost without interruption by the Kenyan government after independence (Oxby, 1975; Bennett, 1988). Attempts at restructuring pastoral production systems to increase their economic self-sufficiency and contribution to the national economy have almost always failed. Instead, such attempts have affected traditional land-use practices, have led to permanent differential access to basic productive resources, and to a substantial increase in income disparities among pastoral households (Hogg, 1986; Mayer et al., 1986).

Another, more serious influence on pastoral production systems relates to the constantly increasing human population in the semi-arid rangeland areas of Kenya, and to the accompanying conflict over land resources. The rapid population growth in neighbouring agricultural communities has led to an expansion of cultivation onto semi-arid rangelands. In particular the small pockets of high potential pastoral land which formerly served as dry season grazing reserves, are increasingly being occupied by agriculturalists. These losses have been aggravated by the establishment of commercial ranches and national parks on pastoral land (Schwartz & Schwartz, 1985). The increased competition for resources among different land use systems was also paralleled by a steady growth of the pastoral population, although at a slower rate than in other groups in Kenya, which further decreased the per capita availability of land for grazing (Swift, 1982).

The combined forces of demographic pressure, steady loss of rangeland to other sectors, and development interventions in pastoral economies have contributed to a rapid decrease in the mobility of pastoral herds and settlements throughout Kenyan rangelands (Grandin, 1988; Fratkin, 1992; Schwartz et al., 1995; Roth, 1996). Sedentary pastoralism in the vicinity of small towns, trade centres, famine relief stations, and mechanized water sources has become a widespread practice, especially among impoverished pastoral households (O'Leary, 1990). The transition to settled life, for long a primary objective of development policies aimed at pastoralists in Kenya, is likely to cause substantial ecological and economical problems. Schwartz et al. (1995), for instance, note that concentration areas are marked by severe and spreading degradation of vegetation and soils. This, in turn, lowers herd productivity, increases herd sizes required to meet household needs, and thus further accelerates environmental degradation and the likelihood of destitution.

To illustrate these complex processes some results of a case study on Rendille pastoralism carried out over many years in Marsabit District of Kenya are presented.

2. A CASE STUDY REVISITED—THE RENDILLE PASTORAL PRODUCTION SYSTEM

The Rendille of Marsabit District in Northern Kenya, a small ethnic group, numbering not more than 25,000 individuals, occupy a home range of approximately 15,000 km^2 in the Western half of the District. They practice a form of opportunistic, horizontal nomadic pastoralism with dromedaries, goats, sheep, and cattle. The small size of the group and the limited and well-defined home range were seen as positive conditions to carry out a study of the factors determining land use and migration patterns in a long-term perspective.

2.1. Ecological Potential of the Rendille Home Range

Mean annual precipitation and the number of month per year without effective rainfall are the two most significant parameters determining the ecological potential. Others, such as potential evapo-transpiration, mean temperatures, the sum of sunshine hours per year, the annual and seasonal variation of rainfall and the risk of drought are closely correlated to the two key figures. With increasing aridity, i.e., with decreasing mean annual precipitation, seasonality becomes more pronounced as the periods without effective rainfall grow longer; the growing period for the vegetation becomes increasingly limited and the seasonal and annual variation of rainfall increases sharply. The reliability of rainfall events declines with a simultaneous increase of the risk of drought. If a drought is defined as the complete failure of one rainy season, it occurs two to three times in ten years. If it is defined as the complete failure of two consecutive rainy seasons it occurs two to three times in thirty years. Complete failure of three consecutive rainy seasons might occur once in thirty years, i.e., once during the economically active life span of a pastoralist.

Rainfall, all other factors being equal, determines plant growth. Table 1 gives some estimates of forage production of different vegetation components together with a value for the permissible off-take of the biomass produced. Permissible off-take has been defined as the proportion of the total biomass produced, which is useable as animal feed, if range deterioration or degradation is to be avoided (Schwartz, 1991).

Biomass production in the herblayer, which is the major source of forage for cattle, sheep and donkeys, ranges from close to 500 kg/ha to just over 3000 kg/ha at mean annual rainfall values of 100 to 500 mm. The shrub layer, which is the preferred forage source of

Table 1. Rainfall and Estimated Forage Production in Marsabit District [kg DM*/ha/year].

| Rainfall [mm/year] | DM Production | | Permissible off-take [%] | Growing period [days/year] |
	Herblayer	Herblayer + shrubs		
100	450	600	25	35–65
200	1080	1600	30	55–85
300	1710	2600	40	70–120
400	2340	3600	50	125–175
500	3160	4600	50	150–220

* = Dry Matter.
Source: Schwartz, 1993, in Baumann, Janzen und Schwartz, 1993.

(a)

(b)

Figure 1. Dwarf Shrub *Indigofera spinosa*, Normal Growth and with Heavy Grazing. (a) Normal growth form of a small (60 cm) multi-stemmed dwarf shrub *Indigofera spinosa*, a common leguminous plant in the dry lowlands of Marsabit District. Because of the high protein content of leaves, fruits and young shoots it is a preferred browse species for goats and camels. (b) Growth form of the same species *Indigofera spinosa* after very frequent defoliation by livestock near a permanent watering place. The multi-stemmed shrub has transformed into a tight crown of short secondary and tertiary shoots which protect the remaining foliage from herbivores. Erosion, facilitated by reduced ground cover, has lowered soil surfaces by 35 cm since the shrub has germinated approximately 12 years prior to the date of the photograph. (Photos by H.J. Schwartz).

goats and camels contributes another 30% of the amounts produced in the herblayer. Permissible off-take increases with increasing rainfall from 25 to 50% of the annually produced biomass. Actual off-take through overstocking, however, often exceeds these values by far, leading to impaired vitality of the range vegetation, to shifts in the number and composition of desirable and undesirable species, to long-term reduction of biomass production and ultimately to soil degradation and erosion (see Figures 1a and 1b).

2.2. Materials and Methods

Different types of data sources were used for the study. Historical information dating back as far as the turn of the century, the results of a two-year series of aerial surveys, the results of a large scale ecological survey and mapping exercise which produced a number of biophysical and ecological maps of the District and some satellite images of the area, notably LANDSAT and NOAA images.

2.2.1. Historical Information

A historical study of Rendille migration patterns was conducted by N. W. Sobania in 1980. Some of the raw data of this study, consisting of oral histories of the migration of seven clan oriented settlements (gobs) and using the Rendille traditional calendar and vernacular place names as references, were analysed to produce maps of annual and seasonal occupation densities during the time period between 1941 and 1978. The estimates were based on a map grid of 10 by 10 km.

2.2.2. Aerial Surveys

In an effort to determine short term land use and migration patterns in the Rendille home range 12 aerial surveys were carried out at approximately two-monthly intervals over two years in 1979 to1980. Site and size of settlements as well as numbers of households and numbers of domestic livestock present in the settlements were recorded. This involved total photographic cover of all settlements in the survey area from a low flying aircraft during the hour immediately after sunrise, when all stock was expected to be retained in the night enclosures. All counts were originally accumulated within a map grid of 5 by 5 km squares. For the purpose of this presentation a 10 by 10 km grid was used matching the one used in the historical survey.

2.2.3. Ecological Survey and Mapping Exercise

From 1988 to 1997 the Government of Kenya with the support of the German Agency for Technical Cooperation (GTZ) was conducting a large scale ecological survey and mapping exercise in ten arid and semi-arid districts in Northern Kenya. Each district was covered by 20 to 24 thematic maps on climate, soils, vegetation, hydrology, geomorphology, erosion status, range condition etc. Marsabit was the first district to be finished and all relevant information was available in 1991 (Schwartz, Shaabani, Walther, 1991). These maps were used to interpret spatial preferences recorded in the historical and the aerial survey in relation to permanent and seasonal physical and ecological features found in the district.

2.3. Results

2.3.1. Historical Land Use Patterns

Table 2 summarizes a few results of the analysis of the historical study. Although some of the oral histories could be followed back as far as 1904, it was only in 1941 when all seven settlement records could be matched. 1950 saw the construction of the first two deep boreholes in the area, 1963 independence from colonial rule brought a certain breakdown of internal security (tribal fighting) and 1971 saw the establishment of the first mission station in Rendille country.

The reduction of the home range utilization from 8100 km^2 to 3500 km^2 during this period is obvious, just as the reduction in the number of moves over significant distances, i.e. entrance into a grid square from outside. Clustering of settlement moves increased with time around mechanized boreholes and, during the last period, around two mission stations, which again were placed close to the boreholes.

2.3.2. Present Land Use Patterns (Established by Aerial Survey)

The grid net in Figures 2 and 3 delineates roughly the Southwest quarter of Marsabit District as shown in the map of the present home range of the Rendille (Figure 4). Figure 2 shows the cumulative occupation density by households calculated as household days per grid square for 24 months. It is evident that extreme clustering occurs in two locations, Kargi (grid reference 3/9) in the North and Korr (grid references 8/7 and 8/8) in the centre of the home range. In both locations are permanent settlements with trading centres, dispensaries, primary schools, and mission stations. Cumulative household and also total livestock densities are highest in the vicinity of the two centres. The inverse picture is shown in Figure 3, which represents livestock numbers per household. These are highest at the greatest distance from permanent water sources and human population centres. This indicative of the fact that wealthier pastoral households are more mobile and tend to graze their stock at some distance from highly frequented locations.

Table 3 summarizes some evidence that spatial preferences for settlement site selection change with the season. In survey 3 at the end of a rainy season, extreme dispersal of settlements can be noted, 35 grid squares are occupied. Site preference is given by range condition class [4] whereas in survey 7 at the end of a dry season extreme contraction of settlements has occurred, only 14 grid squares are occupied, and site preference is given to the availability of permanent water.

Table 2. Number of Grid Squares [10 * 10 km] Occupied by Seven Clan Settlements and Number of Entrances by Settlements into Grid Squares in Four Distinct Periods Since 1941.

Time period	No grid squares occupied*	Total no of entrances	No of entrances within the present home range	
			Total/period	Mean annual total
1941–49	81 (+30)	589	499	55.4
1950–62	70 (+24)	564	506	38.9
1963–70	61 (+8)	324	318	39.7
1971–78	35 (+1)	248	237	29.6

*   Figures in brackets indicate number of movements outside the present home range.

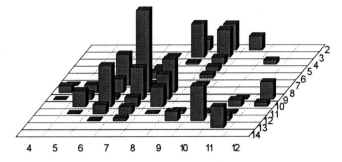

Figure 2. Cumulative occupation density of Rendille home range.
Cumulative occupation density [household days/grid square/24 months] observed in 12 aerial surveys of the Rendille home range. Counted household numbers were multiplied with calculated length a stay in days within one grid square (10 × 10 km).

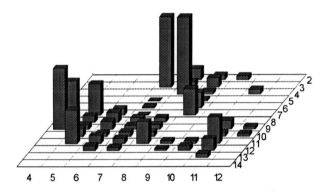

Figure 3. Mean number of livestock [TLU] per household within grid square observed in 12 aerial surveys of the Rendille home range (10 × 10 km). (TLU = Tropical Livestock Unit = 250 kg live weight; 1 camel = 0.7 TLU, 10 sheep or goats = 1 TLU).

Table 3. Occupied Area [Number of Grids], Range Condition Score and Water Availability Score for Nomadic Settlements at Two Different Seasons.

| | Survey 3 | | | Survey 7 | | |
| | End of rains | | | End of dry season | | |
Grid occupancy score	Grid squares occupied	Range condition score	Water availability score	Grid squares occupied	Range condition score	Water availability score
1	10	.76	0	1	.8	1
2	3	.77	.45	—	—	—
3	10	.76	.45	5	.78	.8
4	1	.66	.57	1	1	.75
5	5	.66	.75	2	.55	.87
6	1	.58	1	1	.33	1
7	5	.58	1	4	.28	1
8	1	.36	1	1	.28	1
	total 36	mean .67	mean .44	total 15	mean .58	mean .83

A principal components analysis was carried out using 11 key variables related to site preference such as survey results and data from the mapping exercise. Only about 40 percent of the variance in site preference could be accorded to the variables median rainfall, range condition, vegetation cover index and availability of permanent water. There is strong evidence that non-ecological factors play a major role. In survey 11 a moderate contraction of settlements accompanied by highest livestock numbers per household were recorded in the middle of a dry season. The reason for this and the resulting site selection was a major ritual occasion, the male circumcision, which takes place only once in 14 years. An aseasonal contraction of settlements and livestock was noticeable in survey 8, following an outbreak of banditry in the Western half of the Rendille home range, presumably instigated by Somali tribesmen (*shifta*).

2.3.3. Ecological Significance of the Recorded Land Use Patterns

Figure 4a shows the location of permanent water sources in the south-western quarter of Marsabit District (Bake, 1991). The 10×10 km grid covers the present home range of the Rendille, approximately 15,000 km^2. In Figure 4b, an overlay with the standardized cumulative occupation density by pastoral households illustrates that one of the most important factors for settlement site selection for the majority of pastoral households is availability of permanent water in a short distance (Schwartz et al., 1995).

Total livestock occupation (TLU days) follows the same distribution pattern. In contrast to this livestock numbers per pastoral household show an inverse distribution pattern. Highest numbers of livestock per household are found at larger distances from water (Figures 2 and 3).

Figure 5a shows the livestock to household ratio against the background of a NOAA satellite image, whereas in Figure 5b the background is a map of range condition (Herlocker, 1991). It is obvious that preferred settlement sites for wealthy households are characterized by availability of green vegetation at the beginning of the dry season and by fair to good range condition.

The opposite is the case with animal-poor households. One of the major factors is the availability or the lack of pack animals for household water transport (Schwartz, 1986). Range condition and green biomass availability are clearly a function of distance to permanent water and/or to high human occupation densities. In grid 8/8 (Korr) the highest cumulative occupation density was observed with 224,600 household days in 24 months within one grid square of 100 km^2. Considering the firewood consumption at 2.5 kg/household/ day this alone amounts to an annual off-take of woody biomass of approximately 280 metric tons, not accounting for the off-take for building materials and fences and the amount eaten by browsing livestock. Degradation of vegetation and eventually soils is inevitable at such land use intensity. In consequence one also has to realize that poor households, due to their lower mobility will, usually graze their stock on poorer and further deteriorating pastures (Figures 6a and b). On the other hand rather large areas of the available range at greater distance from permanent will remain under-utilized and in good condition.

2.4. Conclusions from the Case Study

There is a long term trend to increased sedentarization, characterized by reduced number of significant movements of settlements, a contraction of the home range as a whole and particularly during the dry season. The long term trend is masked by seasonal expansion, or rather a strong seasonal fluctuation of home range size. Seasonal trends, however, are again masked by other factors, such as ritual events, security breakdowns or others.

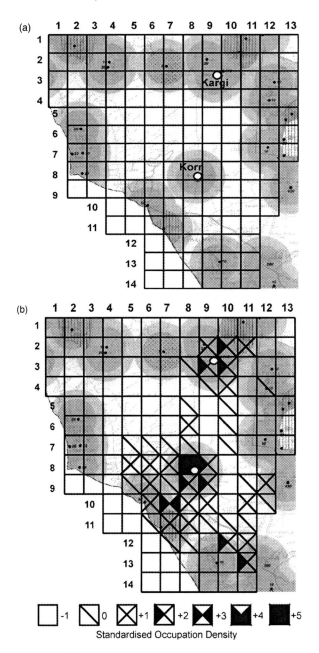

Figure 4. (a) Location of permanent water sources in the Rendille home range. The circles indicate 10 and 15 km distance to the permanent water, the small figures give the estimated water yields in m^3/day. In the vicinity of the two settlements Korr (8/8) and Kargi (9/3) are deep mechanised boreholes which yield 100 and 110 m^2/day respectively. All other water sources within the grid are traditional shallow wells with yields between 10 and 20 m^2/day. (b) Overlay of the standardised occupation density on permanent water distribution. (Standardised occupation density is calculated as mean number of recorded household days per grid square plus or minus standard deviation units. 0 reflects the mean density, -1 densities below the mean, $+5$ signify the highest observed densities.) Highest occupation density is restricted to the 10 km radius around the high yielding boreholes at Korr and Kargi. The highest density was recorded at Korr (grid reference 8/8) with 224,600 household days within the 24 months of the aerial survey.

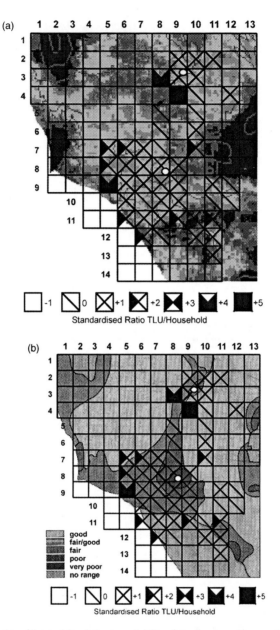

Figure 5. (a) Distribution of livestock in relation to availability of standing green biomass at the beginning of the dry season. Green biomass is shown in shades of green in a modified NOAA satellite image. The standardised ratio TLU/ household has been calculated as mean number of TLU/house-hold within grid square plus or minus standard deviation units. 0 reflects the mean ratio, −1 densities below the mean, +5 signifies the highest observed ratio. Highest numbers of livestock (TLU) per household have been found in areas where green biomass was available at the beginning of the dry season. (b) Distribution of livestock in relation to recorded range condition. (The standardised ratio TLU/household has been calculated as mean number of TLU/house-hold within grid square plus or minus standard deviation units. 0 reflects the mean ratio, −1 densities below the mean, +5 signifies the highest observed ratio.) Highest numbers of livestock (TLU) per household have been found in areas with fair to good range condition. Condition was poor to very poor in the vicinity of boreholes and settlement centres.

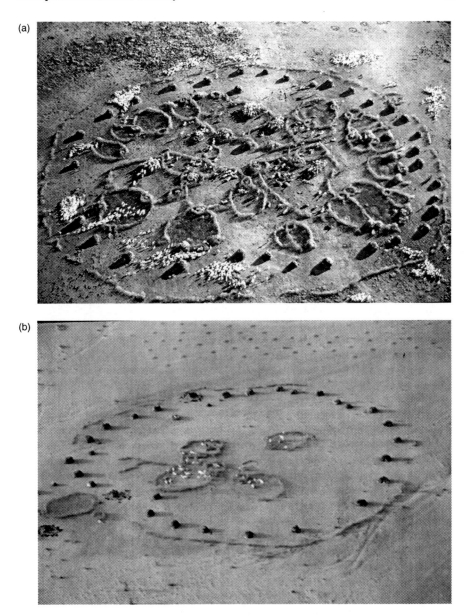

Figure 6. Rendille settlements, with livestock and impoverished. (a) Aerial photograph of a wealthy Rendille Gob (village) settling at grid reference 7/5 approximately 28 km from the nearest permanent water source. There are 32 huts built in a circle around the animal enclosures. The small white dots are sheep and goats; the larger dark ones are camels. The 32 households have available to them at the time an average of 17 TLU/household. (b) Aerial view of an impoverished Rendille Gob (village) settling at grid reference 10/2 at approximately 1.5 km from a deep borehole and approximately 2 km from Kargi trading centre. The 28 households have available to them at the time 35 camels, which translates into 1.8 TLU/household. (Photos by H.J. Schwartz).

Reduced mobility of households is probably compensated for, at least partially, by retaining or increasing the mobility of the herds. The strong fluctuation of livestock numbers per household between surveys can be taken as evidence supporting this statement.

Contraction of nomadic settlements around permanent water sources which have developed into centres of social attraction such as Korr and Kargi are indicative for the fact that subsistence production is giving way to market oriented production supported by wage labour (money returns from migrant workers). Fratkin (1989) found strong evidence for this in Korr.

The long term effect which can be expected is an accelerating depopulation of marginal areas and population concentration around a few commercial centres. This will certainly cause severe environmental degradation, however, only on a very limited scale within the region as documented in the recent range condition map.

3. CONSTRAINTS TO LIVESTOCK PRODUCTIVITY AND TRADITIONAL ADAPTIVE STRATEGIES

Constraints to livestock productivity in the traditional pastoral production systems can be divided into three different categories: normal constraints, disasters, and long term, irreversible changes such as increasing population pressure and constant loss of pastoral lands. The first two have always been part of the systems and adaptive strategies have developed to compensate for their effects. The third group is of more recent origin and largely beyond the control of the pastoralists.

Normal constraints are seasonal, annual, and spatial variation of rainfall and, accordingly, seasonal, annual, and spatial variability of quantity and quality of the available forage. Other normal constraints are endemic diseases, helminth burdens, external parasites and losses through predators and stock theft. Normal constraints can reach disastrous proportions from time to time. Rainfall variability can turn into drought, endemic diseases into epidemics and stock theft into tribal or civil war, which in turn can result in catastrophic stock losses for individual stockowners or even whole groups of pastoralists.

As an insurance against such events pastoralists strive to increase stock numbers, in order to provide security in case of losses, to leave a remainder of feasible size, to rebuild their herds. Thus, the expansion of herd sizes in "normal" times, not stricken by drought, disease or unrest, is a rational strategy and not a projection of prestige, social status, and wealth. Although it is true, that parallel to increased numbers of animals, an increased social standing for the owner will develop, this has to be seen as a favourable by-product of an effort to safeguard future survival.

Traditionally, risk-reducing adaptive strategies are herd diversification and herd dispersion. Herd diversification is practiced as an insurance against major disease outbreaks since the different domestic species are generally not susceptible to the same pathogens. Beside this, the different dietary preferences of the various domestic species also allow for a better utilization of pastures that may not be suited for one or the other domestic herbivore species. Herd dispersion is a second risk-reducing strategy, which is frequently practiced in traditional systems. Stockowners separate their herds and have them herded in areas sometimes up to several hundred kilometres apart; this is primarily a measure against forage shortages and raiding. If the family is large enough, the different herding units are managed by its members, and family reunions and rearrangements of the different stock sections take place either during the rainy season or during certain ritual occasions.

A related form of dispersion, although of a different significance is the formation of stock alliances and stock patronages that is independent of family size and social status. Individual animals or small groups of animals are given out to other stock owners who are either needy or in some way entitled to compensatory claims. Often the animals are never recovered by the original owner, but in times of hardship the son or even grandson might reclaim some or even all of the loaned stock from the recipient's heirs. This risk reducing strategy is common among all pastoralists whose social organization is based on clan and age-set structures and should be regarded as a system of mutual social security rather than an actual management tool.

The most conspicuous adaptive strategy of migratory pastoral production systems was, and still is, the mobility of households and herds. The migrations which are dictated by the availability of forage and water can follow various patterns but are always characterized by the combination of individual stock ownership and communal land use. This combination does not usually promote sustained-yield resource exploitation whenever land becomes scarce, and in particular when dry-season grazing reserves are no longer accessible. If confined to rainy season pastures throughout the year, the mobility of pastoral households and herds will be reduced to only minor moves, for hygienic or ritual reasons, since energy expenditure for a major move is not compensated for by a significant improvement of pastures.

4. RECENT IRREVERSIBLE CHANGES IN PASTORAL LIVESTOCK PRODUCTION AND MODERN ADAPTIVE STRATEGIES

Traditional pastoral production systems have remained stable for a long time, particularly through flexible responses to short-term variations of the climatic conditions. Today, however, numerous demographic and economic changes of long-term nature occur which trigger adaptive changes likely to transform this system significantly. The most salient feature is an emerging precedence of market oriented production over the traditional subsistence production. The major changes in the system are as follows:

- *Increased population pressure*: Pastoral populations are increasing steadily in the whole region. These increases at slower rates than in agricultural and urban groups, but may reach as much as 2% per year, which is no longer compatible with the human support capacity of the land.
- *Losses of pastoral lands*: A constant loss of pastoral land has to be noted. Competing land use systems such as commercial ranches, rainfed and irrigated agriculture and National Parks and Reserves occupy increasingly the small pockets of high potential land within the pastoral areas.
- *Reduced mobility*: Increases of the pastoral population and simultaneous losses of communal pastures are leading to a reduced mobility. Deterioration of the internal and external security aggravates this.
- *Environmental degradation*: The major effect of these developments is a general, but locally often severe environmental degradation, particularly around permanent settlements, mechanized water sources, mission stations etc. Although it can be generally stated that semi-arid to arid pastures show a remarkable recuperative potential in times of good rainfall, it cannot be denied that irreversible destruction of range vegetation has occurred and is spreading.

The change from a long ranging and highly mobile herding system to a short-range and semi-sedentary one bears the potential for both negative and positive effects. Amongst the most obvious negative effects are:

- the increase of environmental degradation,
- increased production risks for the individual herd owner as well as for the industry as a whole due to the disappearance of traditional adaptive management strategies,
- and the accelerated breakdown of social structures which previously served as a form of social security system within herding communities.

The emerging trends toward short-range herding systems have definitely deleterious effects on range vegetation and soils. Severest impacts are found around permanent wells and boreholes and in the immediate vicinity of permanent settlements. They are, however, limited spatially to a small section of the total range. Range enclosures and privatization on the other hand may lead to more widely spread damage. Grazing pressure on the residual open range is becoming exhaustive, migrations have to be rerouted, some migration routes may be closed permanently, thus increasing pressure on others, and, since areas with higher potential are usually enclosed first, the residual open range areas possess lower support capacities and are prone to faster degradation.

Enclosed range areas, although generally of slightly higher potential than the open range, are not immune against diminishing range condition, since stocking densities are rarely matched to the support capacity of the range but rather to the needs and demands of the stock owners, which often results in overstocking. Additionally, erratic spatial rainfall distribution during certain seasons or years may reduce forage growth in some enclosed areas and lead to temporary but severe overstocking and irreversible degradation. This may be aggravated if the breakdown of traditional resource-sharing attitudes in the pastoral system prevents emigration of herds from private lands.

Dry land farming, which is expanding within the agro-pastoral context, has adverse effects on range condition and soils. Land clearance and the sparse and temporary ground cover provided by annual crops favour increased erosion and often lead to irreversible degradation of range areas (desertification).

The consequence of all these effects, beside the inevitable range degradation, is a slow decline of herd productivity, reduced size of individual livestock holdings and productive land, and an increasing drought susceptibility of the whole system.

4.1. Economic Consequences of Sedentarization

The sedentarization of pastoral households is intrinsically related to the incorporation of pastoral economies into regional and national markets. Whether the observed emergence of market oriented production over traditional subsistence production is an additional cause or merely a consequence of sedentarization remains ambiguous; both processes are generally so intertwined that it is difficult to distinguish cause and effect (Sikina et al., 1993). Nevertheless, both processes entail a gradual change in pastoral herd management and species composition, and, ultimately, a redefinition of production goals. Many studies in pastoral systems have documented changes in species composition in pastoral live-stock herds arising from population increase, reduced mobility, and commercialization of production. Wealthy Bedouins in Saudi Arabia have turned from camel-rearing to sheep-raising (Abdalla et al., 2001). Likewise wealthy Rendille pastoralists have exchanged camels for cattle because of their higher retail value on local markets (Hary, 2000), and wealthy Boran herders in Southern Ethiopia have switched from cattle to camels to produce milk for sale in urban markets like Moyale or Guba (Younan, 2002).

Poor pastoral households, in contrast, concentrate on sheep and goats. The rapid rate of reproduction of small stock makes them a major means of post-drought recovery. Moreover, accumulation of small stock is a sensible strategy for poor households since they are easily converted to cash for household needs, and can be utilized as a means of acquiring large stock. Data presented by McCabe (1987) for the Ngisonyoka Turkana suggest a shift from a pastoral subsistence based on cattle to one based on small stock in a situation where livestock holdings per capita are decreasing. In contrast to sheep and goats, cattle have the disadvantage of being large indivisible units, such that substantial amount of the herder's wealth is stored in only a few animals. Poor households are therefore less vulnerable to livestock losses when concentrating on small stock, since here, the capital accumulated in each animal is minimal.

Patterns of small stock and cattle or camel utilization vary according to the scale of the operation unit. Poor pastoral producers achieve a greater efficiency of utilization of their small stock in terms of milk, meat and especially live animal sales than wealthy producers. Due to their low livestock to human ratio and pressing consumption needs, poor households have to engage in market exchange in order to convert their livestock products to foodstuffs with higher energetic value (Grandin, 1988; Sikina et al., 1993). With respect to cattle, poor herd owners tend to rely on an intensive extraction of milk from their herds, whereas richer herders deliberately forego some of the potential milk output in favour of calves and derive higher levels of income from live animal trade. In general, poor producers attempt to offset diseconomies of scale by intensive methods of extracting value from animals (Behnke, 1984).

The case of the Maasai group ranches suggests that the growth in small stock holdings may also be linked to a change in rangeland vegetation induced by an increase in land use pressure (Njoka, 1979). Goats in particular have a wider dietary range and lower water requirements than cattle, and are better adapted to cope with drought and poor grazing conditions such as often occurs in the vicinity of permanent settlements. Thus, the increase of small stock holdings especially among stock-poor households could also be interpreted as an adaptation to a degrading habitat. In general, it can be expected that those households that diversify their herds by keeping a significant proportion of sheep and goats may likely adapt more readily and securely to a sedentary lifestyle (Hary, 2000).

5. CONSEQUENCES FOR THE FUTURE OF PASTORAL ECONOMIES

From the foregoing, a number of consequences for the future of pastoral economies in Kenya can be anticipated. First, the rapid integration of pastoral economies in regional and national markets will continue to impact on management strategies and production goals of pastoral producers. The process of commercialization is, however, not neutral to scale, and will therefore affect large and small subsistence operations differently. Large herd owners can more readily accommodate the shift from in-kind milk and meat production to market oriented meat production, since the reduction in overall biological productivity implied by this shift is potentially more than offset by the higher economic profitability achieved through live animal sales (Behnke, 1984). Small operation units, in contrast, can not afford to abandon traditional subsistence forms of animal use. The potential off-take rate of live animals for sale from their herds is too low to meet the household expenses required to substitute the majority of subsistence products for non-pastoral foodstuffs. Small herd owners will therefore have to retain a predominant subsistence oriented mode of production in order to increase herd sizes for an eventual shift towards commercial production. On the other hand, the process of commercialization has been observed to entail a reduction in the number of surplus livestock from wealthier households available for redistribution through animal loans, gifts and other transfers, thus depriving small herd owners of an important source of livestock to build up their herds. In addition, those producers who successfully make the shift to commercial production will attempt to reinforce their economic superiority by acquiring private use rights to land, a widely spread phenomenon in North and East Africa (Bounejmate et al., 2001). Together, these developments will further undermine the viability of small production units and increase income disparities in pastoral societies (Bennett, 1988).

A second concern relates to the tendency of livestock herds to grow in commercialising pastoral economies, a trend which will aggravate the pressure on natural resources already exerted by the growing and sedentarizing pastoral population. This is because the decline in biological productivity per animal unit in commercial operations may force herd owners to increase their herd sizes in order to maintain overall household income levels. On the other hand, herd owners can also benefit from economies of scale by increasing their herds, since larger herds tend to have lower per unit operating costs. Whereas under the traditional, labour intensive subsistence mode of production, herd growth is restricted by labour availability, this limitation is relaxed in a commercialising operation, thus encouraging producers to increase their operations.

Lastly, the economic polarization of pastoral societies will promote the exclusion of poor, subsistence oriented households from the pastoral sector. Households falling below the minimum livestock per capita ratio required to insure self-sufficiency have a limited set of alternative strategies to choose from in order to complement household incomes. They may either realize additional income from herding for other, wealthier households, seek for wage labour in non-pastoral activities, or abandon the pastoral economy altogether and migrate to small towns or large cities in search for employment (Fratkin & Roth, 1990). Yet, the capacity of pastoralism to absorb surplus labour from households which lack sufficient stock to support themselves independently is very limited. In most African countries, employment opportunities outside the pastoral sector are few and pastoralists are at a comparative disadvantage in competing against the often better educated agriculturalists. Under these conditions, the likelihood of an increase in the number of displaced and destitute pastoralists is high.

In light of the complexity of the problems with which pastoral economies are confronted today, it would seem difficult to isolate a single development strategy that is capable of simultaneously alleviating the above mentioned social, economic, and environmental concerns. The present demographic and macroeconomic conditions in the African drylands prevent the relocation of large parts of the pastoral population to other sectors of the national economy. However, the exclusion of marginal operation units from the pastoral economy is largely unavoidable in the ongoing process of commercialization.

REFERENCES

Abdalla, S.H., Hajooj, A., and Simir, A., 2001, Economic analysis of nomadic livestock operations in northern Saudi Arabia. In: Squires, V.R., Sidahmed, A.E. (eds.). *Drylands. Sustainable use of rangelands into the twenty-first century.* IFAD Series: Technical Reports. Rome: IFAD. Reproduced at www.odi.org.uk/pdn/drought.

Ahmad, Y., 2001, The socio-economics of pastoralism: a commentary on changing techniques and strategies for livestock management. In: Squires, V.R., Sidahmed, A.E. (eds.). *Drylands. Sustainable use of rangelands into the twenty-first century.* IFAD Series: Technical Reports. Rome: IFAD. Reproduced at www.odi.org.uk/pdn/drought.

Bake, G., 1991, Water sources in Marsabit District. In: Schwartz, H.J., Shaabani, S., Walther, D. (eds.). *Range Management Handbook of Kenya,* Volume II, 1, Marsabit District. Republic of Kenya, Ministry of Livestock Development. Nairobi.

Behnke, R.H., 1984, Fenced and open range ranching: The commercialization of pastoral land and livestock in Africa. In *Livestock Development in Subsaharan Africa: Constraints, Prospects, Policy,* edited by J.R. Simpson and P. Evangelou, pp. 261–284. Boulder: Westview Press.

Bennett, J.W., 1988, The political ecology and economic development of migratory pastoralist societies in Eastern Africa. In *Power and Poverty: Development and Development Projects in the Third World,* edited by D.W. Attwood, T.C. Bruneau, and J.G. Galaty, pp. 31–60. Boulder: Westview Press.

Bounejmate, M., Mahyou, H., and Bechchari, A., 2001, Rangeland degradation in Morocco: A concern for all. *ICARDA Caravan* No 15, pp. 33–36.

Coughenor, M.B., J.E. Ellis, D.M. Swift, D.L. Coppock, K. Galvin, J.T. McCabe, and T.C. Hart, 1985, Energy extraction and use in a nomadic pastoral system. *Science* 230:619–625

Ellis, J.E., and D.M. Swift, 1988, Stability of African pastoral ecosystems: Alternate paradigms and implications for development. *Journal of Range Management* 41:450–459.

Fratkin, E., 1991, *Surviving drought and development—Ariaal pastoralists of Northern Kenya.* Boulder: Westview Press.

Fratkin, E., 1992, Drought and development in Marsabit District. *Disasters* 16:119–130.

Fratkin, E., and Roth, E.A., 1990, Drought and economic differentiation among Ariaal pastoralists of Kenya. *Human Ecology* 18:385–402.

Grandin, B.E., 1988, Wealth and pastoral dairy production: a case study from Maasailand. *Human Ecology* 16:1–21

Herlocker, D., 1991, Range condition in Marsabit District. In *Range Management Handbook of Kenya, Volume II,1, Marsabit District,* edited by H.J. Schwartz, S. Shaabani, D. Walther. Republic of Kenya, Ministry of Livestock Development. Nairobi.

Hary, I., 2000, *Effects of seasonality on the productivity of pastoral goat herds in northern Kenya.* Ph.D. Thesis. Humboldt-Universität zu Berlin.

Hogg, R., 1986, The new pastoralism: poverty and dependency in Northern Kenya. *Africa,* 56:319–333.

Mayer, H., R.A.B. Abdel, and B.B. Bös, 1986, The pastoral nomadic subsistence sector in Africa. Sources and consequences of economic and social transformation processes. *Economics* 33:98–110.

McCabe, J.T., 1987, *The importance of goats in recovery from drought among East African pastoralists.* Proceedings IV International Conference On Goats, March 8–13 1987, p. 1531. Brasilia.

O'Leary, M.F., 1990, Drought and change amongst Northern Kenya nomadic pastoralists: the case of the Rendille and Gabbra. In *From Water to World Making. African Models and Arid Lands,* edited by G. Pálsson, pp. 151–174. Uppsala: Scandinavian Institute of African Studies.

Njoka, J.T., 1979, *Ecological and socio-cultural trends of Kaputei group ranches in Kenya.* Ph.D. Thesis, University of California, Berkeley.

Oxby, C., 1975, *Pastoral nomads and development.* London: International African Institute.

Roth, E.A., 1996, Traditional pastoral strategies in a modern world: an example from Northern Kenya. *Human Organization* 55:83–92.

Schwartz, H.J., 1991, Range Unit Inventory. In *Range Management Handbook of Kenya, Volume II, 1, Marsabit District,* edited by H.J. Schwartz, S. Shaabani, D. Walther. Republic of Kenya, Ministry of Livestock Development. Nairobi.

Schwartz, H.J., 1993, Pastoral production systems in the dry lowlands of Eastern Africa. In: M.P.O. Baumann, J. Janzen, H.J. Schwartz, Pastoral production in Central Somalia. *Schriftenreihe der GTZ* 237, TZ-Verlag, Roßdorf.

Schwartz, H.J., C. Mosler, I. Hary, and V. Pielert, 1995, Factors affecting spatial preferences in settlement site selection n migratory pastoralism *Environmetrics* 6:485–490

Schwartz, H.J., S. Shaabani, and D. Walther, (eds.), 1991, *Range Management Handbook of Kenya, Volume II, 1, Marsabit District.* Republic of Kenya, Ministry of Livestock Development. Nairobi.

Schwartz, S., and H.J. Schwartz, 1985, Nomadic pastoralism in Kenya—still a viable production system? *Quarterly Journal of International Agriculture* 24:5–21.

Sikina, P.M., C.K. Kerven, and R.H. Behnke, 1993, *From Subsistence to Specialised Commodity Production: Commercialisation and Pastoral Dairying in Africa.* Pastoral Development Network Paper No. 34d, Overseas Development Institute, London.

Sobania, N.W., 1980, *A Historical Study of Rendille Migration Patterns.* A consultancy report to the UNESCO/FRG Project on Traditional Livestock Management. Nairobi: UNESCO Regional Office.

Swift, J., 1982, The future of African hunter-gatherer and pastoral peoples. *Development and Change* 13:159–181.

Western, D., 1982, The environment and ecology of pastoralists in arid savannas. *Development and Change* 13:183–211.

Younan, M., 2002, Personal communication.

Chapter 5

Cursed If You Do, Cursed If You Don't

The Contradictory Processes of Pastoral Sedentarization in Northern Kenya

JOHN MCPEAK AND PETER D. LITTLE

1. INTRODUCTION

The increased sedentarization of pastoral populations has characterized most arid and semi-arid regions of the world during the past two millennia. The origins of many towns in the Middle East, North Africa, and the Sudan stem in part from the historical process of pastoral sedentarization, whereby segments of mobile herders sought refuge or economic opportunity from settled life. The accelerated settlement of herder populations in Sub-Saharan Africa during the past century provoked claims of 'an end to pastoralism' as diversification into town-based activities was seen as a departure from pastoralism, rather than a supplement or support to it (Government of Kenya, 1980; Snow and Morris, 1984). Northern Kenya is unique in this respect since most settlements and towns have only arisen in the past 50 or so years, and mobile pastoralism still characterizes large parts of the region. The increased sedentarization in the area also reflects a series of external influences, such as the widespread proliferation of food aid and other forms of development assistance, which complicates an understanding of longer-term trends toward settlement. It also questions whether or not increased sedentarization among herders in the region really reflects an enduring commitment away from pastoralism.

JOHN McPEAK • Department of Public Administration, Syracuse University, Syracuse, New York 13244.
PETER D. LITTLE • Department of Anthropology, University of Kentucky, Lexington, Kentucky 40506-0024.

This chapter presents some preliminary results from a three-year research effort examining pastoral risk management in northern Kenya and southern Ethiopia. The materials here are only from the Kenyan research sites and are supplemented by the authors' earlier studies in the area: McPeak studied the Gabra of Marsabit District during 1997–1998 and Little the Chamus of Baringo District, 1980–1998. One main finding of the current study is that what is often called 'sedentarization' does not necessarily reflect a full-time departure from pastoralism, nor does it always jeopardize pastoral production. When one explores intra-family and intra-household dynamics, a changing pattern of diverse combinations of sedentary-like occupations and pastoral activities are revealed. We find that sedentarization does not imply a lack of access to livestock, nor always a lack of mobility for livestock owned by settled households. In addition, we identify a great deal of diversification into non-pastoral activities in areas where households remain involved in pastoral production, even while members are engaged in waged labor and other sedentary activities. The historical process of declining per capita livestock holdings noted by other studies (Fratkin, 1991; Little et al., 2001) has led households to respond by having certain family members leave the system to allow remaining members to pursue herding. Those who depart from the pastoral sector can enter economic niches that generate resources for the pastoral economy through remittances and other transfers. Even agriculture can support pastoralism by generating grains that reduce a family's need to sell off livestock to finance cereal purchases, a pattern that becomes very important in post-drought periods when livestock holdings are reduced (Little, 1983, 1992).

While this appears to be an important trend revealed in our data, our findings provide important counter-examples to this general pattern. In some cases we observe impoverished households that leave the system cluster around towns and earn incomes insufficient to transfer back to the pastoral sector. It is this later vulnerable group that has attracted much of the attention of scholars and development practitioners, as well as those who claim an imminent end to pastoralism. We also find households which fared relatively well in one of our more arid sites (Kargi) where highly mobile pastoral production was practiced during the most recent drought. In contrast, for one of our most sedentary sites (Dirib Gumbo), we find diversification into agriculture and investment in education did not provide households with many benefits during the drought period. Finally, we suggest that the relationship between sedentarization and vulnerability in livestock wealth and sedentarization and vulnerability in food security need to be carefully distinguished. It is clear in our data that sedentarization does make households more vulnerable to livestock losses in a drought, but that the relationship between food security and sedentarization varies between our sites in complicated ways that we will elaborate later in the chapter.

2. BACKGROUND TO RESEARCH REGION AND DESCRIPTION OF THE STUDY SITES

The chapter draws on research conducted by the Pastoral Risk Management Project (PARIMA) of the Global Livestock Collaborative Research Support Program.[1] The study area covers approximately 10,000 square km and encompasses parts of the rangelands of southern Ethiopia and northern Kenya. The study region is bounded by the towns of Hagre Mariam and Negelle in Ethiopia and Isiolo and Marigat in Kenya (see Figure 1) and includes Boran, Gabra, and Guji of Ethiopia, and Ariaal, Boran, Il Chamus, Gabra, Rendille, Samburu, and Tugen peoples of Kenya.

The data presented in this study draw on six northern Kenya study sites, where 30 households were randomly selected in each site. Four sites are in Marsabit district, one is in Samburu district, and one is in Baringo district. Sites were chosen to represent diversity in ethnicity, mean rainfall, and market access as described in Table 1 and elaborated upon below. The sites are noted on the map in Figure 1.[2]

A unique aspect of the study is that we used an areal sampling framework based on the Kenyan administrative unit, the Location. This means that our sampling methodology did not distinguish between pastoral and sedentary households. Thus, the study includes households and individuals residing in or near towns, as well as mobile pastoral households residing away from towns. Because the sample was randomly selected within each of the

Table 1. Sites Where Data Were Gathered.

Site	District	Predominant ethnic group	Average annual rainfall	Market access
Dirib Gumbo	Marsabit	Boran	650	Medium
Ngambo	Baringo	Il Chamus	650	High
Sugata Marmar	Samburu	Samburu	500	High
Logologo	Marsabit	Ariaal	250	Medium
Kargi	Marsabit	Rendille	200	Low
North Horr	Marsabit	Gabra	150	Low

Survey Sites in
Southern Ethiopia and Northern Kenya

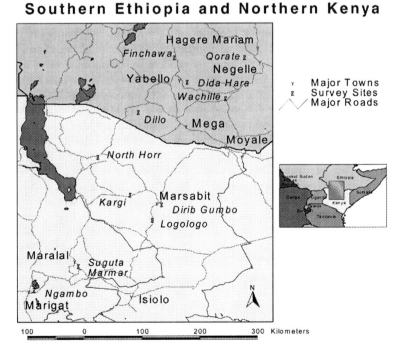

Figure 1. Pastoral risk management project study area.

six locations, the degree of sedentarization reflected in the sample can be taken to be representative of the degree of sedentarization of the location level population.

Households were interviewed with a baseline survey in March 2000 and were re-interviewed at three-month intervals following this baseline. The data presented in this chapter draws on the baseline and the first six repeated surveys, covering the period March 2000 to September 2001; and on a series of qualitative, structured interviews conducted during July to October 2001. Table 1 summarizes the study sites, and is followed by brief descriptions of the six sites ordered from highest rainfall to lowest rainfall.[3]

Dirib Gumbo is a Boran settlement approximately 10 km from Marsabit town.[4] The majority of the market activity undertaken by Dirib Gumbo residents takes place in Marsabit town. Most the residents of this area reside on the upper slopes of Marsabit Mountain, and practice rain fed cultivation. Many households in this area also keep livestock. In some cases these animals are used to plow fields. Very little large-scale migration of animals takes place from this location, both due to relatively small herd sizes and because nearby pastures are controlled by other ethnic groups. Herders instead rely on crop residues, forest products, or pasture on the lower slopes of Marsabit Mountain to feed their animals.

Ngambo is an Il Chamus settlement approximately 10 km east of Marigat town (see Little 1992). Marigat town is located 100 km north of Nakuru on an all-weather road. Marigat town is the major market center used by Ngambo residents. Marigat is particularly lively during the twice-monthly livestock auction held just outside town. Ngambo is located near the Pekerra irrigation scheme, and a large number of households of this location either grow crops in this scheme themselves or work as laborers in these fields. Their form of pastoralism is markedly sedentary, but does entail seasonal herd movements of 20–30 km, during the dry season. The majority of family members rarely move during the year.

Sugata Marmar is a Samburu settlement on the Laikipia–Samburu District border, approximately 50 km south of Maralal on the Maralal–Rumuruti road. Significant populations of impoverished Turkana and Pokot are resident in this location as well. Sugata Marmar has a large weekly livestock market offering households the opportunity for alternative income sources and a place to sell animals. Some rain-fed cultivation is practiced in this area, particularly in the higher elevation areas towards Maralal town, the administrative center of Samburu District. Pastoralism in the area is moderately mobile and cattle can be moved distances of 50+ km during harsh dry seasons or droughts. Rather than the whole family moving with the herds, households mainly rely on a combination of satellite camps of young men (16 years and older) to care for and migrate with the animals and, for polygamous households, moving animals between households of different wives that are established in different areas. From interviews it seems that families themselves used to migrate with the herds more frequently before the 1980s.

Logologo is an Ariaal settlement approximately 40 km south of Marsabit town on the main Isiolo–Marsabit road. Ariaal are a group that mixes elements of Samburu and Rendille culture (see Fratkin, 1991). Logologo residents utilize markets in both Marsabit town and in Logologo town. Rain-fed agriculture is possible in the higher areas of this location, and a very small amount of small-scale irrigation is practiced in town. Most households in Logologo settled there in the 1970s following a series of poor rainfall years and herd losses. Like the Samburu mentioned above, they no longer move the whole family with their animals. Instead, they keep small herds in the area around town and send the majority of their animals to satellite camps in the surrounding rangelands.

Kargi is a Rendille settlement approximately 75 km to the west of Marsabit town in a flat, arid basin. Kargi residents mostly conduct market activity in Kargi town, although they make occasional use of Marsabit markets. No cultivation is practiced in this area. Over the past 20 years, formerly nomadic Rendille have settled around the town center in clan groupings. Rendille in the Kargi area keep small herds in the area around town and rely on young men to stay with the remainder of the herd in highly mobile satellite camps. They keep relatively large numbers of camels and goats and it is not unusual for their camps to move several times during a season.

North Horr is a Gabra settlement approximately 200 km west of Marsabit town on the northern edge of the Chabi desert. Similar to Kargi, most market activity takes place in North Horr town, although residents do make occasional marketing trips to Marsabit town. No cultivation is practiced here. Many Gabra are nomadic in the traditional sense, as households move their house and household belongings to new areas with their animals with some frequency. However, the time between these moves is becoming longer and the area covered by these moves is becoming smaller as Gabra slowly appear to be moving toward the satellite camp based system of their Rendille neighbors. Gabra also keep relatively large numbers of camels and goats in their herds.

3. PRELIMINARY SURVEY FINDINGS

In this section, we present some of our preliminary findings based on cross-community comparisons. At the time this chapter was prepared, we were still involved in the data gathering process. Because of this, we have not yet prepared the data for analysis at the individual and household levels. However, it is possible to analyze the data set in its current form at the community level, but even in this case the findings should be treated as preliminary. Therefore, we advance these findings as broad cross community comparisons that will be refined by further analysis at the household and individual levels when data gathering is complete.

3.1. Herd Size Change, Mobility, and Sedentarization

The period covered by the survey began with a generalized drought in northern Kenya. Figure 2 presents information on rainfall in our study sites. The main impact of the drought was felt throughout the study area in 2000. Overall, sample households lost an average of around 25% of their herds between March 2000 and March 2001. The overall rate of stock-less households in our sample increased from 7% in March 2000 to 12% in September 2001.

Survey results indicate there are large differences in herd size between sites and in the losses experienced. With regard to herd size, these differences generally follow the pattern that larger household herds are located in the study region's drier areas, such as North Horr and Kargi. The listing of sites from largest median herd to smallest median herd when all households and time period specific observations are included is as follows (Total Livestock Unit values in parentheses[5]): Kargi (21), North Horr (20), Logologo (9), Dirib Gumbo (6), Sugata Marmar (4), and Ngambo (2).[6] However, as seen in Figure 3, there are major differences between locations in the pattern of herd size change over the survey period.

Figure 3 indicates that there are differences in the severity of herd losses when the different areas are compared. We propose that part of the explanation for these differences is found in the degree of livestock mobility. Increased mobility has a relationship with decreased herd loss. This is revealed by the information presented in Table 2. Table 2 presents for each site the overall change in herd size between March 2000 and September 2001, the maximum decline during this time period, the average number of water points used by a household having herds per period, the percent of these points that were described as satellite camps, and the total number of water points used by households in

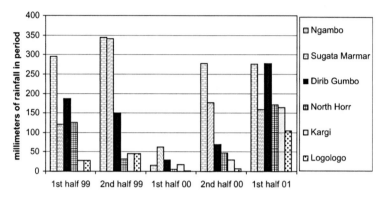

Figure 2. Rainfall in the study sites.

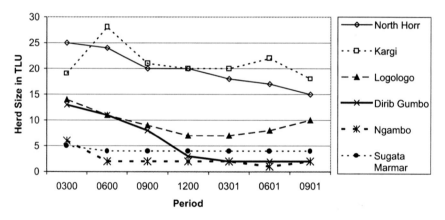

Figure 3. Median herd size by season, March 2000 to September 2001.

Table 2. Herd Size Change and Mobility.

	% Decline 0300 to 0901	Maximum % decline	Average H_2O points used	% Satellite camps	Total H_2O points named
Dirib Gumbo	−85%	−85%	1.1	46%	10
Ngambo	−67%	−83%	1.5	1%	18
North Horr	−40%	−40%	1.7	45%	56
Logologo	−29%	−50%	2.0	81%	54
Sugata Marmar	−20%	−20%	1.3	28%	60
Kargi	−5%	−5%	3.3	88%	40

each location over the March 2000 to September 2001 period. Mobility is measured by the number of water points and reliance on satellite camps by the different communities.

With the exception of Sugata Marmar, the table indicates that higher levels of mobility are associated with lower herd losses. In most sites, the higher the average number of water points used per period, the lower the average herd size decrease. This is also true for satellite camp use. Findings from Sugata Marmar suggest that having many water points to visit may also make a difference. Herders in Sugata mentioned more points than any other area, suggesting they are able to spread out over more area. Households use relatively fewer points per period than in other areas, but use more water points overall than in other areas.

When we turn to the issue of herd size change within areas, the data suggest that herd accumulation is an effective strategy for ensuring a viable post-drought herd size. Figure 4 presents (natural log transformed) herd size in March 2000 compared to (natural log transformed) herd size in March 2001 at the household level. A 45-degree line is added to this graph. In this graph, a dot above the line indicates a household herd increased in size over the period, a dot on the line means the beginning and ending herd were the same size, and a dot below the line means the household herd decreased in size over the study period.

The cloud of dots slightly below the 45-degree line reflects the decreases in average herd size reported in figure one. However, note the pattern of the dots is generally upward sloping. Households that had more animals in early 2000 tend to be the households that have more animals in 2001. The relationship is not perfect, but it does help to explain why herders attempt to maximize herd size in good (pre-drought) years—an attribute that many 'experts' associate with pastoral irrationality. Some households did better than others, and actually realized increased herd size over the one-year interval. Others did much worse. The most extreme examples are found for the households represented by the dots along the lower axis. These dots indicate that households with herds of up to 25 TLU in March 2000 had become stockless only a year later.[7]

While it is early in the herd rebuilding process, our preliminary results indicate herd rebuilding is more a matter of the biological process of animal birth within the family herd

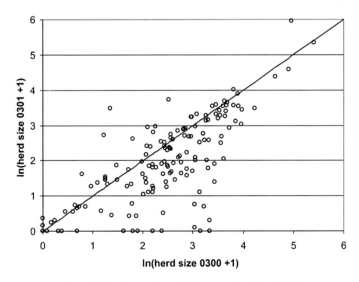

Figure 4. Herd size in March 2000 compared to March 2001.

than restocking through transfers from other herders or restocking through use of the market. From March 2001 through September 2001, the average birth rate from the total herd in TLU terms was an annualized 28 percent, compared to purchases in the markets and receipts of livestock gifts and loans, which account for an annualized rate of 3 percent respectively. In short, if a herder was more mobile and had a larger herd s/he generally fared better during and immediately after a drought than others.

The following three case studies of individuals illustrate the kinds of strategies that herders invoke to cope with and recover from drought.[8]

Case A. William is an Il Chamus from Ngambo who is in his mid-30s and is relatively well-educated with two years of secondary school training. He lost about 75 percent of cattle during the 1999/2000 drought and is attempting to rebuild his herd through several different mechanisms. First, he is relying on small stock, which reproduce faster than cattle. He even recently bought additional goats from the sale of a young bull that survived the drought. Second, he is pursuing agriculture both to produce surplus grain which he can sell locally to buy livestock, and to have sufficient food so that he does not have to sell animals to buy grain. This latter strategy is cited by a number of herders in our study region, but only where cultivation is feasible. Finally, William is rebuilding his herds through the use of remittances from family members working outside Ngambo and from customary livestock gifts and exchanges. In comparing his current herd rebuilding strategies with those following the equally devastating 1991–1992 disaster, he complains that local livestock gifts and transfers were more common in 1992 than now, because widespread poverty has diminished the capacity to help family members and friends. "A destitute person cannot help other destitute persons," he explains. According to William, it took about two years to rebuild his livestock after the 1991–1992 drought, but he thinks it will take longer now.

Case B. Lenapir is an Ariaal Rendille from Logologo, living about 40 km south of Marsabit town. She heads her own household and is engaged in farming and some petty trade around Logologo town. She is a widow with six children. A lion killed her husband in 1992. During the 1999–2000 drought she lost 75 percent of her animals and is now a fairly destitute pastoralist. Her most important strategies for herd recovery are assistance from her brothers (one of whom is relatively wealthy), receiving food aid so that she can meet subsistence needs without having to sell animals, and natural reproduction of her remaining herd. She was expecting that some of her cattle would give birth in late 2001, helping her on her way to recovery. Compared with the drought of 1991–1992 when she lost 60 percent of her cattle, she thinks herd recovery in 2001–2002 will be more difficult. The main reason for this is because her wealthy brother also lost many of his animals and will not be able to help her as much as in the past. In addition, local assistance networks to help impoverished herders are not as salient as they were in the 1970s and 1980s.

Case C. Gondara is a Gabra household head form North Horr who lost more than 50 percent of his livestock during the 1999–2000 disaster. He practices a very mobile form of pastoralism but still lost more than 50 percent of his livestock holdings. In addition to relying on food assistance, he feels that herd reproduction will be the main mechanism to assist herd recovery. He does not receive any remittances and is too far from markets to rely on these sources to rebuild his battered livestock herd. Already he says that recovery is well underway because he is focusing on goats, which have been giving birth in large numbers since the end of the drought. He also says that his productive female camels are already pregnant (October 2001) and will soon give birth. He also has given two goats to a fellow clan member who was particularly affected by the recent drought. He remains optimistic that his herds will recover within a few years.

The three case studies presented above show the different strategies that herders invoke to recover from drought. Those households in areas with good access to markets and employment sources, such as Ngambo, rely more on waged incomes and livestock purchases to recover, than the more remote sites (such as North Horr) where herd recovery is mainly through biological reproduction. Local livestock exchanges and gifts also seem to be more important in areas without favorable access to markets; in both Ngambo and Suguta Marmar local exchange systems were described as playing minimal roles in herd recovery.

3.2. Income and Expenditure

In the preceding section, we considered the issue of herd size and herd size change. This gave us some understanding of household welfare in terms of household assets (especially livestock) and household asset change. In this section, we investigate household welfare in terms of income and expenditure. We use these measures to provide some understanding of household welfare and food security during the study period.

To begin with, we consider the general pattern of income generation revealed in our data set. In Figure 5, we present the relative proportions of income accounted for by different activities in our total sample. The right side of the pie chart represents income generated from livestock or livestock products (livestock sales, milk sales, hide and skin sales). The left side represents non-livestock related income sources (salary labor, trading revenue, wage labor, firewood or charcoal sales, cultivation, craft sales, and water sales). Interestingly, there tends to be highly gendered access to different income sources, with women playing a key role in many of the non-livestock related activities (craft sales, cultivation, and firewood sales) and in the sale of dairy products.[9] In a related study, Nduma et al. (2001) find that there are differences between women that influence which of these activities will be emphasized. We expect that further analysis will deepen our understanding

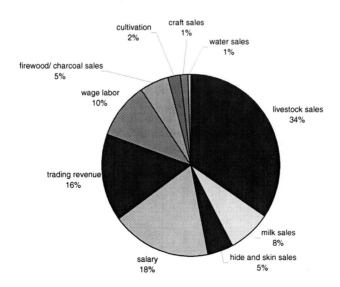

Figure 5. Overall shares of income from different sources in the data set.

of how individual characteristics, particularly gender and wealth, influence activity choice in this area.

Figure 5 indicates that slightly over half (53%) of the total income recorded for our households came from sources other than livestock or livestock product sales. When we turn to the site-specific data, we can see that this overall pattern varies significantly between sites. Figure 6 shows that non-livestock income sources provide the majority of income for three sites: Logologo, Ngambo, and Sugata Marmar. Intriguingly, these three sites have higher income than the sites where income from livestock accounts for the majority of income. It is also interesting to note that the income from livestock sales is less variable across sites than is the variability in income from non-livestock sources.[10] Those households with better access to markets and infrastructure have higher and more diversified incomes. These households also tend to be characterized by a greater degree of sedentarization and decreased mobility, but there are important exceptions to this tendency that are discussed later in the chapter.

We also asked households to report their cash expenditures over a two-week period for a variety of commodity categories.[11] The sum of these expenditures provides a measure of household well being, under the assumption that higher expenditures reflect higher consumption. We present these results as cash expenditures, which often provide a better indicator of well being than reported incomes. This partly is due to difficulties involved in accurately recording total income (see Little, 1997).

One problem with the cash expenditure measure is that it does not include the consumption of non-marketed goods. As milk consumed from household herds can constitute a major proportion of total household consumption in a pastoral setting, a low level of market involvement—hence cash expenditures—may not reflect low welfare if households are meeting their consumption needs by consuming milk from their herds. Milk production at the household level provides a significant contribution to household welfare, as seen in Figure 7, which presents the average amount of milk households reported was available for human consumption in each period.[12]

Recognizing the potential role home consumed milk plays in household well being, a measure of the value of home produced milk was constructed by valuing the daily milk

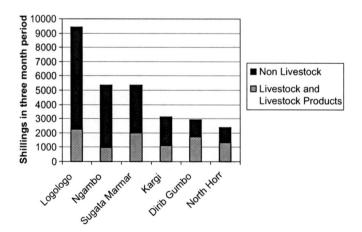

Figure 6. Total income reported over a three-month period.

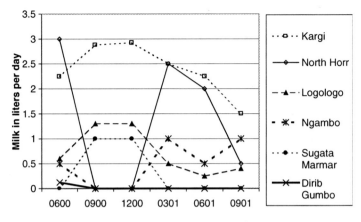

Figure 7. Median daily milk production per household for human consumption.

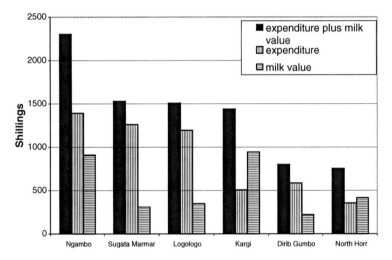

Figure 8. Well-being measures using average expenditure and the value of milk production.

production level reported by the household at the prevailing local market rate and summing this value over a two-week period. The resulting sum was then added to the two-week cash expenditure information to provide a more robust measure of household well-being. Figure 8 presents the average for each site for the following measures: the sum of the cash expenditure and the milk value, the cash expenditure alone, and the milk value alone.

The combined expenditure and milk value data presented in Figure 8 provide a slightly different perspective on household welfare than that provided by the income results presented in Figure 6. Ngambo has the highest welfare by the measure that combines cash expenditure and the value of milk, largely due to the fact that the market value of milk in Ngambo is the second highest of the six sites.[13] In addtion, Kargi appears better off by this measure than by the income measure, as a major part of the diet in Kargi is milk from the herds. Both measures indicate that average welfare in Dirib Gumbo and North Horr is lower than that found in the other sites.

Variation in the expenditure and milk value measures provides one other perspective on household welfare (Figure 9). This is based on the idea that households attempt to smooth consumption over time. A household having food surplus to household needs in one season and then confronted by a food deficit in the following period would almost certainly be better off if they could consume the average of these two extremes in both periods. This would mean that higher variability in the expenditure and milk value measures is associated with lower welfare, as it is assumed that households would be happier if they could avoid such fluctuations in consumption over time.

Again, households in Dirib Gumbo and North Horr are worse off than households in other sites on average. Not only is the average measure of expenditure plus milk value lower than in other sites, it is also relatively more variable. We also find that Ngambo and Logologo are relatively better off by this measure. As shown earlier in Figure 5, salary labor, wage labor, and trading account for a great deal of non-livestock and livestock product income. One explanation for the positive welfare indicators in Ngambo, Logologo and Sugata Marmar is that these three activities account for 61%, 55%, and 41% of income respectively in these sites. In contrast, these three activities account for 30% in North Horr and Dirib Gumbo and 34% in Kargi. Increased welfare, as measured by higher average and less variable expenditure and higher income, are associated with higher levels of salary labor, wage labor, and trading.[14] Conversely, it should be noted that the three more diversified sites suffered severe herd losses during the recent drought (see Figure 3). However, in spite of these losses, it appears that households in more diversified sites were not as exposed to the risk of food insecurity as households in less diversified sites. This leads to the question, what allows one site to become more diversified than another site? Beyond issues of location and infrastructure discussed above, our findings suggest education may play a role in the diversification of household income generation activities. We turn to this topic in the next section.

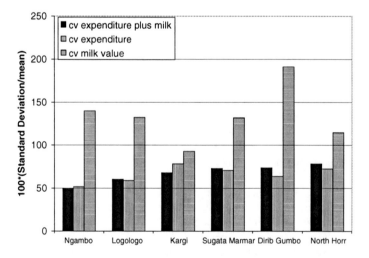

Figure 9. Coefficient of variation in average expenditure and the value of milk production.

3.3. Education

What role does education play in allowing access to the non-pastoral income sources noted in the previous section? To investigate this question, we sum the ages of all household members and the number of years each household member spent in school. This is used to derive the fraction of years spent by household members in a school, which provides a measure of past education. This is presented in Figure 10.

This measure of household education levels is closely related to the measure of the share of income that comes from non-pastoral sources. The areas where households have spent more time in formal schooling are also the areas that derive a higher share of their income from non-pastoral sources. Also, with the exception of Dirib Gumbo, the areas in which households have higher human capital stock of education are better off in terms of the income and expenditure measures of Section II and in terms of food insecurity.

We can again draw on our case study material to elaborate on this issue. The case of Letamara of Ngambo is illustrative of the positive role that education can play in food

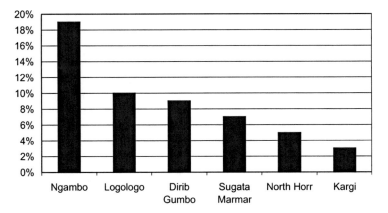

Figure 10. Percent of years household members spent in schooling.

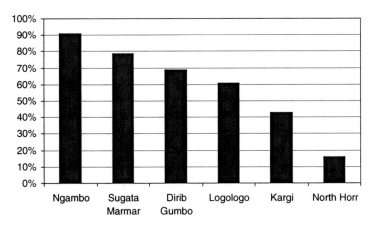

Figure 11. Enrollment of eligible children in school in 2000.

insecurity, even when livestock have been devastated by drought. Even before food relief came to Baringo in 2000, Letamara's household was able to purchase adequate food because he held a government job in Marigat and because his brother was employed in Nakuru town. This occurred despite the fact that he lost almost 80 percent of his livestock holdings and there was virtually no milk available from his herd for household members. Both Letamara and his brother had graduated from secondary school and in the case of the brother he had attained a college degree. Not only did remittances and wages allow Letamara's household to fare better than other households—even those with more livestock—but they also were able to help numerous other family members and relatives with food purchases.

We also present the enrollment rates for each of our sites for the 2000 school year. This provides a measure of current education levels. This information is presented in Figure 11.

The data indicate that enrollment of school age children differs drastically by site. When we compare these findings with the information on mobility presented in table two, it appears there is an inverse relationship between mobility and enrollment, a pattern that is especially apparent by contrasting Letamara's Ngambo area and Gondara's North Horr area. Sites where there is higher mobility have lower enrollment rates and vice versa. Thus, while Ngambo is the least mobile site in our study region (i.e., the most sedentarized), it has the highest levels of education and is among the most economically diversified communities in the area. Moreover, as we have shown above it has been able to parlay education into jobs and to avoid severe food insecurity by using wages and remittances to compensate for herd losses and low herd productivity. Gondara, by way of contrast, has never been to school and has none of his children in school. He finds it difficult to reconcile remaining near towns with the nomadic lifestyle demanded by his herds. However, he is optimistic that the mobile pastoral strategy will allow him to rebuild his herd quickly and thus avoid future food insecurity.

3.4. Comparisons of Different Communities

As this chapter has shown, there are considerable differences in the degree of sedentarization and diversification among the different research sites in our study region. Table 3 provides a summary of some of these findings as they relate to the general themes of sedentarization and pastoral welfare.

In spite losing over 80% of their herds, the average household in **Ngambo** fared relatively well during the recent drought, as shown by their relatively high mean milk value

Table 3. Summary of the Order of Sites for Different Measures (Ranked in Order of Importance).

Higher mean milk value + expenditure	More stable milk value + expenditure	Higher mean income	Higher non pastoral income %	Higher average herd size	Lower maximum herd loss	Higher water points used	Higher enrollment in 2000
Ngambo	Ngambo	Logologo	Ngambo	Kargi	Kargi	Kargi	Ngambo
Sugata M.	Logologo	Sugata M.	Logologo	North Horr	Sugata M.	Logologo	Sugata M.
Logologo	Kargi	Ngambo	Sugata M.	Logologo	North Horr	North Horr	Dirib G.
Kargi	Sugata M.	Kargi	Dirib G.	Dirib G.	Logologo	Ngambo	Logologo
Dirib G.	Dirib G.	North Horr	North Horr	Sugata M.	Ngambo	Sugata M.	Kargi
North Horr	North Horr	Dirib G.	Kargi	Ngambo	Dirib G.	Dirib G.	North Horr

plus expenditure value[15]. As we have demonstrated, Ngambo households have access to work opportunities in the nearby town of Marigat, which has a lively market and is connected by an all-weather road to Nakuru, and can also find work in the local irrigation scheme. Salary, wage labor, and trading account for over 60% of household income. Past education levels and current enrollment rates are the highest of any site in our study region.

In contrast, the average **Dirib Gumbo** household lost over 80% of their herd and their well-being does appear to have been negatively impacted by the drought.[16] Although Dirib Gumbo is not distant from the market town of Marsabit and Dirib Gumbo households have relatively high education levels, they have a relatively low share of their income from non-pastoral sources.[17] Salary, wage labor, and trading account for only 30% of household income. This is probably because most Dirib Gumbo households rely on rain fed agriculture in normal years. When the rains failed during the drought period covered by the study, many of the households sold livestock to meet consumption needs, which may explain why the share of income from livestock and livestock products is relatively high.

The average household in **Sugata Marmar** was not severely impacted by the drought in terms of herd loss or well-being. Partially, this may reflect the fact that the rainfall data from the area (see Figure 2) suggests the drought was less severe in Sugata Marmar than in other areas. It may also reflect the fact that households in this location have access to income generating opportunities arising from the large weekly market held in this town. Households in Sugata Marmar earn a relatively higher share of their income from trading (25%), than is found in any other site, and combined with income from selling their own livestock and livestock products (45% of income), they earn considerable revenues from trade. The mobility of livestock in this area differs from that in other areas perhaps because there are more water points available in the Samburu grazing lands. By relying on satellite herd camps and multiple established households, Sugata Marmar families appear to have found a compromise between mobility and education, as is seen by the relatively high enrollment rates.

The average household in **Logologo** lost roughly half their herd in the recent drought, but this does not appear to have severely impacted welfare as measured by mean income, mean expenditure, or variability in expenditure. Logologo is the only site where the income share from salary (42%) outweighs the income share from livestock and livestock products (35%). Just over half the salary earners work outside the area and are employed by NGO's, the police, the army, the wildlife service or work as watchmen. Schools, government departments, and the police employ local salary earners. Households in Logologo have established links to the larger national economy that allowed their welfare levels to be relatively unaffected during the recent drought.

The **Kargi** results provide an interesting nuance to our understanding of the process of sedentarization. Although the households in Kargi have settled, their animals remain highly mobile. The Kargi results show that pastoral production remains a viable production strategy in some areas. Kargi herders are relatively well off in terms of the mean and variance of the expenditure plus milk value measure, and they lost a relatively small percentage of their herd. Their isolation from market forces actually seems to have allowed them to pursue a form of mobile pastoralism well suited to their environment.

Contrasting the **North Horr** results with the Kargi results provides a fuller understanding of these points. North Horr households are more mobile than Kargi households, as many households still shift their entire household to a new area in search of pasture while in Kargi only the animals are sent. However, results show that Kargi animals are more mobile than North Horr animals. With regard to the viability of pastoral production, it should be noted that the main difference in the welfare measures between North Horr and Kargi is the larger and less variable milk production in the latter site. Although it is not well reflected in the rainfall data for 1999–2001, the spatial distribution of rainfall observed in this area during the study period appeared to create more abundant pasture in key pasture areas used by Kargi herds compared to those used by North Horr herds.[18]

4. CONCLUSIONS

The causes and consequences of sedentarization are complex. In this chapter we have presented information drawn from household level surveys and interviews conducted during 2000 and 2001 in six different sites in northern Kenya. The period of data collection covered the onset of a drought and continues through the early stages of a recovery from the drought.

With regard to livestock wealth, we have shown that larger herds tend to be located in drier areas, where herders derive a higher share of their income from livestock and livestock products and also have more milk available for home consumption. With regard to changes to herd wealth in the drought, we find that areas where herds were more mobile suffered lower losses. We also have noted that households with larger herds before the drought tended to have larger herds after the drought, showing that herd accumulation at the household level provides a self-insurance role.

In contrast to our findings for livestock wealth, we find that areas with higher share of income from non-pastoral sources have higher welfare in terms of higher income, higher expenditure, and lower variability in the measure of milk value plus expenditure. In some cases they also are more food secure because they convert wages into food purchases. In this respect, we find that education seems to play an important role in how households earn their income and cope with food insecurity. Areas where household members have spent more time in formal education have higher shares of their income from non-pastoral sources and tend to have higher incomes and expenditure levels, including on food. We also find there is an inverse relationship between enrollment in school and mobility.

The findings in this chapter corroborate earlier work on pastoralism that suggests sedentarization attracts both poor and relatively wealthy herders (Barth, 1964; Little, 1985). The latter group appears 'blessed' in the kinds of opportunities they can pursue and the degree of support that they can provide the pastoral sector and their mobile relatives and family members. In contrast, the poor appear 'cursed' in the kinds of unremunerative activities they engage in and the extent to which they are caught in a vicious cycle of low incomes, low mobility, and high food insecurity.

Where this chapter departs from these and other studies of sedentarization in Africa and elsewhere is by showing how the process does not necessarily equate to less herd mobility. By utilizing mobile satellite camps, certain members of pastoral households can be sedentary and pursue activities usually associated with sedentarization (waged employment, agriculture, and/or education) while their animals continue to move opportunistically according to climate and resource conditions. These novel forms of adaptation show that while serious development and food security problems still confront pastoral communities of northern Kenya, the pursuit of non-pastoral, sedentary-like activities does not forecast an end to pastoralism. Indeed, we have argued that the types of non-pastoral ('supplemental') activities discussed in this chapter may be what will allow mobile pastoralism to continue in the area for the foreseeable future.

NOTES

1. The PARIMA project is a collaborative effort of Utah State University, the University of Kentucky, Cornell University, Egerton University (Kenya), and the International Livestock Research Institute (ILRI). It addresses the causes and consequences of different types of risk among pastoralists; the means by which herders manage—economically, environmentally, and culturally—endemic and periodic risks; and the grassroots initiatives by herders to address the difficulties associated with high levels of risk. This paper has benefited from discussions with our project colleagues: Abdillahi Aboud, Christopher Barrett, D. Layne Coppock, Cheryl Doss, Getachew Gebru, and Hussein Mahmoud. PARIMA is supported by the Global Livestock Collaborative Research Support Program, funded by the Office of Agriculture and Food Security, Global Bureau, USAID, under grants DAN-1328-G-00-0046-00 and PCE-G-98-00036-00. The opinions expressed do not necessarily reflect the views of the U.S. Agency for International Development.
2. This map was prepared by Ingrid Rhinehart.
3. An interesting variable that generally correlates both with rainfall and the extent of sedentarization is ownership of poultry (chickens). From highest rainfall to lowest: the presence of poultry in the study area was: Dirib Gumbo (57 percent own chickens); Ngambo (73 Percent); Suguta Marmar (40 percent); Logologo (60 percent); Kargi (20 percent); and North Horr (6 percent). As will become more evident below, generally the more mobile the community is, the less important are poultry.
4. Although we describe each settlement by noting the majority ethnic group present in the location, it is important to note that in each site, there are minority populations from other ethnic groups. Given our areal sampling method, members of these minority groups are often represented in our data.
5. Herd size is measured in Total Livestock Units (TLU) following the weighting of the Range Management Handbook of Kenya, where 1 head of cattle = 0.7 camels = 10 sheep = 11 goats.
6. Herd size per capita follows roughly the same pattern, as the average household size is 6 in Kargi and North Horr, 7 in Dirib Gumbo, and 8 in Logologo, Sugata Marmar, and Ngambo.
7. By September 2001 stockless households accounted for 25% of sample households in Dirib Gumbo, 16% in North Horr, 11% in Sugata Marmar and Ngambo, and 3% in Kargi and Logologo.
8. Peter Little developed the interview guidelines for these case histories and the interviews were conduced by Hussein Mahmoud and translated by local research enumerators.
9. One contradictory aspect of sedentarization this leads to is that while sedentarization often decreases household welfare, it frequently opens up new income opportunities for women (Fratkin and Smith, 1995). The gendered dimensions of pastoral diversification in our study region are more fully reported in Little et al., 2001.
10. With regard to diversification of income sources, it is important to distinguish between a given household diversifying into different activities and diversification of different households in a given community into different activities. To make this distinction, we construct a measure of activity concentration that sums the square of the square of the percentage income from each activity. At the household level, concentration of income sources from highest to lowest is as follows: Dirib Gumbo (.90), Kargi (.89), Logologo (.88), Ngambo (.82), North Horr (.76), and Sugata Marmar (.75). At the community level, concentration of income sources from highest to lowest is as follows: Logologo (.28), Dirib Gumbo (.27), Kargi (.24), North Horr (.21), Sugata Marmar (.21), and Ngambo (.20). This indicates there is a great deal more diversification between households than there is within households.
11. The categories are: grains/ flour, sugar / honey, tea / coffee, cooking oil / fat, beans, vegetables / onions / potatoes, meat / milk, tobacco / snuff / miraa, clothes / shoes for self, clothes / shoes for others, beads / jewelry.
12. The ordering of overall milk production is largely reflective of the overall ordering of herd size (median liters of milk per day in parentheses): Kargi (2.5), North Horr (1.5), Logologo (0.6), Ngambo (0.5), Sugata Marmar (0.2), and Dirib Gumbo (0).
13. The cash value of milk is roughly 20 shillings per liter in Kargi and North Horr, 30 shillings per liter in Logologo and Sugata Marmar, 50 shillings per liter in Ngambo, and 60 shillings per liter in Dirib Gumbo.
14. Recall that this conclusion is drawn based on community level averages. Analysis at the household level will provide a more nuanced understanding of this relationship.
15. This assessment is relative to other sites and should not imply that Ngambo households, especially the poorer units, did not suffer during the recent drought.
16. While we do not have firm data to confirm this, it is likely that at least part of the explanation for the severe herd losses experienced in Dirib Gumbo are related to ongoing political conflict over territorial claims and tension between ethnic groups on Marsabit mountain. In contrast to other groups, the Boran of

Dirib Gumbo do not have easy access to lowland pastures surrounding Marsabit mountain due to ongoing struggles over land and water claims among the different groups in Marsabit District.

17. An intriguing topic for further research is an investigation of what causes, if any, can be identified for the relatively low rates of households with access to remittance income when compared to other sites around Marsabit Mountain.

18. This fact does not seem to have escaped the attention of Gabra and Rendille herders. During this period, a relative peace has held between the Gabra and Rendille. They have reached an agreement harmonizing penalties for murder and other personal liabilities, and have used each other's grazing land.

REFERENCES

Barth, F., 1964, 'Capital Investment and the Social Structure of a Pastoral Nomadic Group in South Persia.' In *Capital, Savings and Credit in Peasant Societies*, edited by R. Firth and B.S. Yamey, pp. 415–425. London: Allen and Unwin.

Fratkin, E., 1991, *Surviving Drought and Development: Ariaal Pastoralists of Northern Kenya.* Boulder: Westview Press.

Fratkin, E. and K. Smith, 1995, Women's Changing Economic Roles and Pastoral Sedentarization: Varying Strategies in Alternative Rendille Communities. *Human Ecology* 23 (4): 433–454.

Little, P.D., 1985, 'Social Differentiation and Pastoralist Sedentarization in Northern Kenya.' *Africa* 55 (3): 243–261.

Little, P.D., 1992, *The Elusive Granary: Herder, Farmer, and State in Northern Kenya.* Cambridge: Cambridge University Press.

Little, P.D., 1997, *Income and Assets as Impact Indicators.* Washington, DC: Management Systems International.

Little, P.D., K. Smith, B.A. Cellarius, D.L. Coppock, and C.B. Barrett, 2001, Avoiding Disaster: Diversification and risk management among East African herders. *Development and Change* 32 (3): 401–433.

Kenya, Government of, 1980, *District Development Plan, Baringo District, 1979–1983.* Nairobi: Government Printers.

Nduma, I., P. Kristjanson, and J. McPeak, 2001, Diversity in Income-Generating Activities for Sedentarized Pastoral Women in Northern Kenya. *Human Organization* 60 (4): 319–325.

Snow, R. and J. Morris, 1984, Do Relief Efforts Beget Famine? *Cultural Survival Quarterly* 8 (1): 51–53.

Chapter 6

Once Nomads Settle

Assessing the Process, Motives and Welfare Changes of Settlements on Mount Marsabit

WARIO R. ADANO AND KAREN WITSENBURG

In the recent past, there has been a growing doubt about how future East African pastoral peoples will be able to meet increases in per capita demand for livestock food products. This fear stems from concerns about population growth, large herd losses to catastrophic events and losses of former rangelands to competing uses. When faced with heavy asset losses nomadic pastoralists settle down in order to earn a living and meet basic needs. A typical example of this process takes place on Marsabit Mountain. As a result the mountain has become a meeting locale for different ethnic groups to experiment with farming. Settling down of nomads is associated with new opportunities and experiences. This chapter assesses the sedentarisation process on Marsabit Mountain and the resource endowments of settled households.

The chapter is structured as follows. First, a brief overview of rangeland-livestock debate is presented (Section 1.0). Past experiences of pastoral intervention efforts aimed at providing economic alternatives to impoverished pastoralists are also reviewed (Section 2.0). Section 3.0 gives the history of settlement on Marsabit Mountain from both secondary information and our survey data. The subsequent sections (Sections 4 and 5) report the motives of settlements and identify wealth indicators of settled households on the mountain. The distinguishing characteristics of the households settled through development assistance (planned scheme settlement), and those who settled without support of development assistance (non-scheme settlement), and on gender relations are the core of this analysis. Perception of individuals on changes in their way of life upon settling, are also presented. These responses highlight linkages between the economic mode of production and general measures of well-being. The chapter's concluding remarks raise policy

WARIO R. ADANO AND KAREN WITSENBURG • Amsterdam Research Institute for Global Issues and Development Studies, University of Amsterdam, 1018 VZ Amsterdam, The Netherlands.

issues for supporting the rural poor (Section 6.0), and finds that there is ample opportunity to rethink future development deliverance and to address rural poverty.

1. ATTITUDES TOWARDS PASTORALISTS AND THEIR MOBILITY

There have always been ardent opponents or proponents of sedentarisation of East African pastoralists. The arguments were often environmental in nature. In the 1970s, pastoralists received attention because they were partly held responsible for the desertification process as a result of overgrazing (Lamprey and Yussuf, 1981; Lamprey, 1983; Sinclair and Fryxell, 1985). It was thought legitimate and viable to provide an alternative production system for pastoralists rather than leaving them 'roaming about with unlimited numbers of poor quality herds' (Brown, 1963). Until 1980, most governments, NGOs and missionaries criticised pastoralism as an irrational, ecologically destructive and economically inefficient production system and encouraged sedentarisation as well as advocated the replacement of the traditional system (Helland, 1980; Galaty, 1992; Robert, 1992; Homewood, 1995; Nunow, 2000). Numerous conferences and reports on desertification based on 'tragedy of the commons' (Hardin, 1968, p. 1243–1248) arguments and carrying capacity models were used to advocate the settling down of nomadic pastoralists (Lamprey and Yussuf, 1981; NES-UNCCD, 1999), and the influential Brundtland Report[1] of 1987 (WCED, 1987; UNCED, 1992). At the same time, nomadic pastoralists themselves underwent numerous setbacks in the period after the 1960s. Particularly in Kenya, pastoralists not only lost the relative autonomy they had enjoyed in the colonial time (Dietz and Salih, 1997), they were further marginalized by warfare, droughts, livestock diseases and loss of wetlands to agriculture. There was an increase in the number of projects in which impoverished nomadic pastoralists were enrolled in programmes providing alternative production systems. Examples are the fishery and irrigation projects in Turkana (Hogg, 1988; Galaty, 1992) and group ranching projects in Maasai areas (Helland, 1980; Rutten, 1992). None of these projects was as successful as initially anticipated and evaluation reports showed a depressing state of affairs.

After 1980, a new attitude towards nomadic pastoralism emerged, marked by a conference entitled 'The Future of Pastoral Peoples' (Dietz, 1987, p.13) held in Nairobi in 1980. Some scholars, among them anthropologists and range ecologists, influenced the perception of pastoralism as being a rational and efficient production system in its adaptation to risks and uncertainties in a disequilibrium environment (Baxter, 1975; Salzman, 1980; Galaty, 1981; Behnke, Scoones and Kerven, 1993; Scoones, 1995; Oba, 1996; Fratkin, 1997). Since 1980, the almost rhetoric question "Must nomads settle?" (Aronson, 1980) has become relevant now that scholars and extension officers doubt the rationale of settlement policies. This period indeed witnessed broader advocacy on pastoral development (Salih, 1991). However, it is true that as a result of the perceived negative livestock impacts on the environment development funding and efforts to assist poor pastoralists generally declined since the late 1980s (Morton and Meadows, 2000); apart from relief food operations.

It seems evident that among scholars and extension workers the perception of the viability of pastoralism is inspired by the rangeland debate. When desertification of the range was not as obvious as hypothesized and degradation could no longer exclusively be explained by excessive livestock numbers (Livingstone, 1991), restocking projects and livestock programmes became popular (Mace, 1989; Hogg, 1992). Localized environmental degradation around permanent water points and human settlements in the drylands

(Oba, 1996), apparently due to decreased mobility, are now being used as an argument to discourage sedentarisation. 'Keep nomads mobile' seems the answer to ecological problems in the drylands.

However, despite the initial efforts to encourage settlement prior to 1980 and the later efforts to discourage settlement after 1980, the situation in Marsabit District and in the arid and semi-arid lands (ASALs) in general is one of decreased mobility and increased sedentarisation due to internal and external problems relating to the pastoral production system. As this chapter shows in detail later, our research[2] reveals that 80 to 90 % of the rural settled households on Marsabit Mountain came to the mountain with less than 10 cattle per household or with no livestock at all. At present (see Section 5.2), the settled households have neither better economic nor nutritional position than their mobile relatives (Nathan et al., 1996), and more than 90 % of the households in the District live below the poverty line of one dollar per person per day (Sisule, 2001). Even though settled people involuntarily diversified their production system and incorporated non-pastoral activities, some now perceive a sedentary life to be advantageous, not least because of improved access to a market, education, health and relief services. In addition, especially for the impoverished pastoralists, a settled existence in agricultural areas provides new opportunities to become economically independent, which is widely felt as an important factor in restoring self-respect. The question whether nomads must settle or not has become less relevant today, because a large portion of pastoral households has already settled. The failure of the past settlement projects has brought about a negative attitude among development scholars and non-governments towards sedentarisation (Morton and Meadows, 2000; Oba, Stenseth, and Lusigi, 2000), and this in turn has resulted in neglect of the poorest and most vulnerable group in the pastoral society. However, today it is worth noting that not only observers (politicians, scholars, NGOs and extension workers) have changed their ideas on sedentarisation, but also that the attitude among pastoralists towards sedentarisation is not as negative any more as in the past.

Moreover, following Baxter (1994) and Morton and Meadows (2000, p. 7) 'pastoralism' should not be defined as a production system where people derive 50% or more of their revenues from livestock (also referred to as pure pastoralism), but should also include individuals or households who, although they currently derive their income from other activities, have a pastoralist background, who ethnically belong to a group of people who subsist on livestock keeping, or who see pastoralism as a vocation in future rather than their present occupation. From this perspective, sedentarisation can be perceived as a form of livelihood diversification, either voluntary or involuntary. We advocate a new agenda for policy and analysis concerning pastoral development, which acknowledges pastoralism in the drylands as a form of sustainable land use, but which could and should be strengthened by diversification to non-livestock based activities.

2. SEDENTARISATION IN MARSABIT DISTRICT

2.1. Common Arguments against Sedentarisation

The successful dissemination of the views resulting from 'the new ecology' approaches (Scoones, 1995) and the 'mobility paradigm' (Niamir-Fuller, 1999), backed by new insights on indigenous knowledge and skills of pastoralists, was partly evoked by negative accounts of external development interventions deploying sedentarisation efforts.

These accounts show how (colonial) government policies, missionaries or NGOs have deliberately encouraged the settlement of nomadic people and in doing so disrupted the traditional pastoral setting, causing more poverty, dependency and environmental degradation.

The problems of sedentarisation which are often addressed (for example see Little, 1985; Nathan et al., 1996; Mitchell, 1999; Nunow, 2000) can be briefly summarized as follows:

1. Sedentarisation causes environmental degradation because of overgrazing and removal of wood vegetation for fuel.
2. Sedentarisation increases land use and tenure conflicts, when communal land undergoes a process of privatisation.
3. Sedentarisation closes off areas formally used by pastoralists, and can reduce accessibility of resources for herders.
4. Sedentarisation competes for space with pastoralists.
5. Living conditions in settlements are poor, unhealthy and there are more children malnourished.
6. Agricultural settlement schemes failed because people reinvested in livestock, and move away abandoning settlement projects.

We do not deny that these problems exist, apart from the last one (we rather consider any developing programme among settled pastoralists successful if people are able to reinvest in livestock), but it is necessary to first look at what gives rise to sedentarisation. From our survey it seems that sedentarisation is a result of poverty (Section 4.1). Settled people are the victims of adverse processes in the pastoral production system, as they have lost their herds and could not be helped by the traditional safety net. If they, or the settlements, are again blamed for obstructing the pastoral lifestyle, the arguments loose direction in causal explanations.

It is necessary to deal with the above-mentioned problems instead of ignoring them. Obviously, for individual herders who have no livestock the advantages of sedentarisation at the household level outweigh the negative consequences at the village or district level.

2.2. Forms of Sedentarisation

Sedentarisation is a broad concept. The word has been used to refer to several forms of pastoralists reduced mobility. If we draw an imaginary line on which pastoral households can score on their degree of sedentarisation or reduced mobility, we have at one extreme the highly mobile, purely pastoral households[3] which move with their herds every season, and at the other extreme permanently settled households with a mixed livelihood profile. In each region in sub-Saharan Africa there is a mix of households who would score differently on this line. Reduced mobility results in different forms of settlement. In addition, each household can change with regard to its score on 'degree of sedentarisation' over time, while individuals within a household can be less or more mobile than other members of their family. Broadly speaking, one could try to identify mobile, semi-sedentary and permanent villages across different ethnic groups. However, this typology would not cover all the nuances of reduced mobility within a household. The fact that a household is permanently settled does not mean that its livestock is not mobile (see O'Leary, 1994). In fact, reduced mobility of a household can result in increased mobility of its animals. Arrangements by which households employ a herdsman, or send their animals with relatives, or which entail the son himself herding the animals in distant satellite (*fora*) camps are widespread.

Furthermore, one needs to realise that settlements in pastoral areas have their own dynamics, and are not necessarily permanently inhabited all the time by the same people, nor are they end stations for impoverished pastoralists as is shown by literature on transhumance. While sedentarisation is only sometimes a process in which people simply suddenly decide to stay, or have to stay, in one place and form a village, it is more often a gradual process evolving in stages. Some settlements seem permanent for years but gradually disappear, especially the drought-caused settlements, while others seem only temporal but appear to stay in the same place for years.

Fratkin and Smith (1995) differentiate communities according to their economic strategies and identify four types of communities among the Rendille. These are the mobile camel pastoralists, the sedentary cattle pastoralists, the sedentary agriculturalists and the urban wage workers. This typology is, for the time being, adequate for our purpose because it also allows us to locate villages at different ecological zones, as long as we keep in mind that the sedentary status of a household does not say much about the sedentary or mobile state of all its members and its herds. We have therefore opted to identify settlements according to their economic strategies and ecological differences, but with due attention for the dynamics and internal differences in households and villages. It might not be possible to classify each household, each village or each pastoral group on their 'degree of sedentarisation', but it is however necessary to understand the motives that gave rise to certain forms of sedentarisation.

In Marsabit District, we can discern a continuum from reduced mobility to complete sedentarisation around water points in the desert and around mission centres in the lowlands like Korr and Kargi in Korole and Kaisut desert, and Maikona in Chalbi desert, in Marsabit Town and in the agricultural villages on Marsabit Mountain. In this chapter we will describe the latter: the form of settlement which gave rise to the rural villages on Marsabit Mountain, where people are able to combine livestock keeping with arable farming and other income-generation activities. Although the 'sedentarisation paths' on Marsabit Mountain are different for each household, they all experienced a sudden and traumatic change due to the Somali secessionist war and the devastating droughts in the early 1970s and, before they settled in the rural villages. Especially the drought of 1973, that caused about 40 percent loss of livestock in the district (Hogg, 1986), left many people destitute and therefore paralysed the indigenous social networks and the traditional herd redistribution mechanisms in the society. Impoverished pastoralists started to live in the slums around Marsabit Township and Laisamis surviving largely from begging and emergency relief handouts. In a (partly) joint effort of the government, the National Christian Churches of Kenya (NCCK), the African Inland Church (AIC), CARE (K) and the Marsabit Catholic Mission, settlement scheme projects for impoverished pastoralists were started in 1973 and in 1976 and at least 5 villages were established on the mountain where housing, farm plots, tools and seeds were provided (Marsabit Development Plan, 1979; Adano, 2000).

The following section briefly describes how Marsabit Mountain became permanently inhabited, followed by a description of how 4 specific settlement schemes established on the mountain 24 years ago, through development assistance efforts, are presently faring. Scheme villages settled with external assistance are compared with nearby villages where people settled on their own efforts, depending on their own social network, and without much help from external organizations. In addition, Badassa refugee scheme, which is a settlement established by United Nations High Commission for Refugees (UNHCR) for refugees from the Ethiopian civil war in the 1970s to the early 1980s, is included. We explicitly want to know whether planned scheme households, which were established over twenty-five years ago, today stay poor, get well or worse off, compared to non-scheme settlement households. A range of wealth indicators is used to address this question, by investigating characteristics that differentiate these households.

3. THE HISTORY OF EARLY SETTLEMENTS ON MARSABIT MOUNTAIN

Marsabit Mountain is a high *inselberg* of volcanic origin, rising to an altitude of 1700 m out of the surrounding semi desert at an altitude of 400 metres. A dense mist forest grows on its peaks, and is responsible for the cool and sub-humid climate on the mountain. The average rainfall on the Mountain ranges from 800 mm to 1000 mm annually, dispersed over a bimodal rainfall pattern. The continuous wind direction from Northeast to Southwest causes a relatively humid ecology on the eastern side of the mountain and a dry ecology on the western side of the mountain. On average, rainfall drops to 200 mm in the surrounding lowlands. Vegetation growth in large parts of the District is scanty due to high salinity of soils and water resources.

So far, little is known about the people who originally lived on Marsabit Mountain. All pastoral groups in the District (the major groups are Rendille, Boran, Gabra and Samburu) were highly mobile before 1900, and moved according to season, rainfall, range and security conditions. Pastoral villages moved in small conglomerations of families, using different resources at different times. Flexible movement was accommodated for by 'permeable' ethnic boundaries; individual households, families and whole clans could join neighbouring groups and adopt their way of life (Schlee, 1989). Several pastoral groups used Marsabit Mountain only as a fall-back area in times of drought and in dry seasons.

No source reports the presence of settlements on Marsabit Mountain before 1900. No signs of prehistoric occupation were found within the Forest Reserve (Phillipson and Gifford, 1981). The first reported "permanent' construction" was a little mud–and-wattle house at the edge of the crater-lake by the Boma Trading Company in 1907 (Brown, 1989, p. 323). The adventurers-cum-traders Riddell and Hornyold established a small base on Marsabit Mountain in the same year, which was supported by the colonial administration and heralded a new era of permanent habitation on Mount Marsabit (Brown, 1989). The "Boma Trading Company" disappeared in 1909, while the colonial administration came to stay in Marsabit for the next 50 years. By 1911, the officer Geoffrey Archer had erected an extensive station of mud and thatch residences, stores, offices and staff lines, and a solid wooden-block house. The first clearing of land for planting was undertaken in the same year (Brown, 1989, p. 342). Later the post moved to "Delameres Njiro" (present Aite well), where the settlement started to expand. The settlement, consisting of government servants only, grew steadily, in an area in the forest that is still called Karantina (after "quarantine" of animals for trade, introduced by the colonial government). Construction in the forest was later forbidden and the small colonial settlement moved out. However, the foundations of the houses and offices in the forest are still there today.

In 1921, the Northern Frontier District (NFD), of which Marsabit was a part, came under the military rule of the Kings African Rifles (KARs), and Marsabit District was renamed Gabra District (Marsabit Handing Over Report, 1922). Large police offices, officers' residences and lines were built. The development even included the construction of the first piped water supply. In 1925, the area was organised under civilian rule and many more administrative posts were created (Tablino, 1999, p. 230). British District Commissioners, District Officers and Goan clerks came to live on the Mountain. There were a total of 262 people living a settled life on Marsabit Mountain in 1931. The population mainly consisted of police, administrators and, some Somali and Indian shopkeepers. The sudden increase of Burji and Konso farmers after 1931 was owing to the fact that the English colonials needed farmers to grow their food (Marsabit District Annual Report,

1931:7; MDAR for brevity), and called them over from Moyale. This migration resulted in doubling of the total population in a few years time, and in 1935, the total settled population amounted to 664 people (MDAR, 1935).

Until the Second World War the British administrators discouraged settlement by the three larger pastoral groups: the Boran, the Gabra and the Rendille on the mountain, mainly for fear of overgrazing and pressure on limited water resources. They only wanted people to settle if they took up farm plots and started farming, which could make a clear contribution to the economic on the mountain, as long as they were not Somali and not bringing too many animals.

> "With a view to encourage the local people, the number of Alien Somali traders from outside the province was strictly controlled ... The tendency for the Township population to increase (is) owing to the advent of detribalised natives from local tribes. Unless there was some adequate reason for their remaining, they were sent back to their tribal areas (MDAR, 1937:20–1)."

Additional efforts to control population and livestock numbers on the mountain were effected through a 'one man one *shamba*' policy, restricting movements of agro-pastoralists, with absentee landlordism not allowed. Colonial administrators also permitted settled people to own no more than 20 cattle in the township and 10 cattle for farmers (MDAR, 1937, p. 22). People with more animals were sent off the mountain. Although it is not sure whether people were obeying the rules, *if* they obeyed then people could only survive on the mountain by farming.

3.1. Population Trends

As shown in Figure 1, Marsabit District had an estimated population of 10,399 people by 1929, with Marsabit Township making up less than 2% of the district's population (MDAR, 1929). By 1958, Marsabit Mountain had 2038 inhabitants, accounting for about 7% of the total population of the district. By 1959, the township had grown in size (MDARs, 1958, 1959), despite the restrictive policies of the colonial government. The mountain population in 1969 had tripled in 10 years.

The share of the mountain population in the district progressively increased by the weighted averages of about 10% in the 1960s, 15% in the 1970s, 21% in the 1980s, and about 21% in the 1990s. The results further reveal fast growth rates of the mountain population between the 1960s and 1970s compared to the period between the 1980s and 1990s (GoK, 1969, 1979, 1994, 2001). The population increase in the 1970s coincides with the droughts and massive livestock losses in the region, which subsequently resulted in the establishment of a number of pastoral settlement schemes on the mountain. The population increase on the mountain of the 1970s did not only coincide with the drought and massive stock losses, but the civil war in Ethiopia also caused a flow of migrants from Ethiopia to the Mountain. The slow-down in both district and mountain populations during the last two decades, compared to the previous ones, is true also for Kenya as a whole according to a recent UN-Population Fund assessment report.[4] Overall, between 1958 and 1999, whereas the district's population grew nearly six folds, the population on the mountain grew slightly more than 18 times. This is evidence of a more proportionate growth of the mountain population in relation to the district as a whole. Today we see the colonial government had started a process they had actually wanted to prevent, namely sedentarisation of pastoralists and increasing population pressure on Marsabit Mountain.

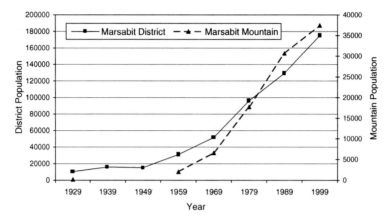

Figure 1. Human population trends in Marsabit District and Marsabit Mountain.
Sources: Data compiled from Handing Over Report 1922; MDARs 1931, 1935; 1937, and GoK, various years.

3.2. Settlement Schemes on the Mountain

Northern Kenya experienced a series of severe droughts in 1960s and in the 1970s that caused large herd losses. In response to such events many organisations joined hands in establishing settlement schemes from 1970 to 1980. These projects mainly targeted impoverished pastoralists and refugees from Ethiopia to become self-supporting. In 1973, the NCCK started a settlement scheme on the mountain for impoverished pastoralists from Laisamis in Songa, Nasikakwe (Karare) and Kituruni on the mountain, where the AIC missionary Anderson was mainly involved in the practical implementation. Around 1976–77, the government, through the Department of Social Services, and with assistance from the Catholic Mission designed two settlement schemes in Manyatta Jillo and Sagante (MDAR, 1977).

The creation of settlement schemes on the mountain started with demonstration farms where nomadic families were shown how to cultivate. In the settlement projects many settled families were allocated an iron-sheet house, some animals and farm plots, ranging from two to five acres (Marsabit Development Plan, 1979). Each scheme was offered communally shared oxen to be used for ploughing. Houses and farm plots are valuable assets owned by the settled families today. In the following section, we will elaborate on the past and present of settled households.

4. THE HOUSEHOLD SURVEY

A survey of eight villages on Marsabit Mountain was undertaken in 1998 by the authors. The villages were Manyatta Jillo, Dirib Gombo, Badassa, Daka-baricha, Sagante, Kituruni, Hula Hula and Karare (see Map 1, Figure 2). Some of these villages have scheme households, as described above, among them. In total, 287 households were surveyed using a stratified sample that covered both scheme and non-scheme households within a village, where applicable. Table 1 illustrates sample sizes of the villages, and those villages with scheme households and without scheme households; non-scheme households.

As we have stated earlier, the scheme households were assisted by various organizations in settling down on the mountain in the 1970s. These households accounted for 23% of the entire sample. Both non-scheme households within scheme villages, and households in non-scheme villages settled as a spontaneous process. However, unlike the planned scheme households, the non-scheme households did not receive assistance at the time of settling down. Today, the missions, NGOs and the line ministries consider all the households in both scheme and non-scheme settlements needy for receiving relief food distributions. In the chapter, we compare the wealth levels of planned scheme villages and non-scheme villages that settled without aid.

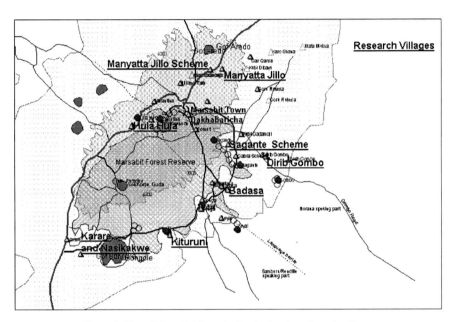

Figure 2. Map of Marsabit Mountain showing location of sample villages.

Table 1. Sample Size and Type of Settlement by Village.

Village	Scheme villages		Non-scheme villages	
	No. of scheme HH	No. of non-scheme HH	Village	H/holds
Manyatta Jillo	17	20	Dirib Gombo	35
Badassa*	16	20	Daka-baricha	35
Sagante	15	20	Hula Hula	38
Kituruni	4	32		
Karare	15	20		
Total	67	112		108

Note:
HH denotes households and * all the scheme households at Badassa are also refugee households, which originally were from Ethiopia.

Source: Authors' survey, 1998.

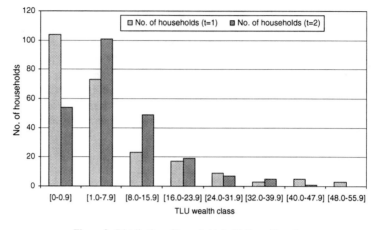

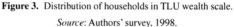

Figure 3. Distribution of households in TLU wealth scale.
Source: Authors' survey, 1998.

4.1. Main Characteristics of Currently Settled Households in Marsabit Mountain

More than 87% of the households in the sample are first generation migrants, of which about 84% belong to the major pastoral groups (Boran 33%, Rendille 21%, Samburu 19%, Gabra 10%). Of those who do not belong to the major pastoral groups, the majority have farming backgrounds and belon to the Burji, Konso and Amharic ethnic groups. The majority (82%) of the settled people who belong to the major pastoral groups owned 12 tropical livestock units (TLU)[5] or less per household at the time (t = 1) they came to the mountain, and 79% still owns 12 TLU per household or less (t = 2) (Figure 3). It is important to realise that a mobile pastoralist way of life is hardly possible for households who own 12 TLU or less. Additionally, the data reveals that virtually 80% of the migrants had 8 TLU per household or less, and owned only 16% of the total TLU at the time of settling on the mountain. Even at the time of settling high disparity in herd wealth existed, since only 3% of the households owned nearly 26% of the total herd wealth, in TLU measure.

A striking point is that 35% of the households said that the male head of the migrated family, or the deceased husband in the case of a female-headed household, was a first-born son of the family. In pastoral societies in Northern Kenya the eldest son inherits most of the family herds. The eldest son is responsible for the future of his brothers and sisters and carries a heavy responsibility. First-born sons in stock poor families hardly have the opportunity to herd other people's herd and rebuild his own herd, like the younger siblings in poor pastoral families. A first-born son is not supposed to depend on other families, and he will therefore opt to try out farming rather than to stay a poor pastoral labourer.

4.2. Time of Arrival, Origin and Motives[6] of Settled Pastoralists

Figure 4 shows, an increase in settlement since 1950, with a peak in 1970–1974, and a steep decrease after 1984. The graph shows that for more than a decade there was continuous immigration of new settlers to the mountain from 1970 until 1984. This is a clear

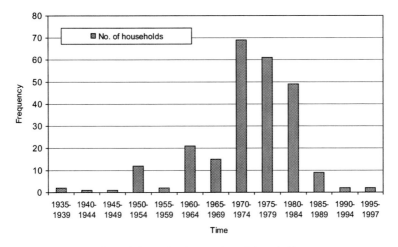

Figure 4. Time of settlement of sample households.

Note: A total of 37 second-generation respondents, and 4 missing cases are excluded from this figure.

Source: Authors' survey, 1998.

Table 2. Time of Arrival on the Mountain and Causes of Animal Loss to Settle.[a]

Animal loss due to:	1950– 1954	1955– 1959	1960– 1964	1965– 1969	1970– 1974	1975– 1979	1980– 1984	1985– 1989	1990– 1995	Total
Drought	4		4	8	39	26	20	4	1	106
Raids/shifta war	1	1	9	5	15	10	7			48
Disease					1	2				3
Total responses	5	1	13	13	55	38	27	4	1	157

Note:

a   Sample size is 242 people. The people who are born on the mountain and 8 missing cases are excluded.

Source: Authors' survey, 1998.

indication that settlement was largely a result of droughts and war after 1970. This was also widely reflected in the reasons mentioned by respondents. The majority of the people in the schemes, except for the refugee scheme in Badassa, mentioned animal losses as a motive to have come to Marsabit, and this corresponds to the original aim of the settlement schemes which was to give impoverished pastoralists an alternative livelihood option. In Manyatta Jillo and Sagante settlement schemes, about 88% and 67% of the people, respectively, stated they had came there because of animal losses. In Kituruni only one person of the four remaining in the settlement scheme said he/she had come there because of animal losses, while in Karare nearly 47% mentioned animal losses as a reason for coming to Marsabit Mountain. Animals were lost in droughts, in raids and *shifta* war[7] (Table 2).

Drought (about 68%); raids (about 31%), diseases or a combination of these factors are some of the reasons people mentioned as giving rise to loss of animals prior to settlement on the mountain (Table 2). Nearly 56% of the first generation respondents said that they settled because of animal losses. Nearly one-third (i.e., 30%) of those born on the mountain knew that their parents settled because of animal losses. Table 3 shows how often other motives were mentioned.

Table 3. Reasons for Settling of Respondents, Apart from Animal Loss, on the Mountain.

Reason for coming	1935– 1959	1960– 1964	1965– 1969	1970– 1974	1975– 1979	1980– 1984	1985– 1995	Total
Civil war in Ethiopia	3	3	1	3	18	10	1	39
Looked for better grazing land	1	2		3	12	8	1	27
Looked for wage/paid work	4	2		3	4	5	2	20
Settled by missionary/ NGO/government				12	2			14
Came to farm	1	1	1	2	1	2	1	9
Social differences with relatives				1	3	2		6
Relatives called me		1		1	2	1		5
Transferred job/retired from job	1		1		2		1	5
Husband passed away				2	2		4	
Stayed after ceremony (age-set, marriage)					2			2
Divorce				1			1	2
As first-born I did not inherit		1					1	
Total	10	10	3	28	48	28	7	134

Note: Columns 2 and 8 cover longer period than other 5-interval years due to scanty responses.

Source: Authors' survey, 1998.

Some motives mentioned for coming to the mountain do not exclude loss of animals (Table 3). For instance if someone said, "I came to the mountain to look for wage/paid work' or I came to farm, or I was settled by mission", it is probable that loss of animals could have been a reason why person started searching for work. For example, as Figure 3 shows, although 40% of the respondents from the major pastoral groups had no livestock at all, 60% brought a few animals with them to the mountain.

Marsabit Mountain has always had the function of a 'fall-back' reserve for dry season grazing and in times of drought, especially for Samburu, Gabra and Rendille cattle keepers. In the early 1970s, Marsabit also provided more security against the Somali secessionists war that lasted from 1964 to 1967. A new situation that arose though was that people did not move away from the mountain anymore once circumstances improved, but decided to stay on the mountain for various reasons, including security, schooling, relief food and medical services.

4.3. The Origin of Settlers on the Mountain

In the survey, heads of the household were asked about their place of birth, or the region of origin of their parents and their fathers' parents if they were second or third generation settlers. Women whose husbands were absent or died were asked about their husband's place of birth and his parents and his father's parents' region of origin for purpose of comparison. More than 23% of all sample households migrated from Ethiopia, and more than one third (38.4%) have grandparents on their father's side who lived in Ethiopia. Nearly 19% of the male heads were born on the mountain, while very few grandparents originally came from the mountain area. Another important area of origin is the Korr/Kargi area, where one quarter of the respondents were born.

Our survey reveals that over 40% of people from all origins came in the 1970s, and that people came to the mountain from all the places in the district and beyond. Almost half of the Ethiopian immigrants, amounting to about 46% (mainly Boran, Burji and Konso), arrived in the late 1970s and the flow continued until 1984. Over 50% of the people from

Moyale areas, and almost half of people from Laisamis/Maralal (Samburu/Rendille), as well as immigrants from Chalbi (mainly Gabra) area, arrived on the mountain between 1970 and 1974. People from the Korr/Kargi (Rendille/Samburu) area arrived slightly later, as nearly 87% of them came between 1974 and 1984.

The present-day ethnical map of Marsabit Mountain shows a clear division between the Samburu/Rendille-speaking communities on the southern part and Borana-speaking communities on the northern part of the Mountain. Since independence, many (violent political) conflicts have occurred between both sides and these contributed to the present day ethnic segregation on the mountain. The Samburu/ Rendille side of the mountain is ethnically more homogenous than the Borana-speaking part, which is inhabited by a mixture of Boran, Gabra, Waata, Konso and Burji, Sakuye, Garri and Sidam. The area occupied by these groups has relatively high human population densities (GoK, 1994, 2001) compared to the other side of the mountain – almost three times as high as the southern side. For example, the recent 1999 Population and Housing Census indicates that, overall, the eastern side of the mountain had human population density of about 20 people and the southern side about 7 people a kilometre (GoK, 2001).

5. SOCIO-ECONOMIC PROFILE OF THE HOUSEHOLDS

In the previous Section 4.1, we showed that more than 80% of the households had very few animals to live a pure pastoral life. People had to find other sources of food or income to survive. One of the options is farming, especially on the wetter eastern and south-eastern sides of the mountain. The initiation of farming on Marsabit Mountain is a process, which developed in stages over the years. Some people came to the mountain with their herds and at one stage decided to fence off a piece of land, while others were forced by circumstances to immediately start farming. The time of arrival on the mountain and when people started living a settled life, as farmers, are different for each village (Table 4).

Table 4. Settlement Process per Village (Mean Values) on the Mountain.

Village	No. of years since people arrived	No. of years since people own land	No. of years people started to cultivate
I. Scheme			
Manyatta Jillo	35.5	25.0	24.8
Badassa	18.8	18.5	17.2
Sagante	23.9	19.7	21.0
Kituruni	25.8	25.8	24.3
Karare	27.3	26.3	25.3
Mean (std. dev.)	*26.5 (8.0)*	*22.9 (4.0)*	*22.3 (22.3)*
II. Non-scheme			
Manyatta Jillo	29.7	20.3	21.2
Dirib Gombo	25.2	15.3	14.8
Badassa	19.2	9.1	15.0
Daka-baricha	32.1	23.5	22.7
Sagante	23.2	14.4	14.8
Kituruni	25.4	15.6	16.2
Hula Hula	20.6	18.6	15.6
Karare	17.6	16.4	16.5
Mean (std. dev.)	*24.0 (10.0)*	*16.7 (10.0)*	*17.5 (8.3)*

Source: Authors' survey, 1998.

People in Manyatta Jillo settlement scheme, for instance, arrived nearly 36 years ago, on average. They got land only 10 years later, when they were settled in the scheme, and they started farming not long after they were settled.

In Sagante settlement scheme, people started farming almost 3 years after they arrived, but they acquired their own piece land after another 2 years when they were enrolled in the settlement scheme. In Kituruni, there is hardly any time gap between the time people settled, the time they acquired their own piece of land, and the time cultivation started. The following section reports comparative average herd sizes when people settled and at the time of our survey (in 1998).

5.1. Changes in Household Herd Sizes

Livestock is the chief indicator of wealth and a measure of economic performance among pastoral households in the past. Table 5 below compares livestock herd sizes of scheme and non-scheme villages at the time of settling and at time of our survey, in 1998 (Table 5). The villages seem to have rebuilt livestock wealth, to varying degrees, and these changes in herd sizes are clearly evident when assessed using a combined TLU[8] measure.

At the village levels, except at Manyatta Jillo, Badassa, Daka-baricha and Kituruni, all the other sites in non-scheme villages got relatively richer in livestock in the course of settled life. In the non-scheme villages, Dirib Gombo and Hula Hula achieved the highest increases in herd sizes per household. Among the scheme villages, households at Kituruni and particularly significantly at Sagante, households' herd holdings (both cattle and small stock) increased compared to scheme villages. It is mainly Badassa and Karare among the scheme villages that appear to be herd losers over the years of farming on the mountain. Overall, the non-scheme households reveal a constant mean livestock holding for the period under review, the scheme households achieved an average of herd size increase of 19%.

Table 5. Sample Size; Mean TLU Size in Year of Settlement in 1998, and Change in TLU.

Village	Sample size	TLU size when settled	TLU size (in 1998)	Absolute change in TLU (%)
I. Scheme				
Manyatta Jillo	15	1.4	2.0	(+) 42.9
Badassa	16	8.1	1.3	(−) 84.0
Sagante	15	0.3	10.0	(+) 3.2 × 10³
Kituruni	4	4.2	10.5	(+) 150
Karare (Nasikakwe)	15	6.7	5.9	(−) 11.9
Mean (std. dev.)		*4.2 (8.8)*	*5.0 (6.9)*	*(+) 19.0*
II. Non-scheme				
Manyatta Jillo	19	8.6	7.3	(−) 15.1
Dirib Gombo	35	5.8	10.1	(+) 74.1
Badassa	20	12.8	7.2	(−) 43.8
Daka-baricha	29	8.1	1.9	(−) 76.5
Sagante	19	0.3	6.6	(+) 21.1 × 10³
Kituruni	32	13.8	7.7	(−) 44.2
Hula Hula	38	3.2	7.5	(+) 134.3
Karare	20	7.4	9.6	(+) 29.7
Mean (std. dev.)		*74.4 (12.1)*	*7.4 (7.9)*	*(Constant)*

Source: Authors' survey, 1998.

The general increase in herd size for scheme villages may indicate a high propensity to reinvest in and restock herds by initially impoverished pastoral households after settling on the mountain. However, although we see the gap being closed in the average livestock wealth between the scheme and non-scheme households, the non-scheme households still own higher livestock wealth compared to the scheme ones at t = 1 (i.e., $t_{0.5, 267}$ = 2.19, $p < 0.05$) and $t = 2$ (i.e., $t_{0.1, 268}$ = 1.97; $p < 0.10$).

Fratkin and Roth (1990) find, among other results, that events such as droughts often cause a wider economic differentiation household herd wealth in rural economies and result in urban migration. Here we find an increase in TLU level of the scheme villages and a stabilisation in TLU holding for the non-scheme villages on average which seem to reduce households herd disparity (reduced coefficient of variation from 2.1/1.64 to 1.38/1.07, respectively). This observation would be reason enough to analyse the changes in TLU holdings between settlements and investigate the mobility of households across the wealth scale.

Many previous studies in the region have pointed to 4.5 TLU per person as a threshold livestock poverty measure required by pastoral households for subsistence production, based on calories requirements of meat and milk (Lusigi, 1983; Fratkin and Roth, 1990, 1996; O'Leary, 1985). Among economies where households derive consumption-production jointly from a range of resources like farming and livestock we expect households to require less than 4.5 TLU wealth for food needs. Oba (1997) offers an alternative TLU per household measure among agro-pastoral Obbu Boran based on local people's perception of livestock wealth ranks.[9] If we take the average 6 persons per household of our survey, this animal wealth status favourably compares with benchmark calculations of 4 TLU per capita level by Dietz et al. (2001). Adopting Oba's wealth scales we assessed household mobility across wealth levels at the time of settling ($t = 1$) and in 1998 ($t = 2$) (Table 6). The household intra-herd wealth mobility is then contrasted with the settlement schemes. The two highest wealth classes are more or less above the subsistence food needs threshold of between 4 and 4.5 TLU per capita, if we assume an average households of size 6.

Table 6. Share of HH (%) and Average Herd Size When Settled and in 1998, per Wealth Rank.

| | % of HH and mean | TLU wealth scale | | | | | Total (%, mean/ (std. dev.) |
		no animal	<8	8.1–16	16.1–40	40.1 and above	
t = 1							
Scheme	%	50	37	3	8	2	100 (65)
	Mean	00	3.1	13.6	25.6	42.0	4.2 (8.8)
Non-scheme	%	36	40	10	11	3	100 (205)
	Mean	00	3.5	12.8	25.5	52.6	7.4 (12.1)
t = 2							
Scheme	%	25	56	16	2	2	100 (64)
	Mean	00	3.7	2.3	23.4	43.1	5.0 (6.9)
Non-scheme	%	16	57	21	12	0	100 (205)
	Mean	00	4.0	12.0	24.2	0	7.4 (7.9)

Notes:
[a] Distinction between scheme and non-scheme villages (and also households) was made in Table 1.
[b] Mean number of years resident on the mountain is used for $t = 1$.
[c] Numbers in the brackets in column 8 are standard deviations, and HH denotes household.
Source: Authors' survey, 1998.

Table 6 importantly permits comparison of scheme and non-scheme households with the same level of livestock holding at the time of settling. In other words, for a sound comparison of the households, it is crucial that households share the same 'initial-conditions' at the time of settling. At the time of settling, 33 scheme households had no cattle or small stock. Out of these, 17 households (about 26% of the total sample) moved out of the abject livestock poverty to a higher wealth rank. This ratio is slightly higher than the 20% of non-scheme households with no livestock at the time of settling down who had escaped herd based-poverty by 1998, and this constitutes transitory households with small herds. We also find that 10 scheme households (15%) and 8 non-scheme (almost 4%) fall in the same wealth class of "less than 0.1 TLU" wealth category at the time of settling and in 1998. This particular class comprises those confronted with chronic entitlement failures that entrap households from forward mobility in livestock-based wealth, yet these households have been settled on the mountain for about 21 years.

Assuming an average household size of 6 people, the threshold of 4 TLU per capita poverty level for subsistence needs corresponds to the 8.1–16 TLU wealth scales (rank 2—column 4). Only 9% of scheme households were in the two highest TLU wealth classes at the time of settling down and 15% of the non-scheme households fall in those two highest wealth classes. These shares reduced to 3% of the scheme households by 1998 and 12% of non-scheme households. Overall, about 48% of the scheme households and about 45% of non-scheme households moved ahead in livestock wealth by 1998. In addition, about 12% of the scheme and 22% of the non-scheme households fell back into poor herd-wealth classes. It is maybe this more proportionate loss of herd wealth status by wealthy households of non-scheme households that smoothens out the livestock wealth differential between the scheme and non-scheme villages during the time under investigation.

A number of reasons are responsible for the changes in livestock wealth in time. Although it is not possible to get precise reasons contributing to such differences, individual responses can give a direction to this inquiry (Table 7). The herd accumulation variables of births, purchases and gifts increase household herd size, while off-take variables of deaths, sales, and slaughters reduce household herd sizes (Table 7). The net effect between herd accumulation and off-take determines the herd growth-path, resulting in positive growth when accumulation exceeds off-take and negative growth when off-take is more than accumulation. Table 7 clearly shows that frequency of drought, raids etc seems to be the main factor restricting herd population and is therefore, in effect, an important factor that constraints household endeavours to realize net herd growth. In other words, drought accounts for a major share of livestock off-take and is responsible for drops in herd holdings. The mention of livestock off-take attributed to deaths corresponds to about 40% for cattle and 28% for small stock by non-scheme villages and about 29% for cattle and 24% for small stock for the households in the scheme villages. Home slaughter and sale responses remain low at both settlement types, indicating a low use of livestock for domestic slaughter and marketed off-take. The result of herd off-take responses (Table 7) emphasizes the fact that livestock deaths accounted for most of the decrease in herd sizes or limiting herd growth.

Based on the reasons for herd accumulation, breeding practices coupled with natural birth and livestock purchases appear to be important means of herd investment by the scheme and non-scheme villages. It is worth noting that whereas births generally determine cattle growth, purchases are frequently mentioned as influencing herd growth as well. This may hint at the use of the market to rebuild herds, especially small stocks, although not conclusively. The means by which households strive to achieve net

Table 7. Household Responses to Reasons for Changes in Herd Size.

Site/variable	Reason	Component	Cattle		Small stock	
			Freq.	%	Freq.	%
I. Scheme						
Accumulation on	Births	Herd growth, tick control or use of veterinary medicine	15	36.6	4	12.0
	Purchases	Charcoal, casual work and farm produce	9	22.0	6	16.0
	'Gifts-in'	Inheritance, dowry, restocking and loan	11	26.8	3	24.0
Off-take	Deaths	Drought, raids, diseases, mistrusts, tick infection and mountain climate	12	29.3	6	24.0
	Sales	School fees, food and construction of house	4	9.8	0	–
	Slaughters	Home consumption	1	2.4	0	–
II. Non-scheme						
Accumulation	Births	Herd growth, tick control or use of veterinary medicine	39	25.9	23	6.3
	Purchases	Charcoal, casual work and farm produce	29	19.1	25	24.2
	'Gifts-in'	Inheritance, dowry, restocking and loan	12	7.9	6	26.3
Off-take	Deaths	Drought, raids, diseases, mistrusts, tick infection and mountain climate	60	39.5	27	28.4
	Sales	School fees, food and construction of house	15	9.9	3	3.2
	Slaughters	Home consumption	1	1	0	–

Note: The answers in this table are listed according to the frequency they were mentioned.
Source: Authors' survey, 1998.

herd increases are quite varied, including sale of natural products and transfers through traditional ceremonies of marriage (Column 3). With reference to herd accumulations, live-stock received seem to be the main explanation for apparent herd growth differentials between the scheme and non-scheme villages, especially for cattle. Herd-gifting practice, in particular, tells something about traditional institutions of livestock exchange among pastoral households. It is such indigenous institutions of livestock gifting and loaning arrangements that usually result in a diverse herd ownership structure and complex rights over animals. The inter-household exchange of animals basically functions to redistribute herds and spread risks within a broader framework of a social security networks. Hence, the relatively high responses of cattle exchange by scheme households, and small stock exchange by the non-scheme households might indicate the existence of the traditional institution of herd sharing by households.

5.2. Other Wealth Indicators

Comparing wealth indicators of the sample households in the settlement schemes and in the non-settlement villages reveals only minor differences, with a slight negative outcome

Table 8. Wealth Indicators of Settlement-Scheme and Non-Scheme Households.

	Settlement scheme	Non-scheme	Refugee scheme	P-value
Income per capita	625.5	627.6	239.1	0.287
TLU owned in 1998	6.0	6.9	1.2	0.008*
Acres owned	6.2	6.6	2.1	0.114
Acres borrowed	0.5	0.4	0.9	0.276
Household size	6.3	6.5	5.9	0.750
Literacy in %	32.2	24.5	19.5	0.012*
Maize production[a]	7.5	8.9	10.1	0.583
Beans production[a]	1.2	1.0	0	0.093
Sample size (n)	51	220	16	

Note:
a   Maize and beans production in 100 kg bags.
*   The mean difference is significant at the 0.05 level.

Source: Authors' survey, 1998.

for the settlement schemes. Table 8 shows that the settlement scheme households score somewhat lower on every indicator except for the literacy level. The literacy level is defined as the percentage of the total household size of members who had more than 4 years of education, where we excluded children aged 0–6 years.

One could question how people who settled because of poverty, especially among the schemes, are able to finance educating their children, and attain high educational levels. The higher literacy level in the settlement schemes is a result of efforts of the Catholic Mission in Marsabit that initially sponsored children in the schemes, although the Mission later widened its scope to sponsorships for other needy and poor families outside the scheme. Survey results show that more children in the schemes of the age between 7 and 18 were schooling in 1998, and more children finished primary and secondary school in the schemes.

These results seem to suggest that there are no significant differences in wealth between scheme and non-scheme households. However, the Scheffé Test for unequal sample sizes, and an F-test for overall significance (Analysis of Variance—ANOVA) reveal difference in means for TLU in 1998 ($F_{2, 233, 0.01} = 4.61$), and for literacy level ($F_{2, 247, 0.05} = 3.00$) only, between the three settlement types. These mean differences, in turn, exist between non-scheme and refugee[10] (significant at $p < 0.01$ level) households for TLU, and between scheme and non-scheme (significant at $p < 0.02$ level) households for literacy level. In other words, there is no significant difference in mean values for TLU between the scheme and non-scheme households.

The population in the schemes and outside the schemes certainly have different characteristics (Table 8). People who settled outside the schemes had more animals at the time of settlement. In addition, the percentage of female-headed households in the schemes is much higher than outside the schemes. Around 47% of the households in the settlement schemes are female-headed, (these are households where the husband died, divorced or works elsewhere), while outside the schemes this proportion is 21.

Table 9 shows that female-headed households have less TLU, less land and a lower maize production, but earn consistently higher cash incomes. The female-headed households (FHH) in the schemes earn less than those outside the schemes, but FHH outside the schemes more often had husbands working elsewhere (we might also refer to these as

Table 9. Average Values of Wealth Indicators for Scheme and Non-Scheme Households.

	Scheme households		Non-scheme households		Badassa (refugee) scheme		P-values	
	Male	Female	Male	Female	Male	Female	Male	Female
Cash income/cap.	576.1	681.0	570.2	842.2	398.1	269.8	0.569	0.348
TLU in 1998	8.2	3.6	7.1	6.1	1.2	0.9	0.018*	0.234
Literacy level	26.1	39.0	25.0	22.7	20.4	15.9	0.675	0.009*
Acres owned	7.8	4.3	7.3	4.1	2.1	2.0	0.129	0.436
Acres borrowed	0.3	0.3	0.5	0.2	0.9	0.7	0.392	0.536
Total maize in kg	890	600	1000	490	1120	300	0.678	0.322
Total beans in kg	140	100	100	90	0	0	0.123	0.667
TLU when settled	2.6	3.2	7.8	6.0	8.1	8	0.129	0.509
Sample size (n)	*27*	*24*	*173*	*47*	*13*	*3*		

Note:
* The mean difference is significant at 0.05 level.

Source: Authors' survey, 1998.

female-managed households) and cash incomes that were almost four times higher due to family remittances. Their cash incomes are generally earned by burning charcoal, brewing of alcohol and the selling of milk, activities hardly ever carried out by male-headed households.

However, there are moderate differences in mean TLU among the male-headed households ($F_{2, 165, 0.05}$ = 3.04). Such differences exist in average TLU (in 1998) between scheme and refugee, and again between non-scheme and refugee within male-headed households (sign. at $p < 0.05$ level). In addition, the means differ for literacy level among the female-headed households ($F_{2, 67, 0.05}$ = 5.112), between the scheme and non-scheme households only (sign. at $p < 0.05$ level). It would therefore seem that, even when the gender of the head of the household is taken into account, there is no evidence of wealth differences any more among scheme and non-scheme households.

At household level there is nevertheless a link between cash income, children's education and TLU wealth. There is a strong correlation between cash income and literacy level ($r = 0.204$, sign. at $p < 0.001$ level, one-tailed) overall. There are two explanations for this: a household with many school children needs a steady cash income to finance education. At the same time, we expect the monetary return once children have been educated to be higher (Section 5.3). The survey data demonstrated a positive correlation between the proportion of the children in a household of 7–18 years old who were receiving education in 1998 and the household's monthly cash income ($r = 0.155; p < 0.016$). In both scheme and non-scheme households this correlation is positive, although correlations are higher and more significant for male-headed households than for female-headed households. The correlation between the literate proportion of a household and monthly income is positive and significant for male-headed households in and outside the schemes ($r = 0.423$; $p < 0.028$ and $r = 0.276$ at $p < 0.001$, respectively), but for female-headed households the correlation is not significant.

The number of TLU a household owns does not correlate directly with the proportion of school-aged children attending school, except in the case of female-headed households in the schemes, where a negative tendency is evident. This suggests that large herd owner-ship does not mean that animals are always herded by the own household members. However, the data show that in the case of households in which a large proportion of the family is engaged in herding, the school enrolment of the children is low. The results show a correlation of -0.312 (at $p < 0.001$) between the proportion of school-aged children at school and the proportion of the household engaged in herding among male-headed house-holds outside the schemes, and -0.509 inside the schemes. For female-headed households this correlation was -0.185 (at $p < 0.234$) and for female-headed households in the schemes this correlation was -0.700 (at $p < 0.001$). These results confirm earlier findings from families among the Rendille in Korr whose children participated in herding and which were less likely to send their children to school (Roth 1991, p. 138).

Alternatively, one may assume that, on the mountain, child labour is used for farming activities instead of herding. However, there seems to be no relationship between farming activities and the school enrolment of children. Maize production and the number of people in the household who are engaged in farming activities do not seem to influence school enrolment of children. This seems to suggest that child labour is more often used for herd-ing activities than for farming activities.

5.3. Literacy, Paid Employment, and Cash Income Levels

Examining educational levels of household members, it appears that households in the scheme villages have, on average, more literate members (4 years of education and above) relative to the non-scheme villages. In the non-scheme villages, 26% of the households consist of illiterate household members. In the scheme villages this is 16%. In addition, 11% of the scheme villages have households where more than 50% of the members are lit-erate, while for the non-scheme villages this proportion is only 8%. This difference is partly accounted for by sponsorships, but also for the period during which the households were settled. Households in the schemes have had a longer history of occupation on the mountain and their members are older people (who might have older children who went to school). The average age of the male head of the household in the scheme villages was 57, while in the non-scheme villages the average age was 48 years.

Of the total population (excluding the 0 to 6 years old children) in the schemes, 37% are literate, while 30% of non-scheme populations are literate. Literacy and wage employ-ment, as a positive outcome of childhood education, result in higher households incomes. There is a significant correlation of $r = 0.27$ (sign. at $p < 0.000$) for the non-scheme and $r = 0.34$ (sign. at $p < 0.003$) for the scheme between level of education and per capita cash income (adjusted for household size). The higher literacy level of the scheme village suggests that their members have an advantage over members of non-scheme households in terms of access to future employment opportunities. The scheme and non-scheme villages both have a positive relationship coefficient of $r = 0.55$ between cash income per person and the share of household members in wage employment being $p < 0.01$ (1-tailed).[11] This is evidence of a relatively high degree of correlation between wage (or salaried) employment and monthly per capita cash income, compared to correspondence between literacy measure and per capita cash income.[12] The type of jobs classified under 'wage employment' such as policemen, game rangers, soldiers, chiefs, councillors and foresters, or pastors, nurses and teachers are virtually all jobs for which a certain level of literacy is required. Wage employed members of the households often live in or commute to Marsabit Town.

Table 10. Food Self Sufficiency or Cash Earnings Above the Absolute Poverty Line.

	Scheme (including refugee)		Non-scheme	
	Male	Female	Male	Female
% of households earning more than 56 shilling/day/cap.	2.5	3.7	3.5	6.5
% of households where harvest is more than 256 kg/cap.	12.5	15.4	20.3	17

Source: Authors' survey, 1998.

5.4. Arable Farming and Cash Incomes

Farming production among male-headed households is not significantly different in the schemes in relation to the non-schemes ($t_{235,\ 0.025} = 1.960$). On average, 890 kg were harvested in the schemes, while outside the schemes the harvest was 1000 kg per household in 1998 for two growing seasons. The harvest was highest in the Badassa refugee scheme, where households harvested an average of 1120 kg maize in one year (in two growing seasons). Despite the favourable agricultural climate of that time (1997 and 1998), the harvest in the whole area remained too small to provide enough food for individual families. If we estimate a calorie intake of 2500 per person (not corrected for adult equivalence scales), and assume that 1 kg of maize provides 3560 calories, the harvest needs to be 253 kg of maize per capita for maize production to be self sufficient. Only 12 to 20% of the sample households on Marsabit Mountain reach that level, but it was not area specific. The non-scheme settlements have a higher proportion of food self-sufficient households, compared to the schemes (Table 10). In both non-scheme and scheme villages, about 3% of the households earn Ksh. 2100 per capita per month or more, which is also the proportion of the households living above the commonly used poverty level of US$1 per person per day.

There is no correlation between per capita income and maize harvest per capita or number of TLU, for none of the categories. There is, however, a strong and significant positive correlation ($r = 0.290$, significant at $p < 0.001$, one-tailed) between maize (the main food crop) harvests and the number of TLU per households in male-headed households in the whole area. However, this correlation does not hold for the female-headed households. Male-headed households seem to concentrate on livestock and maize production, while female-headed households concentrate on either livestock plus cash income or farming plus cash income. We can conclude from this that although male and female-headed households each diversify differently, livestock keeping and crop cultivation reinforce each other especially for the male-headed households.

5.5. Perception on Changes and Differences in Lifestyle

With regard to perception, Stiglitz (2002, p. 175) remarks: "… perceptions are sufficiently widespread to at least suggest that there may be some reality in them, and … the perceptions themselves become part of the reality with which we have to deal". So, what do people show with their perception? Do people express how they perceive reality, or do they express how they want reality to be? This is one of the challenges of the perception research. We asked people about the differences between a pure pastoral and a settled lifestyle and whether they would like their household to return to a purely pastoral lifestyle.

The difficulty of this question is that the majority of the respondents involuntarily settled many years ago after traumatic events and severe livestock losses. It is clear that most people would have preferred a wealthy mobile (semi-mobile) life with animals at the time they came to the mountain. Surprisingly, given the present circumstances, the majority look back on the shift in a positive way and said they appreciated the advantages of a settled life on the mountain.

Only 10% of the respondents belonging to the major pastoral groups said they wanted to return to a mobile pastoral life. For those who preferred a mobile pastoral life, there was no difference between the number of female-headed and male-headed households, although the female-headed households who preferred a mobile life are virtually all Gabra and a few Rendille, that is camel herding pastoral groups. The advantages of settled life most mentioned by women were the accessibility of water, market, education and health facilities, while men valued the farming opportunities. These answers, indeed, represent measures commonly used to calculate poverty levels or assess human welfare (Lipton and Ravallion, 1993; Sisule, 2001). We will list some typical positive, neutral and negative expressions on the settled lifestyle, as many of the respondents perceived them. It is, however, difficult to quantify and value the opinions expressed by the respondents. Usually, a combination of answers was given. However, many of the answers presented in the following subsection seem to represent a general feeling.

    a. The following are positive answers to settling on Mount Marsabit:
      "Having a farm is less risky than animals. All animals can die at once."
      "Settled life gives opportunities especially for the poor. When you work hard, you can become independent."
      "In nomadic life one experiences continuous drought, diseases, raids and war."
      "Settled people don't need to sell animals always, because for family needs they produce food, whereas nomads have to sell an animal whenever they have a little problem."
    b. The following are neutral answers on sedentarisation:
      "Both ways of life are good if you can manage together: settled life is good for young people because farming is hard work. Nomadic life is good for their elders."
      "Both ways of life are good if you can work together. But farming is important for survival."
    c. The answers were given to express a negative attitude towards settling on Mount Marsabit:
      "Nomadic life is better than farming. If you sell one bull, it refunds you the produce from three farming seasons."
      "Mobile life is better. Farm work is too heavy for us."
      "Nomadic life is more healthy, because people eat animal products."
      "We admire life of nomads. Farming is for men. Women lose their main task after settling."

Almost 80% of respondents said they would not like a purely pastoral or nomadic lifestyle. Only 22 people (about 9%) said they would like to go back to a purely nomadic or pastoral life, of which almost half come from Sagante, while 2% expressed preference for combining both ways of life. The social facilities like schools, health centres and clean water are not easily accessible in the lowlands. These facilities are currently highly valued and are a reason for people to stay on the mountain, even in instances that they acquire livestock. Farming has become a feasible as well as a viable activity and complementary to pure pastoralism. The direct comparison of land with animals is revealing. Land is perceived as capital as much as livestock is. Land cannot die and does not become a burden in dry times, like livestock does. One stockowner referred to the 'positive caloric terms of trade' argument. It is profitable to sell one bull when the money the sale generates can be used to buy an amount of maize equivalent to three harvests, although the time and labour investment required to raise a bull can only be profitable if you have a lot.

Many people admit that farming is a labour-intensive way of earning a living. That is why old people, who no longer have school-age children, try to revert to a nomadic, animal-oriented existence. The answer "it is good to stay in one place" represents a bundle of answers which more or less refer to the value of having some "permanence" in life, such as, "It is good to have a permanent house", "good to have a land to stay in", "I feel settled", "Settled life gives you peace of mind".

The feeling that sedentarisation leads to "independence" can be perceived in various ways. Some people miss the social structure and support of the traditional pastoral setting. In some ways, the falling apart of social structures can be seen as a loss of traditional institutions. Old social structures were also considered to be a burden. Once people settle, many regard their some degree of self-reliance as something they have newly acquired, and seek other social relationships, thus create interdependence among the households. How important it is for poor people to be, to some degree, independent is a much-undervalued reason for people to settle and start farming. For instance, as has been said earlier, in cases in which sons do not inherit animals from their father, a first-born son has to find income from somewhere else and is not supposed to depend on others. He cannot work as a herdsman for relatives, whereas the other sons can make use of the "social security net" in the pastoral setting to gradually acquire herds. In the case of impoverished pastoralists, farming provides opportunities to become independent while regaining their self-esteem and dignity. This positive change in people's lives is of course hard to quantify.

Some positive and negative answers are completely opposite. Some people find life in Marsabit cheaper, especially when they grow food, while others find the nomadic way of life less expensive. Some refer to a balanced diet in settled life, while others consider the nomadic diet to be healthier. For some women, settled life is not altogether a constructive change. Although most women find the education for their children important, others state that women lose their main tasks when they stop being mobile. That is certainly true for women who have purely nomadic backgrounds. The most important aspects of nomadic life, which distinguish nomads from pastoralists, are the mobility of the wife and her house. The nomadic hut and the camp camels are the property of the married woman. Loading the camp camels and building the hut after a move are the wife's responsibility. Once a household does not have camp camels, the wife cannot move with the herds anymore. The husband could go alone, but a purely nomadic life is over. If the wife does not have her nomadic hut anymore and the husband has built a (semi) permanent house, she also does not perceive the house as her own property in the same way as a nomadic hut would be. Some former nomadic women say they have lost their authority within the house. Seen in this way, a nomadic lifestyle is defined through the role of women in the pastoral household. She loses that status once the household settles. In the settlements, she is not the owner of the house and has no camp animals she can call her property. However, in permanent settlements a woman owns a few animals which she may regard as her property, and yet she loses her skills and tasks tied to ownership of the traditional hut, like weaving mat and rope making. In this regard, loss of status, creativity and increasing boredom were sometimes mentioned. However, households may consist of members who are able to become increasingly mobile when the family herd grows while the head of the household remains in the settlement. In addition, the idea that part of the household would move, due to parents or herdsmen engaging in (semi-nomadic) pastoralism while their children or wives or brothers guard and exploit the farm, may well be one way in which a lot of people envision in their future.

Mitchell (1999) and Nathan et al. (1996) report crucial findings on effects of sedentarisation. Mitchell illustrates the socio-economic status of settled pastoral women in villages on the dry side of the mountain (Karare and Kijiji, also called Ogicho or Parkichon), who mainly subsist on sales of milk. The position of these women seems to be at the extreme end of poverty compared to other communities. Mitchell shows that the perception of women in these villages on sedentary life is rather negative. It is clear that these women had not chosen a sedentary lifestyle voluntarily. A lack of alternative income-generating activities, an insufficient number of animals to subsist from and a lack of farming or gardening options are signs of a gloomy future for these villages. Kijiji is a spontaneous and very new settlement of impoverished households on the dry side of the mountain (established in around 1994) and we think that their current situation represent a typical first stage in the sedentarisation process. If households like these are not assisted, it will take years for them to restock and to improve on their welfare status.

The negative effects of sedentarisation on people's health by the one-sided nutritional status of the diet (which sometimes lacks animal proteins) is shown by the work of Nathan et al. (1996). In their research, they were able to prove that nomadic children were much better fed and healthier than children in the settlements. Even though the settled families in the villages on Marsabit Mountain already constitute a selection of impoverished pastoralists, and should be compared with the poorest people in the nomadic society, the children of poor nomadic families apparently had a better state of health than children of poor families in agricultural villages (Fratkin, 2001, personal communication). This might partly be explained by the intact social safety net among the mobile pastoralists, where poor children also have access to milk and also poor nomadic families also being in custody of herds of settled kin. However, we think that the *very* poor families who do not have access to this safety net are the ones who settled. If they had had access they might not have settled. It is important to understand that poverty and lack of a social safety net is not caused by sedentarisation, but people settle because they are poor and lack an adequate social safety net. This in itself is a sufficient reason why the sedentarisation process warrants more attention than it presently gets.

5.6. Issues for Future Concern when Sedentarisation Continues

a. *Promotion of agriculture on Marsabit Mountain:* Evidence shows that crop cultivation and livestock keeping reinforce each other. As we have seen, households with a high maize harvest are likely to have more livestock, and productive farmers tend to reinvest in animals. This corresponds with findings of a large-scale research done by Bourn and Wint (1994) who found a high correlation between livestock numbers and agricultural activities in Mali, Niger, Nigeria, Sudan and Chad between 1980 and 1993. Farmers who have livestock are likely to profit from manure and oxen for ploughing, extra protein from milk and have inflation-proof store of cash-surpluses in times of drought or sudden fall in expenditure. Pastoralists with a farm profit from crop cultivation, because they need not deplete their herds when they are in need of grains. Promoting agriculture on Marsabit Mountain will directly promote the pastoral production system as a whole, which is a realistic aim considering the rapid population growth. Development intervention should aim to increase food security through diversification, of which agriculture is just one option.

There is no doubt that the trend towards reduced mobility for a number of people will continue. A crucial question is to what extent can Marsabit Mountain host more impoverished pastoralists in future? Do we assume that the natural resource base of the mountain is a limiting factor, or is the stock of resources dynamic and adaptable? As far as we know, there has been no thorough research done on this question. It is necessary that each component (forest, rangeland, arable land, water) be assessed on its potential for more intensive exploitation. We assume, based on evidence from our own research, that the stock of resources is to a certain extent dynamic and adaptable, and the limiting factors at the moment are capital, labour, unequal access to resources and education.

b. *Inter-ethnic relationships:* Increased sedentarisation has resulted in a concentration of different ethnic groups in Marsabit Mountain. Although there is likelihood that the ethnic groups contend for scarce resources, there is no evidence that violent conflicts in the Mountain area are increasing or are scarcity induced (Witsenburg and Adano, 2004). For example, interethnic sharing of water sources in the height of drought is more common than fights over water. In 2000, Samburu/Rendille, Gabra, Burji, and Boran shared waterholes at Hula Hula, Karantina and Bakato among others. The present vagueness in property rights permits flexibility in access rights to key natural resources on the Mountain. If a process of privatisation replaces the current flexibility and vagueness in resource tenure, then an increase of violent conflicts is likely to emerge.[13]

c. *Effect of sedentarisation on natural resources:* Although sedentarisation should not be an environmental question *per se*, the environment remains an important variable especially in fragile ecologies. Marsabit Mountain has a unique ecology because of the presence of its montane cloud forest. There are indications that until 1940 the mountain was less densely forested than at present. The frequent and intense forest fires in the past century greatly reduced forested areas since the 1970s. In addition, agricultural activities resulted often in planting of live-fences and a range of tree species in areas where there were only dusty plains before.

Tree planting has had positive effects against wind erosion, and resulted in increased agro-diversity. On the other hand, people cleared bush-land for cultivation, and cut trees for fuel and construction. This has a damaging impact on the forest ecology. Despite human and livestock population increases on the mountain and localized gully erosion especially along cattle tracks (Umuro, 2002; Agricultural Officer, personal communication) to-date the settlement has not induced widespread deterioration of the mountain ecology. The main forest conservation concern today is selective depletion of forest products and decline of forest conditions from rapid rates of local exploitation.

d. *Resource complementarities in household production:* Human settlements and economic activities on the mountain are linked to diverse functions of Marsabit Forest. For example, forest products sold on the market alone accounted for 25% of rural households annual income (Adano, 2000). This adds to over 80% of rural and peri-urban households use of fuel wood, largely drawn from the forest, for domestic uses. The current problems of the forest are increasing exploitation by households and commercial enterprises, dry season grazing and watering livestock in the forest, and pressure of forestland conversion to urban settlement. The need for addressing these problems rests on the hydrological services of the forest, its species diversity and its support for farming.

(a) Dry season around 1980

(b) Dry season 2000

Figure 5. Photos of Manyatta Otte, Marsabit Mountain.

(a) Wet season 1987

(b) Wet season 1998

Figure 6. Photos of Marsabit Township.

6. PRÉCIS AND CONCLUSION

There is evidence in the literature that indicates past rangeland development projects met limited success. This may not be surprising given that projects worried about range conditions and focused on the rangeland degradation question, without considering the welfare of pastoral people. Yet the effect of livestock on the range remained a perceived problem rather than empirically proven. The rangeland debate was largely driven by ill will about pastoralism, premised on inappropriate range models and on perceived ecological boom thinking. This view supported the gloomy scenario painted by the Club of Rome in the 1970s. It is also clear that failures of the rangeland projects (for instance IPAL) later on resulted in a scaling-down of international development projects (Morton and Meadows, 2000).

There existed also a fear within donor circles that people enrolled in development programmes would lose their social security network and lack initiatives to tackle their own problems. A question arising from this is whether donor aid for settling nomads is sustainable. In this regard a few observations based on our study are worth pointing out. First, the settlement process on the mountain would have occurred even without droughts and without settlement programmes, but may be over a longer period of time. However, development aid has helped people at a critical time of settling down. Secondly, the scheme villages are seemingly not worse off than non-scheme ones. Moreover, two noticeable differences in favour of the schemes are: (i) large increases in herds, and (ii) higher educational levels, especially among children in female headed households, which is a promising trend for the future. Indeed, the schemes villages show higher educational levels, seem to have rebuilt more herds, and the also show a restored social network. While social security networks cannot withstand shocks beyond certain level, they can be restored in a community after successful intervention efforts.

What differentiates our results from other assessments of settlement schemes is that our research studied development impacts after 25 years, while others report short- to medium-term results. Recipient populations need time to familiarise themselves with new settings and technology of production. Given that effective adoption of new technologies has always taken a long time everywhere (Kanbur, 2001), pastoral aid or range development projects should not be expected to be different. Our findings suggest that other settlement programmes negatively assessed in the past, might show positive effects after a long time. In addition, when households welfare in different settlements are to be compared it is critical that they have similar characteristics at the time of settling.

Governance and administration should deal with the fact that many more nomads are going to live a sedentary or semi-nomadic way of life, and they should deal with the consequences these changes will bring about. That does not mean that the nomadic lifestyle should be further discouraged or made impossible through prohibitive legislation and further privatisation of land. On the contrary, support of the pastoral production system is needed. Sedentarisation is not a last stage in a process of modernization in a society. 'Nomads need not settle to change but will settle if the move (to stop moving!) serves them well' (Aronson, 1980, p. 184). We would like to add that individuals should not be coerced into mobile pastoralism, but would voluntary do so if moving would serve them well.

We plea for a reorientation of policies, based on the lessons learned from pastoral development programmes in the past but aiming at poverty alleviation and livelihood diversification, which would be much more beneficial in an attempt towards self-support. There is need to re-focus the debate away from a concern about environmental conditions of the rangelands to the socio-economic situation of impoverished pastoralists.

ACKNOWLEDGEMENTS. The material on which this chapter is based comes from the authors' Ph.D. research projects. These projects were funded by Netherlands Foundation for the Advancement of Tropical Research (*WOTRO*), and Amsterdam Research Institute for Global Issues and Development, Department of Geography and Planning, University of Amsterdam. The authors would like to kindly thank both organizations for their financial support. We also thank all our field assistants in data collection, and the cooperation of our respondents is very much appreciated. We would also like to kindly thank Eric Roth for his valuable editorial comments, and Elliot Fratkin for offering us the opportunity to contribute this chapter to the book.

NOTES

1. This report also, in particular, spells out the links between poverty and environmental degradation in developing countries.
2. The process of settling down of nomadic or pastoral households and factors that compel them to do so are central to our broader Ph.D. research, from which this chapter heavily draws.
3. Pure pastoral households in Marsabit District can be defined as family units where food and income is completely obtained through livestock production. Pure pastoral households partly feed from their own herds, and exchange animal products on the market for grains and other household needs.
4. See Redfern, Paul (2001) 'Sharp drop in Kenya's population growth rate'. Horizon; Daily Nation, Thursday, November 15, 2001. This report further states that Kenya has at present one of the lowest rates of population growth in sub-Saharan Africa.
5. Tropical livestock unit (TLU) is derived from cattle and small stock holdings only, using 1 cattle = 0.8 TLU, and 1 small stock = 0.1 TLU, a scale commonly used in the District (Lusigi, 1983; Fratkin and Roth, 1996; O'Leary, 1985, 1990).
6. The materials for this part are drawn from two surveys undertaken among settled households on Marsabit Mountain. The first, in 1998 had 287 respondents from 8 villages around Marsabit Forest Reserve. Scheme households and non-scheme households were selected for purposes of comparison. A follow-up survey was held in 2000, among the same households in 6 villages. The follow-up survey had sample size of 203 households.
7. '*Shifta*' means 'bandit' in Amharic and is a common word used for bandits and raiders. The 'shifta' war was a guerrilla war caused by the conflicts between the Kenyan government and Somali secessionists. Somali clans living on Kenyan territory wanted to join Somalia, and wanted Marsabit District to fight with them. Non-Islamic Boran and Gabra had no desire to join an independent Somaliland. Somali fighters retaliated by raiding, fighting, killing, burning houses and cropland. The *shifta* secession war lasted from 1963 to the early 1970s.
8. The scheme and non-scheme villages show little difference in donkey holdings, an average of 0.3 and 0.5 respectively. However, a non-scheme household is one and half times as likely to own chicken as a scheme household. In the past, rearing or consumption of chicken was a taboo among pastoral groups. Today, chickens provide supplementary sources of food, and thus defend present consumption during peak periods of stress and shortfall in food demand and as well as provide an added source of protein and enable households to diversify their diet.
9. These scales are equivalent to wealth status of very poor (scale 1 = 8 TLU), poor (2 = 8.1–16 TLU), moderate level (3 = 16.1–40 TLU) and rich (4 = 40.1 TLU and above) per household. This scale recognises wealth as a relative measure influenced by existing wealth levels, and based on household's consumption needs from livestock sector contingent upon resource complementary supply from other household activities.
10. We analyse Badassa Refugee scheme separately because of its exceptional position in the area. Ethiopian refugees first came in the 1970s during the war. When the war was over many returned, but others who had supported the previous government came. So the village is a mixture of people who did not come at the same time, and who had been enemies in Ethiopia. They have basically no relatives on the mountain, much limited social networks and restricted access to land outside the refugee scheme, except when they rent.

11.   Overall, per capita cash income correlates significantly with the ratio of household members employed ($r = 0.54$) and literacy level ($r = 0.28$) (both at $p < 0.01$, one-tailed). Wage employment and literacy levels correlates positively at $r = 0.32$ ($p < 0.01$, 1-tailed test), and at $r = 0.60$ for the scheme and only 0.20 for the non-scheme households.

12.   In a study on income opportunities and level of diversification among Rendille women in Korr, it appeared that there was a negative relationship between level of education and diversification activities (Nduma et al. 2001). Our data does not allow clear comparisons: in our sample there were only seven educated women who headed a household and they were all engaged in farming activities. Five of them had considerable income from family remittances.

13.   In this respect it is worth mentioning that the recent 'water conflicts' along Tana River between Pokomo farmers and Orma pastoralists started after the government launched a land adjudication programme along the river (Daily Nation Team, 2001, March 10). Where vagueness in property rights along the Tana River existed, the attempt to define territory triggered violent conflicts resulting in at least 60 deaths and numerous injured and displaced families, while it is explained as a typical case of scarcity induced conflicts.

REFERENCES

Adano, W.R., 2000, Costs and Benefits of Protected Areas: Marsabit Forest Reserve, Northern Kenya. In *The Economics of Biodiversity Conservation in sub-Saharan Africa: Mending the Ark*, edited by C. Perrings, pp. 115–158. Edward Elgar, Cheltenham, UK.

Aronson, D.R., 1980, Must Nomads Settle? Some Notes Towards Policy on the Future of Pastoralism. In *When Nomads Settle: Processes of Sedentarisation as Adaptation and Response*, edited by Philip Salzman, pp. 173–184. Praeger Publishers, New York.

Baxter, P.T.W., 1975, Some Consequences of Sedentarisation for Social Relationships. In *Pastoralism in Tropical Africa*, edited by Theodore Monod, ed. pp. 206–228. Oxford University Press, Oxford.

Baxter, P.T.W., 1994, *Pastoralists are People: Why Development for Pastoralists, not the Development of Pastoralism?* The University of Reading, Agricultural Extension and Rural Development Department. Bulletin, No. 4, April 1994: 3–8.

Behnke, R.H., I. Scoones, and C. Kerven, 1993, *Range Ecology at Disequilibrium: New Models of Natural Variability and Pastoral Adaptation in African Savannas*. London: Overseas Development Institute.

Bourn, D. and W. William, 1994, *Livestock, Land Use and Agricultural Intensification in Sub-Saharan Africa*: May 2002; London: Overseas Development Institute, Pastoral Development Network Discussion Paper 37a. (Online): http://www.odi.org.uk/pdn/papers/index/.html

Brown, L., 1963, *The Development of the Semi-arid Areas of Kenya*. Nairobi: Ministry of Agriculture.

Brown, M., 1989, *Where Giants Trod: The Saga of Kenya's Desert Lake*. Quiller Press, London.

Dietz, T., 1987, *Pastoralists in Dire Straits: Survival Strategies and External Interventions in a Semi-arid Region at the Kenya/Uganda Border: Western Pokot, 1900–1986*. Ph.D thesis, Dept. of Human Geography, University of Amsterdam.

Dietz, T., A.A. Nunow; A.W. Roba and F. Zaal, 2001, Pastoral Commercialization: On Caloric Terms of Trade and Related Issues. In *African Pastoralism: Conflict, Institutions and Government*, edited by M. A. Salih, T. Dietz, and G. M. Abdel, pp. 194–234. Pluto Press, London.

Dietz, T. and M. A. Mohamed Salih, 1997, *Pastoral Development in Eastern Africa: Policy Review, Options and Alternatives*. Report For I/C Consult, Zeist (For BILANCE). Amsterdam and The Hague. The Netherlands.

Fratkin, E., 1997, Pastoralism: Governance and Development Issues. *Annual Review of Anthropology*, 26: 235–261.

Fratkin, E. and E.A. Roth, 1990, Drought and Economic Differentiation among Ariaal Pastoralists of Kenya. *Human Ecology*, 18(4): 385–402.

Fratkin, E. and E.A. Roth, 1996, Who Survives Drought? Measuring Winners and Losers among the Ariaal Rendille Pastoralists of Kenya. In *Case Studies in Human Ecology*, edited by Daniel, G. Bates and Susan H. Lees, pp. 159–173. Plenum Press, New York.

Fratkin, E. and K. Smith, 1995, Women's Changing Economic Roles with Pastoral Sedentarisation: Varying Strategies in Alternate Rendille Communities. *Human Ecology*, 23(4): 433–454.

Galaty, J.G., 1981, Introduction: Nomadic Pastoralists and Social Change Processes and Perspectives. In *Change and Development in Nomadic and Pastoral Societies* edited by J.G. and P.C. Salzman, pp. 4–26. International Studies in Sociological and Social Anthology, series 133. Leiden: Brill.

Galaty, J.G.,1992, "The Land is Yours": Social and Economic Factors in the Privatization, Sub-Division and Sale of Maasai Ranches. *Nomadic Peoples*, 30: 26–40.

Government of Kenya, 1969, *Kenya Population Census, 1969*. Nairobi: Government Printer.

Government of Kenya, 1979, *Kenya Population Census*, 1979. Government Printer, Nairobi.

Government of Kenya, 1994, *The 1989 Population and Housing Census*, Volume I. Government Printer, Nairobi.

Government of Kenya, 2001, *The 1999 Population and Housing Census*, Volume I. Government Printer, Nairobi.

Hardin, G., 1968, The Tragedy of the Commons. *Science*, 162: 1243–1246.

Helland, J., 1980, *An Outline of Group Ranching in Pastoral Maasai Areas of Kenya. International Livestock Centre for Africa (ILCA), Kenya*. Working Document No. 17.

Hogg, R., 1986, The New Pastoralism: Poverty and Dependency in Northern Kenya. *Africa*. 56(3): 319–333.

Hogg, R., 1988, Changing Perceptions of Pastoral Development: A Case Study from Turkana District, Kenya. In *Anthropology of Development and Change in East Africa*, edited by D. Brokensha and P. Little, pp. 183–199. Westview Press, Boulder, Co.

Hogg, R., 1992, NGOs, Pastoralists and the Myth of Community: Three Case Studies of Pastoral Development from East Africa. *Nomadic Peoples*, 30: 122–146.

Homewood, K., 1995, Development, Demarcation and Ecological Outcomes of Maasailand. *Africa*, 65(3): 331–350.

Kanbur, R., 2001, Economic Policy, Distribution and Poverty: The Nature of Disagreements. *World Development*, Vol. 29, No. 6, pp. 1083–1094.

Lamprey, H. F., 1983, 'Pastoralism Yesterday and Today: The Overgrazing Problem'. In *Tropical Savannas. Ecosystems of the World*, Vol. 13: 643–666. edited by F. Bouliere, Elsevier Press, Amsterdam.

Lamprey, H. F. and H. Yussuf, 1981, Pastoralism and Desert Encroachment in Northern Kenya. *Ambio*, 10: 131–134.

Lipton, M. and M. Ravallion, 1993, *Poverty and Policy: Policy Research Department, The World Bank*. Working Paper series 1130.

Little, P.D., 1985, Social Differentiation and Pastoralist Sedentarisation in Northern Kenya. *Africa*. 55(3): 243–260.

Livingstone, I., 1991, Livestock Management and "Overgrazing" among Pastoralists. *Ambio*, Vol. 20(2): 80–85.

Lusigi, W.J., 1983, *Integrated Resource Assessment and Management Plan for Western Marsabit District, Northern Kenya*. UNESCO-IPAL. Integrated Resource Assessment Part 1; Technical Report, No. A-6. Nairobi: UNESCO.

Mace, R., 1989, *Gambling with Goats: Variability in Herd Growth Among Restocked Pastoralists in Kenya*. London: Overseas Development Institute (ODI). Paper 28a.

Marsabit District Handing Over Report, 1922, *Marsabit District (Gabra) Handing Over Report*, 1916–1930, PC/NFD2/2/1.

Marsabit District Annual Report (MDAR), 1929, *Marsabit District Annual Report*, 1929, PC/NFD1/2/2, 1930–36.

Marsabit District Annual Report (MDAR), 1931, *Marsabit District Annual Report, 1931*, PC/NFD1/2/2, 1930–36.

Marsabit District Annual Report (MDAR), 1935, *Marsabit District Annual Report, 1935*, PC/NFD1/2/2, 1930–36.

Marsabit District Annual Report (MDAR), 1937, *Marsabit District Annual Report, 1937*, PC/NFD1/2/3, 1937–43.

Marsabit District Annual Report (MDAR), 1958, *Marsabit District Annual Report, 1958*, PC/NFD1/2/5, 1951–60.

Marsabit District Annual Report (MDAR), 1959, *Marsabit District Annual Report, 1959*, PC/NFD1/2/5, 1951–60.

Marsabit District Annual Report (MDAR), 1977, *Marsabit District Annual Report, 1977*. Unpublished report, 1977.

Marsabit District Development Plan, 1979, *Marsabit District Development Plan, 1979–1983*. Government Printer, Nairobi.

Mitchell, J.D., 1999, Pastoral Women and Sedentism: Milk marketing in an Ariaal Rendille Community in Northern Kenya. *Nomadic Peoples*, 3(2): 147–160.

Morton, J. and N. Meadows, 2000, *Pastoralism and Sustainable Livelihoods: An Emerging Agenda*. Policy Series No. 11: Natural Resources Institute University of Greenwich, UK.

Nathan, M.A., E. Fratkin, and E.A. Roth, 1996, Sedentism and Child Health among Rendille Pastoralists of Northern Kenya. *Social Science and Medicine*, 43(4): 503–515.

National Environment Secrétariat—UNCCD, 1999, *National Action Programme to Combat Desertification*. Kenya National Report on the Implementation of the United Nations Convention to Combat Desertification (UNCCD). Nairobi: Ministry of Environment and Conservation.

Nduma, I., P. Kristjanson, and J. McPeak, 2001, Diversity in Income-Generating Activities for Sedentarised Pastoral Women in Northern Kenya. *Human Organization*, 60(4): 319–325.

Niamir-Fuller, M., 1999, *Managing Mobility in African Rangelands. The legitimisation of transhumance*. London: IT publications.

Nunow, A.A., 2000, *Pastoralists and Markets: Livestock Commercialisation and Food Security in North-Eastern Kenya*. Ph.D. thesis, University of Amsterdam.

Oba, G., 1996, The Range Degradation Debate and its Implications for the Drylands of Africa. In *Approaching Nature from Local Communities: Security Perceived and Achieved*, edited by Anders Hjort-af-Ornas, pp. 100–124. EPOS, Research Programme on Environmental Policy and Society Institute of Tema Research, Linköping University, Sweden.

Oba, G., 1997, *Indigenous Strategies for Coping with Droughts: Options and Prospects for Obbu Booran Pastoralists of the Kenya-Ethiopia Borderlands*. Final (Consultancy) report submitted to Marsabit Development Programme/GTZ—Marsabit, 1997.

Oba, G., N.C. Stenseth, and W.J. Lusigi, 2000, New Perspectives on Sustainable Grazing Management in Arid Zones of sub-Saharan Africa. *BioScience*, 50(1): 35–51.

O'Leary, M., 1985, *The Economics of Pastoralism in Northern Kenya: the Rendille and the Gabra*. IPAL Technical Report, No. F-3. Nairobi: UNESCO.

O'Leary, M., 1990, Changing Responses to Drought in Northern Kenya: The Rendille and Gabra Livestock Producers. In *Property, Poverty and People: Changing Rights in Property and Problems of Pastoral Development*, edited by P.T.W. Baxter with R. Hogg, pp. 55–79. Manchester: Department of Social Anthropology and International Development Centre, University of Manchester.

O'Leary, M., 1994, Patterns of Range Use, Nomadism and Sedentarisation: The Case of the Rendille and Gabra of Northern Kenya. In *A River of Blessings: Essays in Honour of Paul Baxter*, edited by D. Brokensha, pp. 99–111. New York: Syracuse, Maxwell School of Citizenship and Public Affairs, Syracuse University.

Philipson, D. and D. Gifford, 1981, Kulchurdo Rock Shelter and the Stone Age of Mount Marsabit. AZANIA: *Journal of the British Institute in Eastern Africa*. Vol. XVI: 167–177.

Robert, H., 1992, *Pastoralism in Africa: Paths to Future. A Review of Mennonite Experience with Africa Pastoralist Communities*. Mennonite Central Committee. Nairobi: Unpublished manuscript.

Redfern, P., 2001, Sharp drop in Kenya's population growth rate. *Horizon; Daily Nation*, Thursday, November 15, 2001.

Roth, E.A., 1991, Education, Tradition, and Household Labour among Rendille Pastoralists of Northern Kenya. *Human Organization*, 50(2): 136–141.

Rutten, M.M.E.M., 1992, *Selling Wealth to Buy Poverty. The process of the individualization of landownership among the Maasai pastoralists of Kajiado District, Kenya, 1890–1990*. NICCOS Studies no. 10. (University of Nijmegen), Saarbrücken Breitenbach Publishers.

Salih, M.A.M., 1991, Livestock Development or Pastoral Development. In *When the Grass is Gone*, edited by P.T.W. Baxter, pp. 37–57. Uppsala, SIAS.

Salzman, P.C., 1980, Introduction: Processes of Sedentarisation as Adaptation and Response. In *When Nomads Settle: Processes of Sedentarisation as Adaptation and Response*, edited by Philip Salzman, pp. 1–19. Praeger Publishers, New York.

Scoones, I., 1995, *Living with Uncertainty: New Directions in Pastoral Development in Africa*. International Institute for Environment and Development. London: IT Publications.

Schlee, G., 1994, *Identities on the Move: Clanship and Pastoralism in Northern Kenya*. Gideon S. Were Press, Nairobi.

Sinclair, A.R.E. and J.M. Fryxell, 1985, The Sahel of Africa: Ecology of a Disaster. *Canadian Journal of Zoology* 63: 987–994.

Sisule, T.P.M., 2001, *Poverty in the Eyes of Poor Kenyans: An insight into the PRSP Process*. Tegemeo Institute, Egerton University (Kenya).

Stiglitz, J., 2002, *Participation and Development: Perspectives from the Comprehensive Development Paradigm*. Review of Development Economics 6(2): 163–182.

Tablino, P., 1999, *The Gabra: Camel Nomads of Northern Kenya*. Paulines Publishers Africa, Nairobi.

United Nations Conference on Environment and Development (UNCED), 1992, *Agenda 21. Report of the United Nations Conference on Environment and Development*. Rio Janeiro: United Nations.

Witsenburg, K. and W.R. Adano, 2004, *Population Growth, Resource Scarcity and Violent Armed Conflict. No Evidence of Causal Relationship*. University of Amsterdam.

World Commission on Environment and Development (WCED), 1987 *Our Common Future*. Oxford University Press, Oxford.

Chapter 7

From Milk to Maize

The Transition to Agriculture for Rendille and Ariaal Pastoralists

KEVIN SMITH

1. INTRODUCTION

Agriculture is becoming an increasingly common, sometimes necessary, subsistence strategy among many East African pastoralists, albeit one that is often considered a poor second choice to animal husbandry by both pastoralists and researchers alike. Rendille and Ariaal of northern Kenya are no exception, particularly since the arid and semi arid lands they occupy are uniquely suited to the rearing of livestock. In the early 1970s, however, a small number of impoverished Rendille and Ariaal began farming in the newly established community of Songa, located in an unusually well watered area that, combined with relatively rich soils, permits the irrigation of fruits and vegetables alongside the dryland farming of maize and beans. A generation later, this number has grown to over 2500 permanent inhabitants, with a continuous flow of new arrivals spurred on by the apparent greater viability of farming over herding as a subsistence strategy, especially for poor pastoralists who have lost so many of their animals to numerous droughts in recent years. This chapter draws upon research among Rendille and Ariaal in Songa and in five pastoral communities to show how Songa's food security, favorable climate, permanent lifestyle, and market opportunities make farming an attractive option. Despite certain reservations about culture change and ethnic conflict, the overwhelmingly positive attitude Songa residents have toward farming indicates that this economic strategy will remain a viable and even preferable option, yet one that is not available to all Rendille and Ariaal.

The pastoral mode of subsistence in East Africa has become particularly difficult to practice in recent years as the result of several exogenous factors. Marginalization,

KEVIN SMITH • United States Agency for International Development, Nairobi, Kenya.

137

impoverishment, and oppression have increased throughout the twentieth century, diminishing rangeland freely, to the point of preventing pastoralists from subsisting as they once did (Baxter, 1991: 8). In the past decade, over fifty million Africans in arid regions of the Sahel, Sudan, and East Africa experienced the problems of drought and shrinking rangeland (Bonte and Galaty, 1991). These pastoralists have also witnessed agricultural expansion and government programs to sedentarize and usurp their group territorial rights (Gilles, 1990).

Loss of animals from diseases, droughts, or raids either forces pastoralists into a subordinate relationship with wealthier pastoralists or makes them rely on alternate subsistence strategies, from foraging to trading livestock for agricultural produce (Sobania, 1988: 222–223). The severe droughts and famines that struck northern Kenya in 1971–73 and 1982–84 greatly impoverished many pastoralists and coincided with the introduction of churches and development projects that attempted to settle these people. Catholic and Protestant missions, along with several development agencies, have led many members of the closely related Rendille and Ariaal ethnic groups to settle around the Catholic mission centers at Kargi, Korr, and Laisamis and at the African Inland Church (AIC) establishments at Loglogo and Ngrunit. Korr and Kargi used to be dry-season waterholes, but by the late 1970s became permanently settled centers with mechanical boreholes and food aid provided by the Christian missions. During this period, the UNESCO sponsored Integrated Project in Arid Lands (IPAL) encouraged market development of local pastoral populations, to reduce the number of animals on the range (Fratkin, 1991). In addition, government policy in Marsabit District favored settling pastoralists and promoted irrigation schemes for agriculture on the Mountain, rather than strengthening the livestock sector, the mainstay of the district's economy (O'Leary, 1990: 163). This rapid growth of towns, sedentarization of nomads, and increased participation in the market economy through livestock sales and wage jobs introduced new options for pastoral Rendille and Ariaal (Fratkin, 1992: 119). These options, however, are tempered by constraints to mobility and sustainability.

Although Rendille and Ariaal have not practiced farming historically, some members of these ethnic groups have had to turn to it in the past. While agro-pastoralists such as Dassanetch adjusted to the severe livestock epidemics that swept East Africa in the late nineteenth century by concentrating on farming, purely pastoral Rendille, as well as Samburu, were sometimes forced to rely on agro-pastoral neighbors to survive (Sobania, 1991: 136–137). And while inter-group migration was more fluid before colonial authorities set boundaries and delineated ethnic groups, colonialism interfered with the inter-societal means of "insurance" used to ameliorate localized destabilizations (Sobania, 1991: 140). The former reliance Rendille and Ariaal had on agro-pastoral neighbors during times of drought has reoccurred today. This time, however, it is with members of their own ethnic groups who, after settling out of the necessity of poverty, have become local examples of self-sufficiency and market integration.

2. THE COMMUNITY OF SONGA

Songa is an irrigated agricultural community of approximately 2750 inhabitants, located on the southern slopes of Marsabit Mountain, in Marsabit District, Kenya's largest, most arid, and least populated district (Fratkin, 1991: 1), prior to being subdivided into two districts in 1996. Marsabit Mountain is an extinct volcano that rises from the desert floor to an altitude of 1865 meters, the upper elevations of which have rich volcanic soils with high water retention capacity (Republic of Kenya, 1994–96: 21). This soil is very deep and well drained, with a topsoil rich in organic matter (GOK, 1988). Rainfall is high and evaporation low, with many mornings of fog and mist. The southern and eastern faces of this zone include Marsabit Forest Reserve, an area covering 1441 square kilometers of dense forest (DDP, 1994–96: 22–23). It is on the lower reaches of this sub-humid ecological zone where Songa lies.

In response to the Sahelian drought of the early 1970s, an AIC missionary founded Songa, as well as two other mountain communities of Kitiruni and Nasikakwe, in 1972 for destitute Rendille and Ariaal who were living in the desert town of Loglogo at the southern base of Marsabit Mountain. While Rendille (population 25,000) live in the Kaisut Desert with their camels, goats, and sheep, Ariaal (population 10,000) predominantly live on the fringes of the Ndoto Mountains southwest of Marsabit so that they may be nearer to environments suitable for cattle as well as camels and small stock.[1] These two ethnic groups share strong economic and social relationships, not only with each other, but also with Samburu (population 100,000), who live to the south and west. In fact, the intermigration of Rendille and Samburu gave rise to Ariaal in the middle of the nineteenth century (Fratkin, 1991; Spencer, 1973). And it is from Samburu whom Ariaal, and more recently Rendille, originally obtained their cattle. Members of these three ethnic groups continue to intermarry, share pasture, and exchange livestock today, although almost all of the inhabitants of Songa are historically Rendille and Ariaal.

As with pastoral Rendille and Ariaal, households in Songa are matrifocal and minimally include a woman and her children. Most houses have water taps, around which crops that need frequent watering are grown, with *mashamba* (Kiswahili for farms, sg. *shamba*) lying adjacent to the houses. Some people clear land in unoccupied lower reaches of Songa or establish gardens near waterholes in the small valleys within Songa. If a man is polygynous, all of his wives can have their houses on the same plot or, if his plot is not large enough, his wives will live apart from each other and have separate farms.

Farms of married sons are typically on their fathers' plots, although most farms are not large enough to be divided into viable units after two or so generations. Even the relatively large farms of the early settlers have substantially diminished in size, as subplots have been given to sons or friends and relatives who have settled in Songa more recently. Unmarried warrior sons and mature daughters are often given small gardens to cultivate and sell crops, while a married daughter may establish a house and farm on her father's property if her husband is not from Songa and wishes to settle there.

Several crops are grown in Songa, the main ones being *sukuma wiki* (Kiswahili for collard greens), maize, and beans. In addition, many people grow fruits and vegetables such as bananas, mangoes, papayas, oranges, tomatoes, onions, and green peppers. All of these crops may be sold at the district capital of Marsabit town, whose more than 13,000 residents provide opportunities for local farmers and food sellers unparalleled in the whole of Marsabit district. Almost every day women, and sometimes men and girls, make the two-hour journey of approximately 15 Km to Marsabit town's marketplace. Although *sukuma wiki* sales predominate, the sale of milk increases toward the end of the rainy season and beginning of the dry season because *sukuma wiki* is particularly vulnerable to insects and the stem rot virus, both of which flourish during the rainy seasons. It also takes a few months for the new crop to establish itself. For those who do not have large enough vegetable gardens, milk remains the most important item to sell regularly.

Compared to selling vegetables, however, there are two disadvantages to selling milk: smaller profits and competition with women from other Rendille and Ariaal communities on the mountain.

Herd owners in Songa, like pastoral Rendille and Ariaal, keep only a few milking cattle, calves, goats, and sheep within the community. Some men keep a couple oxen for plowing as well, renting them out to those preparing their fields just before the rains begin. The rest of the animals are herded away from Songa, not only for the lack of good pasture, but also because livestock will damage the farms, the climate is cold and wet, and the high tick infestation is particularly harmful to camels and goats. The few people in Songa who have camels keep them permanently with pastoral relatives in the lowlands.

Many Songa warriors herd cattle around the lowland pastoral towns of Loglogo and Laisamis, while others herd them on Marsabit Mountain and come home more often. This latter arrangement enables warrior-age brothers to alternate between herding and farming more easily. Songa boys and girls herd cattle and small stock locally. But since many of them attend school and cattle camps are too distant, they are not as involved in herding as are pastoral boys and girls. If a man has no warrior sons or his sons are otherwise occupied with school or wage jobs, a relative can herd his animals. Elders, women, and *nekerai* (female adolescents of age to be girlfriends of warriors) herd less frequently because their primary work is on the farms (Smith, 1999).

Songa has not developed independent of influence from Western institutions. Various governmental and nongovernmental organizations have left their mark. The National Council of Churches of Kenya (NCCK) built the irrigation system, primary school, and original dispensary. More recently, Food for the Hungry International (FHI) constructed an electric fence around Songa to keep out elephants, expanded the primary school, sponsored children's school fees, and introduced hygienic birth practices to local midwives. The Kenyan government built, staffed, and continues to operate a modern dispensary and a nursery where farmers can obtain tree saplings and some agricultural advice. Finally, the European Union is providing funds to upgrade and expand Songa's irrigation system, a project that is still in its early stages. With regard to agricultural development projects aimed at improving crop productivity or marketing farm products, however, the community has been left relatively untouched.

2.1. Resource and Labor Exchanges

Farming Rendille and Ariaal maintain important ties to the pastoral communities around and below Marsabit Mountain. Friends and relatives herd some of their animals, they exchange crops or animals for livestock, and their identity is maintained through inter-marriage with the pastoralists and through major age-set rituals that are rooted in the pastoral culture. As the emphasis on farming precludes keeping many animals in Songa, those who have more than a few animals will have them herded by relatives living in the lowlands or in the pastoral settlements on Marsabit Mountain. Some Songa warriors may even spend a good deal of their time herding their families' animals with the animals of these pastoral relatives.

With regard to transactions over subsistence items, most of Songa's inhabitants to exchange food with pastoral Rendille and Ariaal. The main items exchanged are maize and beans for cows and goats, although sometimes animals can be exchanged. As with exchanges among pastoral Rendille and Ariaal, these exchanges are usually not simultaneous. Typically, a relative or, less often, a friend from the pastoral area will borrow food from a farmer in Songa and pay him back later. Promises are not always kept, however, a frustration some informants expressed to the author, and the farmers will end up giving maize and beans, sometimes even cows or goats to pastoralists without reciprocation. Regardless of this frustration, or the fact that simultaneous exchanges are uncommon, 80 percent of respondents still give crops or animals to some friends or relatives without expecting anything in return. These gifts are usually initiated by the pastoralists, who come to Songa during the harvest season and bring the crops or animals back to their communities.

In addition to exchanges or gifts of crops or animals, Songa residents on occasion give money or cloths and blankets (used for clothing, the typical style of dress) to their pastoral friends or relatives. Sometimes these items are payments to pastoral warriors for herding their animals, other times they are for poor pastoral Rendille and Ariaal. Farming and herding warriors may also exchange beaded jewelry with each other as friends, a practice that is common between pastoral warriors.

2.2. Advantages and Disadvantages of New Land Titles

Among the trend toward Rendille and Ariaal becoming permanently settled is a current land survey of communities on the Rendille side of Marsabit Mountain. Funded by GTZ, the survey is being conducted by government surveyors, with the plan of giving land titles that officially recognize outright ownership. (Previously, landowners names were only registered with the county council.) So far, the survey has concentrated on Songa and nearby Lpus and Kitiruni because these communities have the most productive land and are thus being contested by neighboring Boran. The District Officer (DO) officially opened the survey on April 25, 1996, with rough sketches of Songa, Kitiruni, and Lpus made shortly thereafter.

Although lack of accommodations for the survey staff was the official reason given for why the survey was delayed (PDR, 1996: 49), Rendille and Ariaal claim it was to give non-Rendille more time to secure land in Songa. A wealthy Indian merchant from Marsabit town, the undersecretary of land, and some Boran families own farms in Leiyai, the closest ridge in Songa to the Boran community of Badasa. In addition, Boran from Badasa have attempted to claim Leiyai as their territory. Just before the land survey was officially opened by the DO, several Boran from around the mountain brought their animals below Leiyai to claim it for Boran. The DO, however, produced a map showing that the large valley between Leiyai and Badasa separates Rendille and Boran territories and Boran were ordered to return their animals to the Badasa side of the mountain. As recently as July 1998, some of the Boran who fled Leiyai during ethnic clashes with Rendille and Ariaal in 1992 have attempted to reclaim their land. But this effort was refused again by Songa residents, with government backing.

By 1997, the survey of Kitiruni was complete, although title deeds had not been issued. The process of mapping farms was still underway in Songa when the author visited the area in 1998. Several people in Songa have not had their farms surveyed either because they did not pay the 300/ = fee ($1 equaled about 57/ = in 1998) or because they are waiting for male relatives with whom they share their land with to be present. Neighbors who dispute their borders will not have their farms surveyed until Songa's' indigenous local land committee settles the matter. A disputant who feels cheated by the committee's decision, however, can file a complaint with the district land committee in Marsabit, chaired by the District Commissioner, which will often refer the dispute back to the local committee.

According to some of Songa's residents, the issuing of title deeds will create a problem for those who are not aware of lands' value. Title deeds give the government land commission jurisdiction to settle all disputes. With no committee of elders to halt or reverse sales, comparatively wealthy outsiders who know the value of farmland could eventually purchase all of the farms in Songa, thereby forcing Rendille and Ariaal off of what was once their land. Rendille and Ariaal are just beginning to realize the value of farming and subsequently investing in their land. But the temptation to sell land at what may appear to be a lot of money may be too great for a poor pastoralist to resist.

Those who stand to lose the most from the new title deeds are women. The introduction of land registration systems in Africa has historically stressed exclusive male ownership, denying women their traditional rights (Bryston, 1981: 33–34). Throughout the developing world, the shift from communal land ownership to Western-style private ownership through colonial policies marginalized women by preventing them from obtaining land (Rogers, 1980: 38). In pre-colonial times, African women often owned land because it belonged to the person who cultivated on it. However, colonialism introduced land privatization and attendant legal rights to men, thereby excluding women from land ownership (Rathgeber, 1989: 22). Colonial administrators failed to register women's assets or usufruct rights, while women's lack of access to cash prevented them from purchasing land.

As with livestock, Rendille and Ariaal elders are acknowledged as owners of the land in Songa, while women have only use rights (Smith, 1998: 463). Because they seem to benefit the most from selling crops, women are in a particularly disadvantaged position with regard to having these use rights abrogated if their husbands decide to sell land against their will. This same problem extends to the land rights of widows when sons or brothers of the deceased husband sell land, spoiling their hopes of gaining some degree of self-sufficiency later in their lives. In fact, widows throughout Africa often have to depend on their adult children for access to land, as well as housing (Potash, 1986: 4–5).

Despite the potential for alienating Rendille and Ariaal from their land, title deeds do secure their territorial rights, which are becoming increasingly valuable as Marsabit Mountain's population expands with the influx of pastoralists from the surrounding desert and refugees from war and famine torn Ethiopia. The Rendille chief of neighboring Karare location commented to the author that Rendille might have given in to Boran pressure to abandon their side of the mountain had they not been taught how to farm and made aware of land's value in recent years.

3. METHODOLOGY

In understanding Songa's place in the realm of economic options for Rendille and Ariaal, it is important to discover who settles, why, and what they think of their life after having done so. That is, what are the pros and cons of living in Songa versus the pastoral areas, and what draws people to settle in Songa? Interviews with 29 elders, 42 women, and 32 warriors from Songa focused on how they compared farming to herding and life in Songa to life in the pastoral areas. Families were selected based on whether there was at least one warrior-aged son in the family. Warriors in their mid-teens or younger were not interviewed because their immaturity made it more difficult for them to answer questions thoughtfully. In addition to those with warrior aged sons, 11 families were chosen that did not have warrior-aged sons so as to capture a wider age range of elders and women.

Informants were not chosen randomly. Instead, assistants contacted households based on the above criteria because the goal of the author was to interview people of various ages who had lived in Songa for different lengths of time in order to include a wide variation of responses to questions. Of the roughly 285 households in Songa in the mid-1990s, the time in which the research was conducted, 42 were selected. Rather than surveying a large number of households, this number was chosen to allow more time for multiple interviews with informants and for other research both in Songa and in several pastoral communities.

One woman from each of the 42 households was interviewed, whereas only 29 elders were interviewed. This discrepancy is related to the fact that some of the women interviewed were widows or were second or third wives whose husbands lived elsewhere. In one case, the husband was herding his animals far from Songa and did not return home while the research was being conducted. Since a first wife is typically fifteen years younger than her husband and it is acceptable for a man to remarry after his wife has died, widowers are extremely rare among Rendille and Ariaal, and the author did not encounter any while living in Songa. In contrast, widows cannot remarry and are often household heads. Because some of the households were too recently established to have sons of warrior age or the warrior sons were away herding, only 32 warriors were interviewed. A return visit to Songa one year after the original fieldwork was conducted included reinterviews with 10 elders, 11 women, and seven warriors to discuss the conclusions drawn from the initial fieldwork.

To place farming and marketing's effects on culture change in a broader perspective, 21 elders, 23 women, and 21 warriors from 23 pastoral Rendille and Ariaal households were also interviewed. The interviews took place in the nomadic settlements of Ndikir and Lewogoso, as well as permanent settlements of Ngrunit, Korr, Kargi, and Karare. All of the settlements are in the Kaisut desert except Karare, which is located on Marsabit Mountain, approximately 18 kilometers by foot from Songa. The questions were similar to those asked in Songa, but were modified to suit the pastoralists' situation (e.g., "Would you consider moving to Songa?" rather than "Why did you move to Songa?"). Reinterviews with six elders, six women, and six warriors were conducted upon the author's return to the area one year later.

4. THE FARMING ALTERNATIVE

One approach to helping destitute pastoralists in East Africa has been to establish small-scale irrigation schemes. While these schemes may have been successful in helping people get back on their feet, permanent settlement did not occur. Instead, participants frequently reinvested in livestock and returned to the pastoral sector, leaving behind poor farmers who, although they still thought of themselves as pastoralists, did not have enough animals to return to pastoralism (Anderson, 1988: 241). What is unique about Songa, however, is that its inhabitants do not want to return to pastoralism: they prefer to invest more in their farms than in their herds. This attitude is understandable because, even though animals are still recognized for their economic value, past experience with herd losses that brought people to Songa cautions them from investing heavily in livestock.

Concentrating on farming, while still keeping animals, is attractive to poor Rendille and Ariaal because they can make a living and even earn cash more easily than through herding exclusively. Although pastoralists do not have to sell animals constantly, instead selling milk if there is a nearby market, they are faced with a dilemma when doing so because animals take years to replace. This is especially problematic for those who have few animals to begin with. But farmers can sell surplus crops and still have just as much food to sell the next season, without affecting the amount of food grown for consumption. In addition, profits from agricultural products are greater than those from milk products because of the larger amounts that can be grown and transported to town. Women in Songa, for example, who sell garden produce make twice as much money as do their neighbors who sell milk (Smith, 1998: 462).

While it is understandable that poor pastoralists migrate to Songa's predominantly agricultural economy out of necessity, it may be surprising to learn that, given their strong pastoral identity, none of the men and women interviewed by the author plan on returning to pastoralism. A generation ago, consensus among pastoral Rendille and Ariaal was that they would never "scratch the earth" like those in Songa were just starting to do. Today, however, many of these people so admire the food and water security and marketing opportunities Songa offers that they would consider moving there themselves if too many of their animals die. Given the risk avoidance behavior of Rendille, Ariaal, and other East African pastoralists (Smith et. al., 2001), this attitude makes sense.

Although agro-pastoralists on Marsabit Mountain have impressive returns from their farms, they invest relatively little in their animals with regard to feed and adequate water storage (Chabari, 1996: 10). Economic surveys of 42 households in Songa reflect this phenomenon. Table 1 summarizes how many herd owners in Songa bought and sold animals within the past year of being interviewed. Only 35 of the 42 households surveyed owned cattle, 22 owned small stock, and just seven owned camels, while the number of livestock sold greatly exceeded the number bought. These figures indicate that herd owners sell livestock, particularly cattle, to raise cash rather than to reinvest in animals, perhaps partly because of the greater expenses related to settling, such as constructing and furnishing permanent houses, paying school fees, and buying medicine and clothes.

Compared to specialized herders, such as Maasai or Samburu, who have at least 25 cattle per household, agro-pastoralists tend to have far fewer animals (Sperling and Galaty 1990: 76). Table 2 shows that only 35 households owned an average of 19 cattle and small numbers of animals are kept in Songa (4.5 cattle and 6.8 small stock), which demonstrates how little households rely on them for subsistence. This pattern exists despite the fact that agro-pastoralists, unlike pastoralists, can sell crops instead of animals, allowing the latter to increase more easily while still meeting the household demand for purchased food. This greater seasonal food security is indicated by residents in Songa pointing out that they receive less food aid than do their pastoral counterparts in the lowlands (Smith, 1997:103).

People give two basic reasons for settling in Songa: the desire to farm, either to change one's lifestyle or to broaden one's economic options; or the need to farm because of

Table 1. Number of Songa herd Owners and Livestock Bought and Sold. Total Number of Households = 42.

Type of animal	Number of herd owners	Number who bought	Animals bought	Number who sold	Animals sold
Cattle	35	4	7	14	30
Camels	7	0	0	0	0
Small stock	22	4	6	4	7

Table 2. Type and Number of Animals Owned by Households. Total
Number of Households = 42.

Households	Camels	Cattle	Goats	Sheep
Own livestock	7	35	22	13
Average number*	4.3	19	18	13
Maximum number	5	100	81	30
Average number in Songa**	0	4.5	6.8	≤ together

*   For those who owned the type of animal in question.
**   The number for goats and sheep is combined because these animals are herded
together in Songa.

having lost too many animals on which to survive in the pastoral economy. However, only
the latter reason ultimately explains why Rendille and Ariaal have turned to farming. When
asked whether they had many animals upon first settling, almost all informants who said
they came to farm admitted that they arrived with few or no animals, or that the many
animals they had brought quickly died off.

Rather than lament their herding past and the prestige of owning animals, Songa
residents overwhelmingly say they prefer the farming life. Greater food and water security
are the reasons most often given. For elders and women, the ability to feed their children
more easily is especially important. Songa's climate is also cooler, and there is no need to
shift residence periodically. Furthermore, the same experiences that brought people to
Songa initially cautions them against investing too heavily in animals that may only be lost
to future droughts. A Songa elder who had lost many animals while living in the pastoral
lowlands best explains the advantage of farming.

"Between farming and animals, it is better to farm because if your animals die you
become very poor and it takes a long time to recover. For me, when my goats were finished
I stayed poor for a long time. But when I came here to farm I got food in one season. I fed
my children and it lasted me to the next season."

A major attraction of farming is the greater ease with which a poor person can regain
self-sufficiency. He or she can come to Songa with few or no animals, as almost all settlers
did, and through hard work can have food and money from selling crops much faster than
one could in the pastoral economy. An elder and a woman from pastoral communities,
the former of whom was in the process of establishing a residence in the nearby farming
community of Kitiruni when the interview was conducted, state the advantage of living in
Songa:

"After visiting the area [Songa] I saw many people I knew who had nothing before
have a *shamba* now. They can sell *sukuma wiki* and do not have to sell animals. They have
more food (pastoral elder)."

"I prefer Songa [to Karare] because people grow many things and do not need to buy
as much food. Also, warriors are helping in the *shamba*. Here, they just look after animals.
Life is harder here [in Karare]. There is a problem of water. You have to go to a far place to
get water. But in Songa, it is in your *shamba*. Because there is no water here, you even
delay cooking (pastoral woman)."

One reason it is easier to make money from agriculture than from pastoralism is
the different value placed on crops and animals. Crops are grown in excess of household
needs for the purpose of selling. Their short lifespan makes storing them useless, except for
beans or the staple crop of maize, the latter of which is easily exchanged for animals.

Excess animals, on the other hand, are not raised exclusively for sale. They are an insurance strategy against future droughts, which are never too far off in arid northern Kenya. Animals also have a higher social value than crops, cementing ties such as marriage or friendship, and playing important roles in all major Rendille and Ariaal rituals.

Despite the economic and social importance of animals, farming is more compatible with the trend toward sedentarization, which provides people with more access to education, jobs, marketing entrepreneurship, and health care (Fratkin and Roth, 1990: 400). Furthermore, land tenure in Kenya has favored wealthy farming ethnic groups expanding into pastoralists' territories (Galaty, 1980; Little, 1992: 102). The Kenyan government, both present day and colonial, notoriously denied land rights to pastoralists because, as the government argues, they were not permanently occupying their land. One of the only ways for today's pastoralists to secure land titles is by farming, an important reason Rendille and Ariaal, as well as other ethnic groups in northern Kenya such as Boran (Baxter, 1975: 223) and Il Chamus (Little, 1992: 102), emphasize farming. Farming also offers a more immediate benefit by allowing people to sell crops instead of animals in order to raise cash. Like Il Chamus (Little, 1992: 9), Rendille and Ariaal can reinvest some of the profit from agricultural sales back into their herds, keeping at least some animals for milk, wealth storage, and ritual obligations, while feeding their families at the same time.

4.1. Farming and Advantages for Women

Numerous studies of African groups have described development projects that target men to the detriment of women (see, for instance, Anderson, 1985; Lele, 1975; Overholt et al., 1985; Papanek, 1979; Rogers, 1980). Introducing large scale cash crops (particularly tea and coffee in Kenya), encouraging men to control the profits from agricultural sales, and giving only men technology such as tractors that actually increases women's workload are but a few examples. But Songa has not been influenced by development projects to this degree. Instead, Rendille and Ariaal have learned to farm virtually by themselves, the result of which has provided a unique advantage to women. As they have been able to make significantly more money from produce and milk sales compared to their pastoral counterparts (Fratkin and Smith, 1995: 445), women in Songa now contribute to the economy in the same way that men do. Not only do they provide food from their gardens, but women generate enough cash to have men relying on them for subsistence (Smith, 1998: 465).

A major difference between crops versus milk women sell is that the former can be grown and transported in higher-valued quantities. The amount of milk a woman sells is limited by the number of milk animals available to her and how much milk she can carry to the market. Another limitation is that fresh milk can only be obtained when animals are milked in the morning. By contrast, vegetables can be harvested at any time, usually the day before marketing, and women can carry (on their backs—although donkeys are used by the few women who have access to them) a greater monetary value of vegetables than milk. This difference between how the types of goods are produced and sold does not affect the resources men control, whether land or animals, but it does enable women to make more money and gain economic autonomy through being viewed as more important economic actors than women in the pastoral economy.

The way pastoral Rendille and Ariaal divide control over resources by sex enables women in Songa to have control over what they grow and sell. While men own the farms, women control *sukuma wiki*, just as men own animals and women control milk in the pastoral economy. Songa's residents have basically transferred their ownership and control system of animals onto their farms and what these produce. Women benefit from this arrangement because the products they control are in high demand. They also control a greater amount of cash because sales of *sukuma wiki* and other produce generate twice as much cash as milk does (averaging 87.11/= for produce versus 43.35/= for milk). One difference, however, is that men in Songa grow and sometimes sell *sukuma wiki* and other vegetables, whereas pastoral men neither milk cows or sell milk.

Since farming and produce marketing are recent occurrences, no precedent has been established in which men exploit women's labor on cash crops without remuneration. While some men claim that women respect elders less because of greater marketing opportunities, they are unwilling to take over the "women's work" of produce and milk selling. But men still benefit from women's incomes supporting the households and from receiving small amounts of money from their wives. These two manifestations of the value women have created from their marketing activities discourage men from trying to usurp their new autonomy. Women now make and control more money from the products they sell, and contribute to the household much in the same way men do in the pastoral economy. By contrast, even though pastoral women living near town can generate income from selling milk, they do not make as much money because of the limited amount of milk they can obtain from their animals, the lower sale price of milk, and the inability to transport large amounts of milk to town.

Another advantage to farming is that women's economic benefits are also not undermined from competing products. In the pastoral economy, the male herd owner has the authority to decide whether to sell animals. In Songa's farming economy, there is no competition over which resources to concentrate on selling because all farmers want to raise crops for sale, and women are the members of society who sell almost all of these crops (see Table 3).

Despite the current marketing advantages for Songa women, the ultimately limited availability of land and the value of produce may lead elders in Songa to increase their control over both farms and crops. Because animals reproduce, a man can allocate portions of his herd among family members over time. By contrast, a man's land only decreases as he divides it among his wife and children. This is especially the case in Songa today, where unclaimed portions of decent farmland no longer exist. Another major difference between the two types of resources is the profitability of crops, which increase the value of land. Since produce, especially *sukuma wiki*, is more profitable than milk, Songa women are already claiming that elders are becoming more concerned with controlling revenue generated from it (Smith, 1998: 463).

Table 3. Songa Marketing Activities. Average Income Earned from
Produce and Milk Sales per Trip.

Status	Total number	Percent of total sellers	Average sales*	Average purchases*	Average trips per week
Elder	19	7.73	106.00	78.61	1.22
Woman	152	71.18	79.61	69.47	1.79
Warrior	4	1.59	150.00	133.34	0.40
Nekerai	31	15.26	64.92	63.85	1.67
Boy	3	1.92	45.00	45.00	0.28
Girl	5	2.13	47.00	47.00	0.23
Total	214	100.0	82.09	72.88	0.93

*   Sales and purchases in Kenyan shillings, 52/ = per $1 in 1995.

4.2. Changes in Female Labor and Nutritional Status

Despite the overwhelmingly positive attitude in Songa about how much better farming is than pastoralism, nutritional studies by Roth et al., Chapter 9, and Fujita et al., Chapter 11, this volume suggest the contrary. Nutritional and anthropometric data show Rendille and Ariaal women in Songa to have a poorer quality diet than in the purely pastoral community of Lewogoso, even though it is less affected by seasonality, a finding echoed by Songa residents when they claim that farming provides greater food security. Table 4 supports these findings by showing that calorically low quality sugar, tea leaves, cooking oil, and flour are the main items purchased from the sale of high quality garden produce and milk.

The authors of the nutritional studies link Songa women's low quality diet to the higher percentage of poor households compared to Lewogoso, and it is poor pastoral Rendille and Ariaal who founded Songa and continue to migrate to it. Their findings reflect the limited potential of available farmland and the relative inexperience people have with farming. A herd, unlike a farm, can multiply over time to a point where a man and his family can have hundreds of animals. By contrast, Rendille and Ariaal willingly admit that they are new to farming, and they have given farmland to friends and relatives rather than keeping large farms, or have settled in Songa after large tracts of land no longer exist. They are also just learning the commercial value of crops and the proper agricultural techniques to have more productive farms. However, differing work profiles also play an important role, as caloric expenditure rises for women when farming and marketing replace herding activities. Combined with a lower quality diet, the fact that women in Songa work longer hours, and in the more arduous task of farming and walking two hours to and from town to sell their produce, contributes to a depressed nutritional status relative to their pastoral counterparts.

Table 4. Number of Individuals (by age and sex status) and Percentages of their Most Commonly Purchased Items.*

Status	Number	Sugar	Tea	Oil	Flour	Beans	Meat	Tobacco	Soap
Elder	19	95	63	42	32	5	0	21	5
Woman	152	100	70	50	37	7	7	3	5
Warrior	4	25	0	25	0	0	0	25	0
Nekerai	31	97	84	19	3	0	0	0	0
Boy	3	33	33	33	33	0	0	33	33
Girl	5	100	60	60	60	0	0	0	0

* Based on six roadside surveys, 1994–96, of individuals returning to Songa from the Marsabit market.

Table 5. Women's work vs. Leisure. Chi-Square test of Agro-pastoral Songa and Pastoral Lewogoso.

Location	Non-Leisure* No. (%)	Leisure No. (%)	Total observations No. (%)	Non-Work** No. (%)	Work No. (%)	Total observations No. (%)
Lewogoso	1,257 (63)	752 (37)	2,009 (100)	831 (41)	1,178 (59)	2,009 (100)
Songa	1,127 (74)	386 (26)	1,513 (100)	455 (30)	1,063 (70)	1,513 (100)

* Chi-Square: 55.38 > 10.82 at .001 ci, 1 df.
** Chi-Square: 49.87 > 10.82 at .001 ci, 1 df.

Table 6. Songa Women's Non-work vs. Work. Chi-Square Test Excluding Farming or Marketing.

	Non-work	Work excluding farming*	Non-work	Work excluding marketing**	Total observations
Community	No. %	No. %	No. %	No. %	No. %
Lewogoso	831 (63)	1,178 (37)	831 (41)	1,178 (59)	2,009 (100)
Songa	602 (40)	911 (60)	638 (42)	875 (58)	1,513 (100)

* Chi-Square: .82 < 3.84 at .05 ci, 1 df.
** Chi-Square: .2 < 3.84 at .05 ci, 1 df.

The author conducted a time allocation study of Songa and compared the results with a time allocation study of the pastoral community of Lewogoso (Fratkin, 1987) to determine whether work and leisure times of males and females changed with the transition to farming. While a chi-square test of Rendille and Ariaal elders from Songa and Lewogoso reveals no difference between their work and leisure times (Smith, 1999: 147), there is a highly significant difference for married women. Table 5 shows both chi-square values of 55.38 and 49.87 to be greater than the distribution table value of 10.82 at the .001 confidence interval and one degree of freedom (Ott, 1988: a7). These tests indicate that Songa women have significantly less leisure time and work significantly more than their pastoral counterparts in Lewogoso.[2]

The increase in women's work is the result of both farming and marketing adding to their labor time. Table 6 shows that when farming or marketing are removed from the activity profile of Songa women, there is no significant difference between their work time and that of Lewogoso women. The chi-squares values of .2 and .82 are both below the distribution table value of 3.84 at the .05 confidence interval and one degree of freedom (Ott, 1988: a7). Women in Songa are engaged in livestock activities less than are Lewogoso women (4.2% versus 13.9%), but they make up for this difference by spending 10.4% of their day performing farming related activities (Smith, 1999: 143, 145). Furthermore, whereas Lewogoso women did no marketing, 12.5% of Songa women were observed selling their produce or milk on any given day.

A major reason women in Songa work more is their time involved in marketing: walking to and from Marsabit town, selling their produce or milk, and buying food and other items for their families. By contrast, marketing opportunities are virtually nonexistent for nomadic women who live in Lewogoso because there are no nearby towns with substantial populations to purchase their milk. Meanwhile, pastoral Rendille and Ariaal women settled around the pastoral town of Ngrunit have a much smaller customer base compared to Songa women, even though they can keep as many milking cows at home as their Lewogoso counterparts. Ngrunit's is a fraction of the 13,000 plus people living in Marsabit town, and many people in Ngrunit do not buy milk, whether they already have access to milk from their own animals or they are too poor to do so.

5. THE FUTURE FOR RENDILLE AND ARIAAL IN SONGA

Agriculture has been noted for enabling pastoralists who live in zones of adequate climate to stabilize their livelihoods (Hogg, 1980; Little, 1992). Such is the case for pastoralists in northern Kenya who engage in farming when they lack enough animals on which to support themselves (Smith et al., 2000). Unlike other East African pastoralists who usually take up farming in relatively arid environments, however, those in Songa live in an area highly suitable for farming. The rich volcanic soil makes for high crop yields. Rainfall is adequate for growing maize and beans, and the irrigation system allows fruits and vegetables to be grown year round. Market demand for Songa's crops further enhances economic opportunities. When asked to compare their pastoral to their farming lives, there is no question that Songa's residents prefer the latter. Many pastoral Rendille and Ariaal even admire those in Songa for being able to pull themselves out of poverty and make a more secure living.

Rendille and Ariaal in Songa, as well as in neighboring settled communities, would find it difficult to return to pastoralism even if they so desired. Several factors beyond their control have entered the economic and political scene in northern Kenya during the 20th century and have particularly intensified since the 1970s. The Kenyan government, churches, and non-governmental organizations have tempted pastoralists into settling by distributing food aid and providing permanent water and medical care (Fratkin, 1991). Desires of the government to control pastoralists and of researchers to minimize environmental degradation have discouraged pastoral mobility that is so essential to maintaining herd viability. Commoditization has not only introduced a desire for Western goods, but has also encouraged people to seek ways to make money. All of these factors were nonexistent until relatively recently, and long after Rendille and Ariaal had created economic systems based on mobility.

Flexibility via wide ranging kin ties and varying economic strategies, including farming and foraging, have contributed to the survival of East African pastoralists (Baxter, 1991: 19). One instance of this flexibility is the symbiotic relationship that has developed between Songa and pastoral Rendille and Ariaal, with farming becoming fully integrated into the permanent range of their subsistence strategies. Exchanging animals for grain between pastoral and agricultural Rendille and Ariaal is facilitated by a common ethnic identity, intermarriage, and shared rituals. Songa is of great use to pastoral Rendille and Ariaal, whether they must obtain food from their farming relatives in times of need or switch to farming entirely. Pastoralism is also considered valuable by those in Songa, who still keep animals for economic, social, and cultural reasons.

The poor diet and nutritional status of women in Songa somewhat contradicts their expressed preference of agriculture over pastoralism. Poverty, related to the limited size of farms and Rendille and Ariaal inexperience with farming, partially explains why these women have low caloric intakes. However, farming and marketing activities in Songa exacerbate poverty's nutritional stress on the body. Of course, a switch from pure pastoralism to predominantly agriculture means that people will have a different diet. What is less obvious, however, is the way in which marketing particularly influences women's work. As the main produce sellers, the frequency with which women go to the market, combined with the two-hour's trek to and from town, increases their workload and only adds to their already limited caloric intake.

Despite all of Songa's benefits, land privatization threatens to distance Rendille and Ariaal in Songa from their pastoral relatives, effectively creating a new and separate society. Among East African pastoralists, the expansion of the farming economy privatizes land ownership and affects social relations, possibly undermining the communal ideological and social bases of pastoralism (Hjort, 1981: 141; Rigby, 1981: 161–62). Communal land rights are crucial to pastoral solidarity, permitting individuals from far and wide access to each others' pastures. But the issuing of land titles currently underway in Songa challenges the solidarity between pastoral and agricultural kin. Whether land privatization ultimately severs ties between farming and herding Rendille and Ariaal is yet to be seen.

One certainty is that Rendille and Ariaal receive economic benefits from farming, provided there is adequate farmland and water and a market for their produce. All these aspects are found in Songa, which may lead one to conclude that the future for pastoralists is as farmers. But there are not enough communities like Songa for all Rendille and Ariaal. As more and more of them migrate to Songa, land and water will become increasingly overtaxed. Furthermore, as good farmland becomes increasingly scarce on Marsabit Mountain, greater competition from wealthy outsiders and the more numerous and politically powerful Boran will make it more difficult for Songa's residents to keep their farms or for their pastoral relatives to obtain land some day. Those in Songa are already aware of these problems and are concerned about the future of their own livelihood as farmers.

ACKNOWLEDGEMENTS. Research for this paper was collected in 1994–95, 1996, 1997, and 1998. I am grateful to the Office of the President, Republic of Kenya for permission to conduct research in Marsabit District, and to my principle assistants, Korea Leala and Rose Galoro, and to Daniel Lemoille, Nuria Herse, and Sammy Leala in Songa. Much thanks must also go to Kawab Bulyar, Annemaria Aliyaro, Lucy Fofen, Raphael Gudere, Margaret Leala, Sophia Lekuton, and Silas Leruk for their assistance in the various pastoral communities. I am also grateful to the five members of my doctoral committee for reviewing major portions of this paper: Drs. Elliot Fratkin, Patricia Draper, Eric Roth, Patricia Lyons Johnson, and David Shapiro. Funding for fieldwork in 1994–95 was provided by a J. William Fulbright scholarship and a Hill Foundation fellowship from Penn State University. Research in 1996 and 1997 was provided by National Science Foundation grant nos. SBR-9400145 and 9696088, administered by Dr. Elliot Fratkin and Dr. Eric Roth. Research in 1998 was provided by USAID grant no. DAN-1382-G00-0046-00, 1998–99 administered by Dr. D. Layne Coppock. Finally, I would like to dedicate this work to Jamila Mohammed, whose kindness and generosity toward me far outlived her brief time on this earth.

NOTES

1. Population estimates are estimated, based on data reported by Fratkin (1998: 4).
2. Activities such as "eating" and "rituals" are excluded because they are neither leisure nor work.

REFERENCES

Anderson, M.B., 1985, Technology Transfer: Implications for Women. In *Gender Roles in Development Projects*, edited by C. Overholt, M. Anderson, K. Cloud, J. Austin. West Hartford: Kumarian Press.

Anderson, D.M., 1988, Cultivating Pastoralists: Ecology and Economy among the Il Chamus of Baringo, 1840–1980. In *The Ecology of Survival: Case Studies from Northeast African History*, edited by D. Johnson and D. Anderson. London: Lester Crook Academic Publishing.

Baxter, P.T.W., 1975. Some Consequences of Sedentarization for Social Relationships. In *Pastoralism in Tropical Africa*, edited by T. Monod, pp. 206–228. London: IAI, Oxford University Press.

Baxter, P.T.W., 1991. Introduction. In *When the Grass is Gone: Development Intervention in African Arid Lands*, edited by P. Baxter, pp. 7–26. Uppsala: Scandinavian Institute of African Studies

Bonte, P. and Galaty, J.G., 1991, Introduction. In *Herders, Warriors, and Traders: Pastoralism in Africa*, edited by J. Galaty and P. Bonte. Boulder: Westview Press.

Bryston, J.C., 1981, Women and Agriculture in Sub-Saharan Africa: Implications for Development (an exploratory study). In *African Women in the Development Process*, edited by N. Nelson. London: Frank Cass and Co. Ltd.

Chabari, F.N., 1996, Livestock Development. In *Project Activity Report*, Marsabit Development Program/GTZ: Marsabit, Kenya.

Fratkin, E., 1987, *The Organization of Labor and Production among the Ariaal Rendille, Nomadic Pastoralists of Northern Kenya*. Ph.D. Dissertation. Washington, DC: The Catholic University.

Fratkin, E., 1991, *Surviving Drought and Development: Ariaal Pastoralists of Northern Kenya*. Boulder Westview Press.

Fratkin, E., 1992, Drought and Development in Marsabit District, Kenya. *Disasters* 16(2): 119–130.

Fratkin, E., 1998, *Ariaal Pastoralists of Kenya: Surviving Drought and Development in Africa's Arid Lands*. Boston: Allyn & Bacon.

Fratkin, E., and Roth, E., 1990, Drought and Economic Differentiation among Ariaal Pastoralists of Kenya. *Human Ecology* 18(4):385–402.

Fratkin, E., and Smith, K., 1995, Women's Changing Economic Roles with Pastoral Sedentarization: Varying Strategies in Alternate Rendille Communities. *Human Ecology* 23(4):433–454.

Galaty, J.G., 1980, The Maasai Group Ranch: Politics and Development in an African Pastoral Society. In *When Nomads Settle: Processes of Sedentarization as Adaptation and Response*, edited by P. Salzman, pp. 157–172. New York: Praeger Publishers.

Gilles, J.L., 1990, Nomads, Ranchers, and the State. In *The World of Pastoralism: Herding Systems in Comparative Perspective*, edited J. Galaty and D. Johnson. New York: The Guilford Press.

Hjort, A., 1981, Herds, Trade and Grain: Pastoralism in a Regional Perspective. In *The Future of Pastoral Peoples*, edited by J. Galaty, D. Aronson, P. Salzman, and A. Chouinard, pp. 135–143. Ottawa: International Development Resource Center.

Hogg, R., 1980, Pastoralism and Impoverishment: The Case of the Isiolo Boran of Northern Kenya. *Disasters* 4(3):299–310.

Lele, U., 1975, *The Design of Rural Development*. Baltimore: The Johns Hopkins University Press.

Little, P.D., 1992, *The Elusive Granary: Herder, Farmer, and State in Northern Kenya*. Cambridge: Cambridge University Press.

O'Leary, M., 1990, Drought and Change amongst Northern Kenya Nomadic Pastoralists: The Case of Rendille and Gabra. In *From Water to World Making: African Models and Arid Lands*, edited by G. Palsson, pp. 151–74. Uppsala: Nordiska Afrikanist

Ott, L., 1988, An Introduction to Statistical Methods and Data Analysis. Boston: PWS-Kent Publishing Co.

Overholt, C., Anderson, M.B., Cloud, K., and Austin, J., 1985, Women in Development: A Framework for Project Analysis. In *Gender Roles in Development Projects*, edited by C. Overholt, M. Anderson, K. Cloud, and J. Austin. West Hartford: Kumarian Press.

Papanek, H., 1979, Development Planning for Women: The Implications of Women's Work. In *Women and Development: Perspectives from South and Southeast Asia*, edited by R. Jahan and H. Papnek. Dacca: The Bangladesh Institute of Law and International Affairs.

Potash, B., 1986, Women in Africa: An Introduction. In *Widows in African Societies: Choices and Constraints*, edited by B. Potash. Stanford: Stanford University Press.

Project Development Report (PDR), 1996, Marsabit Development Program/GTZ: Marsabit, Kenya.

Rathgeber, E.M., 1989, Women and Development: An Overview. In *Women and Development in Africa*, edited J.L. Parpart. Lanham: University Press of America.

Republic of Kenya, 1994–96, District Development Plan: Marsabit, Republic of Kenya: Nairobi: Office of the Vice-President and Ministry of Planning and National Development.

Rigby, P., 1981, Theoretical Implications of Pastoral Development Strategies in East Africa. In *The Future of Pastoral Peoples*, edited by J. Galaty, D. Aronson, P. Salzman, and A. Chouinard. Ottawa: International Development Resource Center.

Rogers, B., 1980, *The Domestication of Women: Discrimination in Developing Societies*. New York: Tavistock Publications.

Smith, K., 1997, *From Livestock to Land: The Effects of Agricultural Sedentarization on Pastoral Rendille and Ariaal of Northern Kenya*. Ph.D. Dissertation, Department of Anthropology, The Pennsylvania State University.

Smith, K., 1998, Sedentarization and Market Integration: New Opportunities for Rendille and Ariaal Women of Northern Kenya. *Human Organization* 57(4):459–468.

Smith, K., 1999, Economic Transformation and Changing Work Roles among Pastoral Rendille and Ariaal of Northern Kenya. *Research in Economic Anthropology* 20:135–161.

Smith, K., Barrett, C.B. and Box, P.W., 2000, Participatory Risk Mapping for Targeting Research and Assistance: With an Example from East African Pastoralists. *World Development* 28(11):1945–1959.

Sobania, N., 1988, Pastoralist Migration and Colonial Policy: A Case Study from Northern Kenya. In *The Ecology of Survival: Case Studies from Northeast African History*, edited by D. Johnson and D. Anderson. London: Lester Crook Academic Publishing.

Sobania, N., 1991, Feasts, Famines, and Friends: Nineteenth Century Exchange and Ethnicity in Eastern Lake Turkana Region. In *Herders, Warriors, and Traders: Pastoralism in Africa*, edited by J. Galaty and P. Bonte, pp. 118–142. Boulder: Westview Press.

Spencer, P., 1973, *Nomads in Alliance*. London: Oxford University Press.

Sperling, L. and Galaty, J.G., 1990, Cattle, Culture, and Economy: Dynamics in East African Pastoralism. In *The World of Pastoralism: Herding Systems in Comparative Perspective*, edited by J. Galaty and D. Johnson, pp. 69–97. New York: The Guilford Press.

Chapter 8

Women's Changing Economic Roles with Pastoral Sedentarization

Varying Strategies in Alternate Rendille Communities

ELLIOT FRATKIN AND KEVIN SMITH

1. INTRODUCTION

The settling of formerly nomadic pastoralists is steadily increasing in northern and eastern Africa. Pastoralists are moving close to towns for a variety of reasons, including loss of herding range and political and economic insecurity, but also by the attractions of town life which offers increased economic opportunities and social security. Grazing lands of many pastoralists have decreased in the past two decades due to privatization of land, growth of both agricultural and pastoral populations, and, in Kenya and Tanzania, the expansion of tourist game parks (Campbell, 1984; Galaty, 1992). In Ethiopia, Somalia, Sudan, and northern Kenya, many pastoral peoples have moved toward towns to escape civil war, armed livestock raiding, or other political insecurities (Clay, 1988; Fratkin, 1992; Shepherd, 1988). In less disrupted regions, pastoralists seek the security of towns and access to improved health care, schools, and markets (Fratkin, 1991; Galaty and Bonte, 1991; Oxby, 1981).

Sedentarization is not a single process, it does not occur in the same way for all pastoralists, nor even in the same way for one pastoral society. People may be attracted to

ELLIOT FRATKIN • Department of Anthropology, Smith College, Northampton, Massachusetts 01063.
KEVIN SMITH • United States Agency for International Development, Nairobi, Kenya.

155

towns to increase the marketing of livestock and dairy products, to care for ill family members in hospitals, to attend school, or to engage in wage labor. Sedentarization is not necessarily a one-way process, as many town residents will return to a mobile livestock economy when conditions and opportunities change. For example, over 75% of Turkana pastoralists moved to mission centers distributing famine-relief during the droughts in northern Kenya in the 1980s; after the droughts passed, over one half of these people returned to their mobile livestock economies (McCabe, 1987). Other pastoralists such as Kenyan Boran, however, faced impoverishment in the 1960s and 1970s by war and drought, where today the majority are settled in towns or farms (Hogg, 1985).

Sedentarization affects all sectors of pastoral society, but town life has a particular impact on the lives of pastoral women and their children. Living near shops, schools and clinics has altered the domestic routine and labor activities of women. Towns enable women to reduce time spent in collecting water or grazing milk animals, but they may also lead to increased labor expenditures in food preparation, child care, and time or income spent finding firewood (Dahl, 1987).

Importantly, town residence contributes to increased participation by pastoral women in the cash economy, particularly by the sale of fresh milk or garden produce (if horticulture is pursued) to regular customers or in the open market. Poorer women with few livestock or garden resources may engage in shop keeping, petty commodity sales (e.g., tobacco, herbal medicines, or the illegal trade of beer and stimulants), or as wage laborers employed as house cleaners, shop clerks, and not infrequently, prostitutes (Dahl, 1987).

This study describes the variety of economic roles pursued by Rendille women of northern Kenya in terms of production tasks practiced in nomadic and town communities, and exchange activities of women participating in the urban marketplace as sellers of dairy products, agricultural produce, petty commodities, or the sale of labor. Specifically, we examine economic strategies pursued by women in four types of Rendille communities— nomadic cattle keeping, sedentary cattle keeping, settled agro-pastoralists, and urban town dwellers—and we link these strategies to differential access to productive resources, proximity to urban markets, and household wealth differences. Comparisons are made using time-allocation data, surveys of women marketing activities, and nutritional indices of children.

2. PREVIOUS STUDIES OF PASTORAL WOMEN AND SEDENTARIZATION

The sedentarization of nomadic pastoralists and its effects on women is a relatively recent phenomenon, but the literature is growing (cf IDA, 1991). Two volumes in particular stand out, Gudrun Dahl's (1987) edited issue of *Ethnos* on Women and Pastoral Society, and Aud Talle's (1988) study of Maasai women, *Women at A Loss*.

Among livestock pastoralists, as in other rural-based societies, women are responsible for feeding household members, either by acquiring and preparing food from their household's subsistence efforts, or purchasing foods with money acquired from the rural economy. Among African pastoralists, women are usually responsible for daily milking of livestock and for purchasing additional foods for household consumption from local shops and trading posts. Furthermore, some pastoral women earn income through the sale of livestock products such as milk, butter, or hides to non-pastoral neighbors, while larger incomes generated from the sale of butchered meat or livestock are almost exclusively controlled by men (Hilarie, 1986). For pastoralists living close to towns, women may earn extra income by selling surplus milk to non-milking urban populations, as among Maasai in Kenya (Grandin, 1988), Hawazma Baggara in western Sudan (Michael, 1987), women of southern Somalia (Little, 1994), and Orma women of northeastern Kenya who sell ghee (processed fat) in town markets (Ensminger, 1987). Women's income generated from the sale of these livestock products usually goes to purchase other types of foods, particularly grains, tea, and sugar, which are used to feed the household group (Hilarie, 1986; Oxby, 1987). Consequently the relationship between women's income and nutritional levels of household members (particularly children) is an important one, perhaps having a larger effect on children's health than income generated by men.

In some cases, sedentarization restricts the ability of women to control milk resources, particularly as dairy marketing gains monetary importance. This occurred among settled Fulani in Nigeria where the responsibility of milking shifted from women to men, and cereals formerly obtained from agriculturalists by women bartering milk are now obtained by cash earned by men selling cattle (Waters-Bayer, 1985). In Omdurman, Sudan, diary marketing is now performed by men as market integration increases, strengthening gender hierarchies and making women more dependent on men for access to cash (Salil, 1985). In Kenya, the shrinking of Maasai territory and the expansion of private land titles have decreased women's economic and social positions as the economy shifts from female-dominated dairy to male-dominated meat production, stimulated by increasing integration in the urban markets (Talle, 1988).

Settled pastoralists are often not able to keep their livestock close by, while many have lost their animals completely and migrate to towns seeking alternative livelihoods. In these situations, women may obtain cash incomes from gathering and selling firewood, making charcoal, selling handicrafts, or finding work employed in housework, concubinage, prostitution or begging (Dahl, 1987). Women with only weak links to male herders—for example, widows without sons or brothers—are among the first to migrate to towns when hard times hit the pastoral economy. In the town of Isiolo, Kenya, for example, single women may combine small-scale trade with prostitution (Hjort, 1990). In nearby Wamba, Samburu women collect and sell firewood, raise chickens for eggs, make and sell beaded jewelry to tourists, and sell individual cups of milk to generate income. Others may work as domestic servants, while some women earn revenue from making local and illegal alcohol (Sperling, 1987: 332).

Although there is little research on pastoral household budgets, studies from rural agricultural households in Africa suggest women use much of their earnings for household domestic needs, particularly food purchases. For example Ghanaian women provide 25% of rural household incomes (Guyer, 1980), although cash contributions by women in pastoral households are probably less and show a high degree of variation. Grandin (1988: 8) estimates for Maasai of Olkarkar group ranch, isolated from town markets, women's milk sales account for only 5% of the total household cash income. Among Nigerian Fulani living closer to large towns, women's earnings from milk sales account for one-third of the total income generated from cattle herds; this income is used mainly for daily household needs (Waters-Bayer, 1985).

3. SEDENTARIZATION AMONG PASTORAL RENDILLE

Until quite recently Rendille subsisted exclusively by camel, cattle, and small stock pastoralism in the very arid Marsabit District of northern Kenya. This district is Kenya's largest and least populated, located between Lake Turkana and Mt. Marsabit and bordering Ethiopia. Marsabit District receives an average of 500 mm of rain per year, with the Chalbi and Kaisut deserts receiving less than 250 mm annually. Nearly all of the district's, 110,000 people, including Rendille (25,000), Gabra (30,000), Boran (30,000), and Ariaal (7000), practiced pastoral livestock economies.

Marsabit's pastoralists began a lengthy sedentarization process in the 1960s when northern Kenya experienced civil war and social dislocation following the Somali led secession movement known as the *shifta* (bandit) war. Severe droughts in the early 1970s and mid 1980s led to large scale mission efforts distributing famine-relief foods, and saw the growth of small towns around food distribution centers. By 1990, over fifty percent of the formerly nomadic Rendille were living near the towns of Korr, Kargi, Laisamis, and Marsabit, subsisting off bought foods acquired largely through selling livestock that were grazed in distant areas by adolescents and members of the warrior age-grade (see Figure 1). This sedentarization process is discussed in detail in Fratkin (1991, 1992).

The towns that developed after the 1960s initially attracted the poorest Rendille—those impoverished and wiped out by both the 1973 and 1984 droughts, as well as a few enterprising shopkeepers and livestock entrepreneurs (mainly Somali). Those Rendille with livestock continued to remain in the pastoral economy. While stock owners

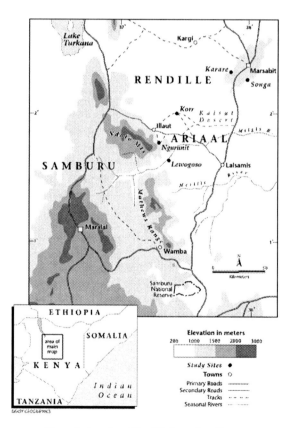

Figure 1. Location of Rendille in northern Kenya.

(i.e., senior males) preferred to keep their animals at some distance from the towns, using traditional methods of mobile livestock camps for non-milking animals, the towns were attractive for women and small children owing to mechanized pumps and education (Fratkin, 1989a).

Camel-keeping Rendille, as well as their related neighbors the Ariaal who keep cattle in highland areas (Fratkin, 1986), developed a variety of economic strategies to cope with both drought and the changing political and economic environment, including the growth of towns around year round water sources and mission centers. Today Rendille live in several types of communities pursuing different economic strategies:

1. Mobile camel pastoralists. The majority of Rendille (total pop. 25,000) live in large semi-nomadic settlements in the Kaisut Desert lowlands between Mt. Marsabit and Lake Turkana. These settlements range from fifteen to fifty houses composed of married adults men and small children subsisting off the milk, meat, and trade of camel herds and small stock flocks. Non-milking stock including cattle and camels are herded in mobile livestock camps in distant pastures, managed by adolescents and members of the warrior age-grade (Fratkin, 1987; Fratkin and Roth, 1990).

2. Sedentary cattle pastoralists. Several large communities of Ariaal are settled permanently in highland locations, particularly on Mt. Marsabit where 2000 people live in one large settlement (Karare) near the district capital of Marsabit. Karare residents are able to graze their cattle permanently in the Marsabit Forest Reserve, and women from Karare have a long association of selling milk in Marsabit town.

3. Sedentary agriculturalists. Since 1973, poor Rendille have settled at several mission-sponsored agriculture projects on Marsabit Mountain, particularly at Songa and Nasikakwe. Songa community (pop. 2000) is a successful farming community 17 kilometers from Marsabit town, where households have gardens irrigated from forest reservoirs and produce market vegetables including kale, onions, peppers, tomatoes as well as corn, beans, and chickens for domestic consumption.

4. Urban wage-workers. In addition to settled pastoralists and horticulturalists, some Rendille have settled in towns including Korr, Ngrunit, and Marsabit, earning their living as shop keepers, traders, livestock brokers, and wage-earners (in construction, herding livestock, or, for women, as household servants cooking, cleaning, and caring for children). These residents may be economically secure (e.g., employed by government or international donor agencies) or economically insecure town residents who live at or below poverty levels. These urban poor obtain their food from low paid casual work (such as latrine digging), selling petty commodities (e.g., tobacco or *mira'a* (the stimulant *catha edulis*, or *khat*), or from illegally working as beer brewers, poachers, or prostitutes (Roth, 1991).

4. METHODS AND MATERIALS

In May and June of 1992, five Rendille communities were surveyed including three previously studied lowland communities of Lewogoso (camel pastoralists, pop. 250), Korr town in the Kaisut Desert (pop. 2000), and Ngrunit town in the Ndoto Mountains (pop. 500), (Fratkin, 1987, 1989, 1991; Fratkin and Roth, 1990; Nathan et al., 1996; Roth, 1990, 1991; Roth and Fratkin, 1990); and two unresearched highland communities on Mt. Marsabit of Karare (sedentary cattle pastoralists, pop. 2000) and Songa (sedentary agriculturalists, pop. 2000).

A time allocation survey was performed in Ngrunit town for comparison to time surveys conducted among Lewogoso camel keepers in 1985; household budgets were obtained from women in Ngrunit, Karare, and Songa; and health and nutritional data obtained from dietary recalls and anthropometric measurements of women and children were obtained in all five communities.

4.1. Time Allocation Surveys

In previous research by Fratkin (1989b), two time surveys were conducted among nomadic camel pastoralists of Lewogoso, one during a wet period in October 1985 and the second during a dry period in January 1986. In the first survey, 286 spot observations were randomly obtained among 39 households over a five-day period between the hours of 0600 and 2000, where every observed individual was recorded for age, gender, and activity type. This survey was aimed at functional rather than structural activities (Hames, 1992), and particularly at specific categories of work roles and amount of rest or leisure time available to each age and sex category. The second survey stratified the community's households for differences in wealth, dependency ratios, and livestock specializations.

In May 1992, a similar time study was performed in Ngrunit town, a small Ariaal Rendille community of 500 people living in the Ndoto Mountains. Many residents of Ngrunit keep some livestock (including camels, cattle, and small stock of goats and sheep) in their compounds, while residents without animals earned livings as shopkeepers, school-teachers, domestic workers, or traders.

This time survey differed from the pastoral survey as observations were made randomly throughout one fourteen hour period (again during the daylight hours of 0600 to 2000 hours), where two observers walked through the town and neighboring pastoral settlements, recording activities of all persons encountered. In all, eight observation walks were made for a total of 463 observations (results listed in Table 1).

4.2. Women's Marketing Budgets

A survey of 39 women living in Songa agricultural community and 61 women from the cattle-keeping community of Karare was obtained in roadside interviews during a three-day period in May 1992. Each day groups of women walk fourteen kilometers to Marsabit town from Karare to sell milk, or seventeen kilometers from Songa to Marsabit to sell agricultural produce, particularly kale, onions, and tomatoes. Interviews conducted with women returning home from the market listed their age, origin, number of children in residence, what type and how much produce was sold, how many times per week they sold produce, and whether they were selling for other women, as well as for themselves. Women also described what commodities they had purchased that day; almost invariably women had transformed their daily earnings from milk and vegetable sales into food purchases, particularly of maize meal, tea, and sugar, and amounts and prices were recorded.

4.3. Household Economic Surveys

Surveys were conducted in each of the five communities of approximately 35 house-holds each, interviewing married women or male elders as household heads. Information was solicited about household income (by annual livestock sales, weekly agricultural sales, wages, remittances, e.g., from wage earning children), and household expenditures (including weekly food or annual livestock purchases). Based on this information, house-holds were stratified into two categories of "food sufficient" or "food insufficient," based on both objective criteria (e.g., total number of livestock or income) and subjective ranking by our hired Rendille assistants, who personally knew members of each community. These rankings were then used in analysis of household marketing budgets, and the health and nutritional data described below.

4.4. Case Histories

Eleven interviews were conducted with women in Ngrunit town who earned their livings by wage labor (usually as domestic servants) or as petty commodity traders (usually selling tobacco, *khat*, or beer). These were conducted with a female interpreter assistant, tape recorded when possible, or otherwise transcribed. Each interview lasted about one hour.

4.5. Anthropometrics

Anthropometric measurements were obtained from 35 women and their under seven-year old children in each of the five communities. Women were selected based on their participation in our previous study in the three lowland communities (Nathan et al., 1996), or by their appearance at medical dispensaries in the highland communities of Songa and Karare in response to our advertisement and payment of 100 Kenyan shillings ($3) per family. Mothers and children were weighed using a CMS hanging scale of children under two years of age, and a standard scale for mothers and children over two. Heights and lengths were measured using a Shorr measuring board; triceps skinfold thicknesses (TSF) were measured with a Holtain caliper, and midarm and head circumferences were obtained

Photo 1. Pastoral women building houses in Lewogoso.

using a Roche disposable measuring tape. Body Mass Indices were calculated by the formula BMI = weight/height squared.

4.6. Dietary Recalls

In addition to anthropometric measurements, dietary recalls were obtained from women about their and their children's diets during the past 24 hours, listing the number of times particular food sources were consumed. It was impossible to measure amounts of food consumed, rather we were interested in the variety of the diet, and the number of times items such as tea, meat, fruit, and vegetables were eaten. An exception was made with regard to liquid milk, where mothers described with reasonable accuracy the number (or portion) of cups of milk consumed the previous day, and where cup referred to a standard container common to the area. Other foods were measured by the number of servings consumed the previous day, as it was not feasible to measure actual amounts of food consumed.

5. RESULTS

Woman have active productive roles in both pastoral and urban settings. In both types of communities, women are responsible for child care, household maintenance, and food preparation. In pastoral communities, women will also milk cattle and small stock (camels are milked by men and boys), handle pack animals (donkeys and camels), provide veterinary care to nursing stock, and herd small stock (goats and sheep) if the household labor force is low. Adolescent girls herd small stock, make long trips to collect water or shop, and occasionally accompany young men for long distance herding, such as grazing cattle from distant highland camps. In addition, pastoral women manufacture many household goods including weaving sisal mats for house roofs, carving wooden milk containers, tanning leather for skirts, and decorative beadwork (Fratkin, 1991).

Figure 2 shows time allocation of all age and gender groups in the nomadic pastoral community of Lewogoso. Married women spent 36.7% of their observed daytime activities in household tasks (cooking, child care), while married men spent only 7.2% of their day in these tasks. Adult men spent 33% of their time in livestock tasks and married women 14% of their day with animals, mainly in milking tasks but also herding small stock and caring for nursing livestock. Adolescent boys and members of the warrior age-grade spent 83% and 71% of their time respectively in livestock related tasks (mainly herding), while adolescent girls spent 44% of their time with livestock. When women are not engaged in household or livestock tasks, they spend time 14% of their time manufacturing household goods, compared to men who were observed spending only 2.3% of their time in these tasks (see Table 1).

Women have significantly less rest or leisure time than the men in the pastoral community, with women spending 35% of their daytime hours resting, compared to men who spend 52.4% of their day in rest or leisure activities, such as playing the *mbau* board game in the men's shade area outside the village.

When the nomadic community was stratified for differences in household wealth, size and dependency ratio, and livestock specialization (whether large stock or small stock), it was found that wealthy women worked less than poorer women, as husbands could hire additional labor, and poorer families concentrated more on small stock (goats and sheep)

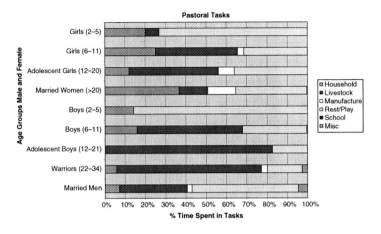

Figure 2. Time allocation of Lewogoso pastoral community, October 1985.

Table 1. Time Allocation Survey in Pastoral and Town Communities. (Lewogoso village, October 1985 and Ngrunit Town, May 1992 (in parenthesis); in Percentage of Observations.)

	Household Pastoral (Town)	Livestock Pastoral (Town)	Manufacture Pastoral (Town)	Rest Pastoral (Town)	(School)
Married Men (>35)	7.2 (10.3)	33.4 (12.1)	2.3 (0.9)	52.4 (72.9)	
Marr. Women (>20)	36.7 (42.9)	14.0 (7.1)	14.0 (13.9)	35.0 (33.6)	(0)
Warriors (22–34)	5.8 (4.6)	71.4 (7.7)	2.9 (0)	17.2 (80.0)	
Adol. boys (12–21)	0 (26.9)	82.6 (30.8)	0 (3.8)	17.3 (23.0)	(15.4)
Adol. Girls (12–20)	12.0 (46.4)	44.0 (14.3)	8.0 (3.6)	36.0 (28.6)	(3.5)
Boys (6–11)	15.9 (21.1)	52.1 (13.5)	0 (0)	31.6 (46.2)	(17.3)
Girls (6–11)	25.0 (32.8)	40.6 (6.2)	3.1 (1.6)	31.2 (53.1)	(6.2)
Boys (2–5)	14.3 (17.5)	0 (0)	0 (0)	85.7 (82.5)	(0)
Girls (2–5)	20.1 (10.6)	6.7 (0)	0 (0)	73.3 (85.1)	(4.3)

which are more labor intensive than large stock raising. Women of families with small children worked harder than women with older children, particularly in livestock tasks, but overall household size did not affect rest time (Fratkin, 1989b).

Time surveys from Ngrunit town showed markedly different work patterns for men and women than experienced in pastoral communities, but with similar inequalities in rest time afforded men and women (see Figure 3).

Married men in towns spent 10.3% of their time in household tasks, 12.1% in livestock tasks, and fully 72.9 of their time in rest and leisure activities. Adolescents spent more time in household and livestock tasks (26.9 and 30.8% for males and 46.4 and 14.3% for girls), while married women spent 42.9% of their time in household tasks, an increase from 36.7% reported for pastoral women in 1985.

Although both men and women had less livestock tasks to perform, women and adolescent girls had no more rest or leisure time than in the pastoral community as their time increased in household tasks including food preparation and child care. Married men and warriors, on the other hand, were able to increase their rest and leisure time to 72.9 and 80% of their day, respectively, an increase from 52.4 and 17.2% in the pastoral

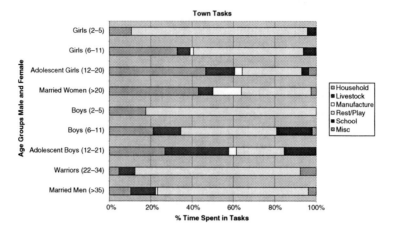

Figure 3. Time allocation of Ngrunit town community, May 1992.

Table 2. Average Weekly Income Earned from Milk and Vegetable Sales,
Marsabit Women, June 1992.

	Number of women	Sales/week	Mean sales K.sh./week	Mean Purchases K.sh./week
Songa (vegetables)	39	3.4	79.9	68
Karare (milk only)	61	3.4	67.3	63

community. However, many of the men observed were only visiting the town to shop or supervise the watering of animals at the wells and were not observed engaged in these tasks, nor in walking the several hours to or from their home settlements.

The important finding here is that Rendille women's labor time does not decrease in the town setting, but remains at 65% of her day time, compared to men whose labor time decreases from 50% in the pastoral community to only 25% in the towns.

Despite their increased labor in household tasks, women have more opportunities to earn money in towns. Women who keep dairy animals can sell the milk surplus, while those women with gardens as in Songa regularly sell vegetables in the Marsabit market, as shown in Table 2.

Women from both agricultural Songa and cattle keeping Karare village average 3.4 selling trips to Marsabit town per week, but Songa women earn twenty percent more from their vegetables (79.9 Kenyan shillings versus 67.3 K.sh.). Women from both communities convert their earnings directly into food purchases, while only a few women reported giving some money to their husbands or saving money for expenses such as children's school fees. This survey did not report additional money that husbands gave to their wives from the sale of livestock, which is the main source of income used to buy food in pastoral communities.

As reported among agricultural communities (e.g., Clark, 1987; Mullings, 1976), the increased income gained by women in the cash market leads directly back to food purchases for the household. Furthermore, the presence of agricultural communities adds variety, particularly in green leafy vegetables, beans, and fruit, to a pastoral diet that is largely milk, maize meal, tea, and sugar. Figure 4 shows the dietary variety between camel

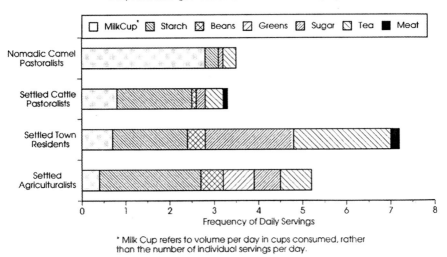

Figure 4. Daily food servings of settled and pastoral Rendille (May 1992).

pastoralists (Lewogoso community), cattle pastoralists (Karare village), town dwellers (Korr town), and agriculturalists at Songa.

Both the camel-keeping Lewogoso and cattle-keeping Karare show high consumption of milk, but Karare residents consumed larger quantities of maize meal, tea, and sugar, owing to their greater participation in the cash market and proximity to Marsabit town. Residents of Korr town show modest consumption of milk and maize meal, but a high level of tea and sugar consumption. There is no single explanation for the high tea and sugar consumption in Korr, although high local temperatures, lack of calories from milk or maize, and relatively more leisure time (as few animals are kept there) are probable causes. Songa agricultural community had the greatest variety of food types in their diet, due to their consumption of garden vegetables and their purchases of maize meal, afforded by high weekly sales of produce in Marsabit town. Songa residents even kept some a few milk animals, and their diet appears to be the most varied and beneficial.

It is not clear if dietary differences (both amounts and types of food consumed) lead to nutritional differences in children. Anthropometric measurements obtained in the one brief wet season sample of the five communities reveal no significant differences in the nutritional statuses of children in these five communities, where the F-ratio for Body Mass Index (weight/height squared) is 1.3 ($p > 0.25$) and for Triceps Skin Fold is 1.33 ($p > 0.25$). (see Tables 3, 4, and 5).

Additional studies of these communities over a long period may show greater variations due to dietary differences, access to store bought foods, and seasonality and food shortages. It is clear that these different Rendille communities vary widely in their access to food sources, with pastoralists having the least variety (few grains and vegetables but high milk and meat, which are calorie poor but protein rich), and Songa agriculturalists having the highest variety, with iron and vitamin rich green vegetables complementing beans and maize. It is also apparent from the data that women's market integration contributes to greater food security in terms of foods bought, particularly high calorie maize meal.

Table 3. Anthropometric Indices (body Mass Index and Triceps Skin Fold) of Children (0–6) in Four Rendille Communities, May 1992.

	N	BMI	S.D.	TSF	S.D.
Camel pastoralists	61	15.2	2.4	7.6	1.7
Cattle pastoralists	62	14.9	2.4	7.6	2.4
Desert town (Korr)	56	14.5	1.9	7.6	1.6
Agricultural (Songa)	61	14.7	1.8	8.2	2.1

BMI = Body Mass Index, Weight/height^2; TSF = Triceps skin fold, in mm.

Table 4. F-test for Analysis of Variance for Anthropometric Means by Location.

	Df	Sum of Squares	Mean Square	F-ratio	Probability
BMI	3	18.35	6.12	1.30	0.275
TSF	3	14.27	4.75	1.33	0.265

BMI = Body Mass Index, Weight/height^2; TSF = Triceps skin fold, in mm.

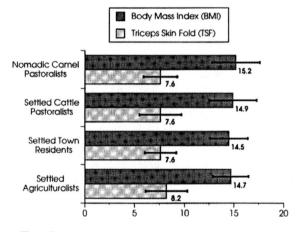

Anthropometric Comparison of Rendille Children (0–7)
Mean Measurement and One Standard Deviation Bars

Figure 5. Anthropometric comparison of Rendille children (0–7).

Songa, Karare, and Ngrunit represent Rendille communities where women obtain cash earnings from the sale of livestock products or farm produce. However, many women living in or near towns have no productive resources of their own, but must sell their labor, petty commodities, or beg to make a living. These women occupy the lower rungs of the economic ladder, and although they are also found in the pastoral setting (particularly as older widows), their situations are quite different in the towns where they often lack large kin networks for support.

The household economic surveys and selected life histories conducted in Ngrunit town show that approximately one half of the women are "food sufficient", as measured by weekly expenditures in household goods. These women are usually married to employed men, or employed themselves as professionals or shopkeepers. The other half are poor or

Photo 2. Songa women walking to Marsabit market with gourds of milk and vegetables.

Table 5. Income and Food Purchases of Women Differentiated for Occupation, Ngrunit Town, May 1992.

	Number	Mean age	No. milk animals	Av. no. children	Daily purchases (K.Sh.)	Daily earnings (K.Sh.)
Pastoral poor	10	35.0	0.25	5.1	8.4	0.5
Pastoral sufficient	11	32.8	1.78	2.4	20.8	3.0
Town poor	16	42.9	0.19	2.9	9.8	0.6
Town sufficient	11	34.4	0.49	3.7	57.9	11.5

"food insufficient" and earn what they can from selling labor (as domestics, house cleaners, baby sitters), selling petty commodities (firewood, tobacco, beer), or for some, by prostitution, illegal beer brewing, or begging.

Several of the poor women in towns were direct immigrants from pastoral society, either widows or women whose husbands were poor; most were older women with grown children and fewer dependents. Several women described their situations in interviews. Gilorit is a widow in her mid-thirties who supports six children. She is dressed poorly in a discarded western dress, barefoot and with no traditional jewelry.

"I sell tobacco now, but I used to have a small garden where I grew maize, beans, and even tobacco to sell. When my husband died I worked for the African Inland Church mission, who paid me 100 shillings a month (about five dollars). It was too little and I asked for a raise, but they refused so I left. I never worked until my husband died. When my animals were finished (by selling them off), I began selling tobacco. You try to offer people something in exchange for food. Some days, you might find food, other days not."

Other women may come to town for increased opportunities. Mairo, aged 50, is the wife of a blind man and works for the local school teacher as cook and baby sitter. She owns some animals (including five milk cows, which are grazed by her sons). She describes her life:

"Before in Longieli (pastoral community), life was not bad, but it is much better here in Ngrunit. I don't have to go far for water or firewood. Also this place is safer for my husband. I can earn some money and help feed my husband. Even when cattle come here, I only have to help water them, I don't have to take them to camp, to herd them. I can keep some livestock here, so if I sell a goat, I am right in town and I don't have to wait for my sons to return some time later. I found work in Ngrunit just by looking. First I worked for a Somali man, I washed his clothes, cooked, cleaned house, and fetched water. I earned 150/= a month, plus food (about seven dollars a month). Then bandits came and tried to steal from the Somali family. They took money, sugar, clothes from the shop—there was nobody to stop them, they were fierce and would kill you, so we all ran away. The Somali family got scared and moved back to Korr. Then I just struggled with my animals and we stayed here. In a short while I heard of another Rendille family here, the schoolteacher, and I saw them making a kitchen. So I came to help, and finished them making the kitchen. Then the teacher's wife said, do you need a job, why don't you stay with me and help me with my children? So I stay now and earn 200/= month, plus food for myself and husband, plus clothing. Some men oppose their wives to work for money; maybe they think we will over look them and become proud, maybe go away with other men and leave them. With some people this will happen. If a man has that problem he does not think he can control his wife. Fortunately I do not have that problem."

6. DISCUSSION

Sedentarization is a process that is occurring rapidly among pastoral populations throughout Africa. This study of Rendille of northern Kenya shows a variety of economic strategies pursued by peoples of different communities, including maintaining a nomadic camel-keeping subsistence economy, living in settled cattle-keeping communities where dairy and livestock may be marketed in urban areas, living in agricultural or agro-pastoral communities where garden produce may be sold, and living in towns as wage earners and entrepreneurs or at lower socio-economic levels engaged as servants, casual labor, prostitutes, or illegal beer brewers.

Town life is attractive to former pastoralists suffering drought, political instability, and economic insecurity. It offers physical security, better health care, and greater educational opportunities for both adults and children. For women in particular, town residence leads to certain improvements in women's lives, particularly in the reduction of drudgerous or dangerous tasks (e.g., water collection or livestock herding). Nevertheless, women's labor expenditures remain high, having no more rest or leisure time in either nomadic or town communities, spending approximately 65% of their day in household, livestock, and manufacturing tasks. This is in sharp contrast to married men, who may spend 50% of their day in nomadic communities and over 70% in towns resting.

Women from rich and poor households show different degrees of integration in town markets. Those living in communities who own their productive resources, such as livestock-raising Karare or agricultural Songa near Marsabit town, are able to sell surplus milk and vegetables to regular customers in town. Wealthier pastoralists may sell surplus milk above their nutritional needs, using remittances to buy agricultural foods. Poor pastoralists however must sell a larger proportion of their milk, using less milk for their household diet and becoming more dependent on store bought grains (Herren, 1990). Ensminger (1987: 39–40) reports for Galole Orma that while both nomadic and settled women sell handicrafts, only wealthier settled women make and sell bread in the local shops because of high start up costs. For both wealthy and poor households, proximity to town markets leads to improvements in child nutrition as milk can be sold to buy maize meal on a regular basis. Although milk is preferred because it is high in protein and vitamins, maize yields five times the caloric value of milk, and is a necessary staple during long dry periods which frequently occur in pastoral regions (Grandin, 1988: 8). Some researchers, however, point out that policies aimed at increasing the economic integration of dairy pastoralists may be harmful as they encourage women to glean too much milk from their stock to purchase costly grain (Hilarie, 1986).

In pastoral communities such as Maasai, poor women may obtain food by selling their labor to richer women by running errands or collecting firewood (Talle, 1990). The Rendille data suggests that town life does not necessarily provide poor women with greater economic opportunities, as possibilities to earn income in town are few. In some cases women may find work as household servants, cooks, or baby sitters, but the supply of women looking for this type of work exceeds the demand, particularly in situations where poor relatives may be directly recruited for such jobs. Some Rendille women turn to selling tobacco or handicrafts, but income from petty commodities brings in less than 200 shillings a month ($6–8). Without livestock, milk, or vegetables to sell, poor women must scrape by in towns, sometimes having no alternative to find food other than church-sponsored programs distributing famine-relief.

Several factors explain pastoral women's differential integration into the market economy: household wealth, sedentarization versus nomadism, control over resources, and proximity to markets where goods or services can be sold. Where women are able to keep milk animals or raise gardens, access to markets increases their ability to earn income, as among Rendille on Marsabit Mountain at Karare and Songa villages. A household's wealth influences a woman's decisions about what she can sell either because wealthy households provide her with more opportunities for selling surplus milk or, conversely, because wealthy households are nomadic and located farther away from market centers. Women in poorer households have less options because they have fewer pastoral resources to sell, particularly if they are settled.

Despite the trend toward sedentarization, many pastoral households are able to keep their herds through sedentary options including agro-pastoralism, individuated ranching, or the separation of livestock into distant livestock camps managed by adolescents and members of the warrior age grade. Furthermore, many town or peri-urban households have access to both town opportunities and pastoralist relatives, and more than a few women are able to keep livestock near town and market their products, particularly milk. But for women and their households who do not have sufficient livestock or access to alternative food resources such as gardening, there are few opportunities to make a living. This especially applies to poor, elderly and widowed women, many of whom are the first to migrate to the towns.

Given its higher integration in the cash market and greater nutritional variety in their diet, the agricultural community at Songa seems a model to emulate. But Songa is an isolated situation particular to the humid conditions of Marsabit Mountain, and is not representative of the very arid resources of Marsabit District. Developers seeking to promote the Songa model need to consider that agricultural schemes accommodate only 10% of the Rendille population, and that agriculture is not a readily available option to most pastoralists. Furthermore, agricultural Rendille on Marsabit Mountain do not have clear title to their farms and may lose their lands if their land tenure remains insecure.

Town residence leads to important changes for pastoral women in terms of work roles and access to food, health care, and social support networks. As this study of varying Rendille communities suggests, the process of pastoral sedentarization is complex and varied and results in different consequences for settled pastoral women.

ACKNOWLEDGEMENTS. Research for this paper was collected in Marsabit District, Kenya in 1985 and 1992. We are grateful to the Office of the President, Republic of Kenya for their permission to conduct research in Marsabit District, and to Larian Aliyaro, Anna Marie Aliyaro, and Patrick Ngoley for their assistance in the field work. Our thanks to John Curry and Rebecca Huss-Ashmore for their invitation to the panel on "Gender and Livestock in Africa" at the American Anthropological Association 1992 Annual Meetings where this paper was presented; to David Reed for his assistance with the statistics and graphics; and to Eric Abella Roth for his suggestions and free advice. Funding for fieldwork in 1985 was provided by the National Geographic Society and the Social Science Research Council; research in 1992 was sponsored by Penn State University and a Mellon Foundation grant to the Population Research Institute at Penn State University. This paper originally published in *Human Ecology* 23 (4): 433–454, 1995.

REFERENCES

Campbell, D., 1984, Responses to Drought Among Farmers and Herders in Southern Kajiado District, Kenya. *Human Ecology* 12(1): 35–64.

Clark, G., 1987, Separation Between Trading and Home for Asante Women in Kumasi Central Market, Ghana. In *The Household Economy: Reconsidering the domestic mode of production*, edited by R. Wilk, pp. 91–118. Boulder, CO: Westview Press.

Clay, J.W., 1988, The Case of Hararghe: The Testimony of Refugees in Somalis. In *The Spoils of Famine*, edited by J.W. Clay, S. Steingraber, and P. Niggl, pp. 157–199. Cambridge MA: Cultural Survival, Inc.

Dahl, G., 1987, Women in Pastoral Production: Some Theoretical Notes on Roles and Resources. *Ethnos*. 52(1–2): 246–279.

Ensminger, J.E., 1987, Economic and Political Differentiation among Galole Orma Women. *Ethnos* 52(1–2): 28–49.

Fratkin, E., 1986, Stability and Resilience in East African Pastoralism: The Ariaal and Rendille of Northern Kenya. *Human Ecology* 14(3): 269–286.

Fratkin, E., 1987, Age-sets, Households and the Organization of Pastoral Production. *Research in Economic Anthropology* 8: 295–314.

Fratkin, E., 1989a, Two Lives for the Ariaal. *Natural History* 98(5): 39–49.

Fratkin, E., 1989b, Household Variation and Gender Inequality in Ariaal Rendille Pastoral Production: Results of a Stratified Time Allocation Survey. *American Anthropologist* 91(2): 45–55.

Fratkin, E., 1991, *Surviving Drought and Development: Ariaal Pastoralists of Northern Kenya*, Boulder: Westview Press.

Fratkin, E., 1992, Drought and Development in Marsabit District. *Disasters* 16(2): 119–130.

Fratkin, E. and E.A. Roth, 1990, Drought and Economic Differentiation Among Ariaal Pastoralists of Kenya. *Human Ecology* 18(4): 385–402.

Galaty, J.G., 1992, 'This Land is Yours': Social and Economic Factors in the Privatization, Subdivision and Sale of Maasai Ranches. *Nomadic Peoples* 30: 26–40.

Galaty, J.G. and P. Bonte, 1991, The Current Realities of African Pastoralists. In *Herders, Warriors, and Traders: Pastoralism in Africa*, edited by J.G. Galaty and P. Bonte (eds.), pp. 267–292. Boulder, CO: Westview Press.

Grandin, B.E., 1988, Wealth and Pastoral Dairy Production: A Case Study from Maasailand. *Human Ecology* 16(1): 1–21.

Guyer, J.I., 1980, *Household Budgets and Women's Incomes*. Working Paper No. 28, African Studies Center, Boston University.

Hames, R., 1992, Time Allocation. In E.A. Smith and B. Winterhalder (eds.), *Evolutionary Ecology and Human Behavior*, pp. 203–235. New York: Aldine de Gruyter.

Herren, U.J., 1990, *The Commercial Sale of Camel Milk from Pastoral Herds in the Mogadishu Hinterland, Somalia*. ODI Pastoral Development Network Paper 30a. London: Overseas Development Institute.

Hilarie, K., 1986, Uncounted Labor: Women as Food Producers in an East African Pastoral Community. In *Proceedings—African Agricultural Development Conference: Technology, Ecology, and Society*, edited by Y.T. Moses, pp. 62–66. Pomona, CA: California State Polytechnic University.

Hill, A.G., 1985, *Population, Health, and Nutrition in the Sahel*. London: Routledge Chapman.

Hjort, A., 1990, Town-Based Pastoralism in Eastern Africa. In *Small Town Africa: Studies in Rural–Urban Interaction*, edited by J. Baker, pp. 143–160. Uppsala, Sweden: Scandinavian Institute of African Studies.

Hogg, R., 1985, The Politics of Drought: The Pauperization of Isiolo Boran. *Disasters* 9(1): 39–43.

IDA, 1991, *Gender Relations of Pastoral/Agropastoral Production: A Bibliography with Annotations*. Binghamton, NY: Institute for Development Anthropology.

Little, P.D., 1994, Maidens and Milk Markets: The Sociology of Dairy Marketing in Southern Somalia. In *African Pastoralist Systems*, edited by E. Fratkin, K. Galvin, and E.A. Roth, pp. 165–184. Boulder, CO: Lynne Rienner Publishers.

McCabe, J.T., 1987, Drought and Recovery: Livestock Dynamics Among the Ngisonyoka Turkana of Kenya. *Human Ecology* 15(4): 371–385.

Michael, B.J., 1987, Milk Production and Sales by the Hawazma (Baggara) of Sudan: Implications for Gender Roles. *Research in Economic Anthropology* 9: 105–141.

Mullings, L., 1976, Women and Economic Change in Africa. In *Women in Africa: Studies in Social and Economic Change*, edited by N.J. Hafkin and E.G. Bay, pp. 239–264. Stanford: Stanford University Press.

Nathan, M.A., E. Fratkin, and E.A. Roth, 1996, Sedentism and Child Health Among Rendille Pastoralists of Northern Kenya. *Social Science and Medicine* 43(4): 503–515.

Oxby, C., 1981, Group Ranches in Africa. *Overseas Development Institute Review* 2: 44–56.

Oxby, C., 1987, Women Unveiled: Class and Gender among Kel Ferwan Twareg. *Ethnos* 52(1–2): 119–136.

Roth, E.A., 1990, Modeling Rendille Household Herd Composition. *Human Ecology* 18(4): 441–455.

Roth, E.A., 1991, Education, Tradition, and Household Labor Among Rendille Pastoralists of N. Kenya. *Human Organization* 50: 136–141.

Roth, E.A. and E. Fratkin, 1990, Rendille Herd Composition and Settlement Patterns. *Nomadic Peoples* 28: 83–92.

Salil, M.A., 1985, *Pastoralists in Town: Some Recent Trends in Pastoralism in the North West of Omdurman District*. ODI Pastoral Development Network Paper 20b. London: Overseas Development Institute.

Shepherd, A., 1988, Case Studies of Famine: Sudan. In *Preventing Famine: Policies and Prospects for Africa*, edited by D. Curtis, M. Hubbard, and A. Shepherd, pp. 28–72. London: Routledge.

Sperling, L., 1987, *The Labor Organization of Samburu Pastoralism*. Unpublished PhD dissertation. Department of Anthropology, McGill University.

Talle, A., 1988, Stockholm: University of Stockholm, Department of Social Anthropology. *Women at a Loss: Changes in Maasai Pastoralism and Their Effects on Gender Relations*.

Talle, A., 1990, Ways of Milk and Meat among the Maasai: Gender Identity and Food Resources in a Pastoral Economy. In *From Water to World-Making: African Models and Arid Lands*, edited by G. Palsson, pp. 73–92. Uppsala, Sweden: Scandinavian Institute of African Studies.

Waters-Bayer, A., 1985, Modernizing Milk Production in Nigeria: Who Benefits? *Ceres*. 19(5): 34–39.

Chapter 9

The Effects of Pastoral Sedentarization on Children's Growth and Nutrition among Ariaal and Rendille in Northern Kenya

ERIC ABELLA ROTH, MARTHA A. NATHAN M. D., AND ELLIOT FRATKIN

1. INTRODUCTION

Despite widespread policy efforts to settle nomadic pastoralists, the costs and benefits of sedentarization are not well understood. Several studies reveal that marketing opportunities may benefit women who are able to sell milk and agricultural products in town (Fratkin and Smith, 1995; Little, 1994; Smith, 1999; Waters-Bayer, 1988; Ensminger, 1992; Sato, 1997; Zaal and Dietz, 1999). Others, however, report negative social and health consequences of pastoral sedentarization: impoverishment and destitution (Hogg, 1986; Little, 1985) particularly affecting women (Talle, 1988); poorer nutrition; inadequate housing; lack of clean drinking water; and higher rates of certain infectious diseases including malaria, bilharzia, syphilis, and AIDS, despite better access of settled populations to formal education and health care (Chabasse et al., 1985; Galvin et al., 1994; Hill, 1985; Klepp et al., 1994; Nathan et al., 1996).

ERIC ABELLA ROTH • Department of Anthropology, University of Victoria, Victoria, British Columbia, Canada V8W 3P5   MARTHA A. NATHAN M. D. • Brightwood Health Center, Springfield Massachusetts 01107 ELLIOT FRATKIN • Department of Anthropology, Smith College, Northampton, Massachusetts, 01063.

This chapter reports the results of a three-year study of pastoral and settled Rendille communities of northern Kenya. Longitudinal research was carried out in northern Kenya among both pastoral and sedentary Rendille communities between 1994–1997, in which we delineated the biological and social consequences of sedentism for these pastoralists. Specifically, we addressed two related research questions:

(1) What are the biosocial consequences of different types of sedentism for these formerly mobile pastoralists? and (2) How can they be measured?

We compared levels of child malnutrition among five different Rendille communities. Analysis of bimonthly dietary recalls, anthropometric measurements, morbidity data, and economic differentiation and specialization among 200 mothers and their 488 children under age 11 revealed large differences in the growth patterns of nomadic vs. settled children. Age-specific height and weight measurements for the nomadic pastoral community are uniformly higher than same-aged measurements of children from the sedentary villages. Differences in child growth are attributed mainly to better nutrition, particularly access to camel's milk, rather than differences in morbidity wrought by infectious diseases. We conclude that international development assistance and government policy should consider the negative effects on child health of sedentarization of pastoral populations and not neglect improvements in livestock production and support of pastoral movements in their work in Africa's arid lands.

2. PREVIOUS RESEARCH ON PASTORAL SEDENTARIZATION AND ITS EFFECT ON HEALTH AND NUTRITION

In the 1980s, Allen Hill (1985) organized a conference on studies of health and nutrition in both sedentary and pastoral groups in Mali, although this was not an integrated research program. At that conference, Chabasse et al. (1985) reported that nomadic groups have higher rates of tuberculosis, brucellosis, syphilis, trachoma, and child mortality (children five and under), which they attributed to differences in access to health care services. Settled agricultural populations, however, had higher rates of bilharzia, intestinal helminths and other parasites, malaria and anemia, which the authors attributed to their stable proximity to water. Hill's studies did not look at the process of sedentarization within a single ethnic group or confined area where health and nutritional outcomes for nomadic versus settled communities could be compared without the complicating variables of ethnicity and distance.

The South Turkana Ecosystem Project of the late 1980s carried out extensive research on ecology, fertility, health, and nutrition of nomadic Turkana of Kenya (Little and Leslie, 1999). It also considered, to a lesser degree, the health and nutrition among settled farming Turkana populations. Researchers found that settled Turkana experienced reduced fertility, increased morbidity (particularly from malaria) and increased child mortality. Settled children under five showed more growth stunting than nomadic children, although settled children over five were heavier, which was attributed to the greater role of carbohydrates in their diets, particularly for children receiving supplemental feeding in schools. Nomadic Turkana women, however, were taller, heavier, and had lower blood pressure than settled women (Brainard, 1990; Campbell et al., 1999; Galvin, 1992).

Diets change when pastoralists settle. Pastoral diets generally include more protein (mainly from milk) but fewer calories than do sedentary diets, and both protein and energy

Photo 1. Milking camel in Ngrunit.

content vary markedly with seasonal rainfall (Galvin, 1985, 1992; Galvin and Little, 1999; Little et al., 1993; Nathan et al., 1996; Nestel, 1986; Sellen, 1996; Shell-Duncan, 1995). The lean seasons for northern Kenya occur at the end of the two dry seasons (November–March and May–August) when livestock water and pasture become scarce in turn limiting availability of drinking water and milk for human consumption. During dry periods, pastoralists are forced to sell their small stock to purchase foods, mostly grains (maize meal called *posho*) and other carbohydrates, including sugar to mix with tea.

The milk-based, high-protein diet of pastoralists, nonetheless, appears to give them a nutritionally adaptive advantage, despite its seasonal fluctuations and overall limited energy content (Galvin and Little, 1999).

On the other hand, in settled communities of former pastoralists, certain families such as those engaged in the commercial livestock economy and those who take up cash-crop agriculture, may have a wider economic resource base. This allows them not only to alleviate seasonal fluctuation of food availability but also to widen the variety of food in their diet. Typically, there are contrasting seasonal patterns of nutritional stress between agriculturists and pastoralists. Critical periods for agriculturists coincide with the food shortage and high labor demand associated with farming and harvesting during the pre-harvesting time (Simondon et al., 1993: 166). Families with sufficient agricultural and/or pastoral resources will be able to even out the seasonal stresses associated with each subsistence mode. By contrast, poorer families who rely on smaller pastoral or agricultural holdings for their subsistence and cash income are more likely to experience seasonal stresses distinct from those of wealthier families.

3. IMPACT ON THOSE AT RISK: REPRODUCTIVE AGE WOMEN AND CHILDREN

The advantages of the high-protein diet may be particularly significant for infants, pregnant women, and lactating mothers. "Children, along with pregnant and lactating women are commonly viewed as vulnerable groups among human communities at risk from poor environments" (Panter-Brick, 1998: 66). Since protein is an indispensable nutrient for reproductively active pastoral women as well as for infants and growing children (Galvin and Little, 1999), the potential protein loss associated with agricultural sedentism may also have a negative impact on maternal nutritional health.

On the other hand, market integration of rural producers in Africa may have both positive and negative consequences on child health and nutrition. Sales of agricultural commodities may diminish child nutrition when they lead to substitution of cheaper, lower-calorie or -protein foods for higher quality ones (Lappé and Collins, 1977). However, other studies report improved child nutrition associated with commercial agriculture when combined with subsistence production, as shown in various production strategies of Taita farmers of Kenya (Fleuret and Fleuret, 1991). Ensminger's (1991) study of the economic transformation of Orma of Kenya found that families who lived in market centers and engaged in agricultural commercialization exhibited improved nutritional markers (weight for height) for adults and male children, but not for female children.

In this perspective the sensitivity of human growth processes to the environment is seen as "one mechanism by which our species adapts" (Johnston and Little, 2000: 40). Poor adaptive responses to environmental change, evidenced by the degree of "growth faltering" of young persons' growth profiles, was originally termed "auxological epidemiology" by Tanner (1981), who traced its origins to studies of the growth and development of British factory children in the 19th century. Today growth faltering is associated with a higher risk of morbidity and mortality on a worldwide basis (cf. Martorell, 1989; Pelletier, 1994). The pattern spirals, with growth patterns deteriorating further as a result of infectious processes. Thus child growth and related maternal health are both sensitive indicators of community health and predictors of overall success of adaptive strategies.

Rendille and Ariaal communities have adopted different subsistence strategies in their transition to sedentism. Some communities now raise vegetables for sale in the main market in Marsabit Town; others sell milk and/or market livestock, while still others combine all these activities. Because of such diverse local economies we chose maternal-child health, measured via morbidity, nutrition and their effects on child growth and development, as the vital currency for appraising the biosocial consequences of sedentism.

In our previous work on these subjects we proposed that sedentarization would affect Ariaal and Rendille child growth in two ways. The first would be via dietary change, with sedentary groups consuming far less milk because they are separated from household livestock herds primarily graze outside sedentary centers in nomadic animal camps called *fora*. The second would be through an increase in density-dependent infectious diseases, as increased population density in sedentary communities would act as reservoirs for infectious pathogens. In outlining these potential pathways of biosocial change we were invoking the well-documented model known as the *nutrition–morbidity synergism* originally described by Scrimshaw and Taylor (1968).

Results of previous cross-sectional analyses partially supported our dual hypotheses (Nathan et al., 1996; Fratkin et al., 1999). Milk intake by sedentary children was markedly lower, but child morbidity patterns were not statistically different. In these previous analyses, we also found that household economic levels had no effect upon levels of childhood malnutrition. These results surprised us, as we predicted that patterns of economic differentiation delineated in earlier livestock-based analyses (cf. Fratkin and Roth, 1990; Roth, 1990, 1996), would translate into varying levels of childhood malnutrition.

Now, armed with longitudinal data we can refocus upon this same nutrition–morbidity framework and examine the ramifications of economic differentiation over time. In addition we can explore possible effects of differential parental investment by gender found in other aspects of Rendille and Ariaal patrilineal, patrilocal cultures (Fratkin, 2004; Roth, 1991, 2000).

The multi-year longitudinal data set that is the basis for analysis is unique for sub-Saharan African pastoralists. Today the standard for longitudinal studies of Sub-Saharan African pastoralist health is the Southern Turkana Ecological Project (Little and Leslie, 1999), which produced two seminal longitudinal health and nutritional studies (Galvin, 1985;

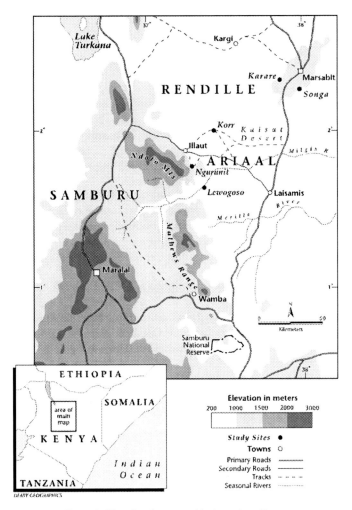

Figure 1. Map of study communities in northern Kenya.

Photo 2. Dr. Marty Nathan and AnnaMarie Aliyaro Weighing Children in Ngrunit.

Shell-Duncan, 1995). Both are based on one year of data collection, and were conducted in drought conditions. In contrast, as shown in Figure 2, mean monthly rainfall data for our multiple year study show normal years for 1995 and 1997, on either side of the 1996 drought year, wherein the spring long rains failed. Thus we can monitor seasonal changes in diet, health, and growth across both wet and dry years.

These data allow us to examine growth patterns for Rendille and Ariaal children in two related manners. We begin with an examination of longitudinal growth patterns based on age-specific anthropometric indicators. This is followed by a study of possible underlying determinants of childhood growth, utilizing dietary and morbidity recall data.

4. MATERIALS AND METHODS

To monitor child growth and health we selected five Rendille/Ariaal communities, four sedentary (Korr, Karare, Ngrunit, and Songa) and one nomadic (Lewogoso) control in Marsabit District, northern Kenya. Their locations are shown in Figure 2. These communities are summarized as follows:

1. *Lewogoso* is a nomadic camel-, cattle-, and small-stock-keeping settlement of approximately 250 people practicing mixed-species husbandry. This community forms a control community for the comparison of the sedentary villages.

2. *Ngrunit* is a sedentary agro-pastoral community of approximately 1200 people located in a forested valley in the Ndoto Mountains. This community has a church, school, and small dispensary but is isolated and not well integrated into marketing activities. Its inhabitants raise vegetables from their gardens and market livestock.

3. *Korr* is a new town in the arid lowlands of the Kaisut Desert below Marsabit Mountain created initially by the Catholic diocese to feed destitute Rendille during the famine of the

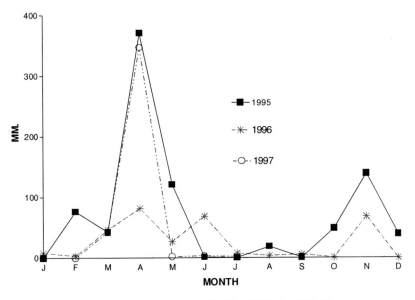

Figure 2. Monthly rainfall in Marsabit District, 1995–97.

1970s. Today, Korr has a sedentary population of about 6000, with semi-nomadic Rendille settlements around it. Korr has poor marketing facilities, although the town provides a local market, mainly represented by small stock sales, for surrounding pastoralists.

4. *Karare* is a settled highland community on Marsabit Mountain about 17 km, from Marsabit Town. Its 2000 residents both keep cattle herds and raise dryland maize. Karare has access to good marketing facilities as well as a large urban population in Marsabit Town and is located on the major truck road from Nairobi to Addis Ababa. Karare women sell milk on a regular basis to Marsabit townspeople.

5. *Songa* is a sedentary highland agricultural community of 2000 people founded by American missionaries from the African Inland Church in 1973 in a forest on Marsabit Mountain. Practicing drip irrigation, Songa's population grows vegetables for sale in Marsabit town.

From September 1994 to June 1997, mothers and their children from these five communities were surveyed bi-monthly for nutrition (foods eaten within last twenty-four hour period) and morbidity (days children ill in last month due to fever, respiratory infections, and diarrheal diseases) via recall methodologies. From each community an availability sample of forty women with children under six years of age was included for repeated measurement. Growth was determined from serial bi-monthly anthropometric measurements: height, weight, mid-arm circumference, and triceps skin-fold. Monthly household expenditures, wages and sales of livestock, milk and/or vegetables, and mother's reproductive status, (pregnant and/or breastfeeding) were the queried socioeconomic variables.

Longitudinal growth data from the five communities produced a total of 5565 measurements from 488 children from birth to age 11. Table 1 groups anthropometric measurements by community and child's year of age, with the final age category lumping children aged 6+ years. Analyzing each age class by the SAS® (1997) PROC NPAR1WAY program for one-way Non-Parametric Analysis of Variance reveals that for all but the last

Photo 3. Measuring triceps skin fold.

Photo 4. Measuring baby's length.

Table 1. Age-Specific Measurements of Five Sample Communities.

Age in months	Lewogoso	Ngrunt	Songa	Karare	Korr	Total
0–11	114	202	110	177	162	765
12–23	184	211	161	181	170	907
24–35	215	203	198	173	212	1001
36–47	187	121	191	161	155	815
48–59	190	109	191	193	191	874
60–71	171	82	146	187	164	750
72+	54	9	108	184	68	423
TOTAL	1115	937	1105	1256	1122	5535

period, constituting a catch-all of ages greater than six years, the samples are statistically non-significant.

The computer program EPI-INFO (Center for Disease Control Centre, 1997) transformed sex-specific measures of weight-by-age, height-by-age, and weight-for-height into standard deviation (Z-scores) according to the formula from Stinson (2000: 443).

$$Z = \frac{\text{Individual subject score-median reference value}}{\text{Standard deviation in reference population}}$$

These are based on growth reference curves developed by the National Center for Health Statistics (Hamill et al., 1979) and recommended by the World Health Organization (1986) for use in Third World countries. Measurements less than two negative standard deviations from the median of the reference population (< -2 S.D.) were classified as mild-to-moderate malnutrition (World Health Organization, 1986), constituting evidence of growth faltering.

5. ANALYSIS AND RESULTS

5.1. Longitudinal Child Growth as Population Health and Adaptation

Examining levels of child malnutrition rates using these measurements, as shown in Figures 3 and 4, reveals large differences in the growth patterns of children for communities. Age-specific height and weight measurements for the nomadic Lewogoso community are uniformly higher than same-aged measurements from the sedentary villages. For the latter, growth faltering, characteristic of many African populations at about the six month range (cf. Eveleth and Tanner, 1990; Little et al., 2000) is notable for both height and weight measures, while this is true only for weight in Lewogoso, and not nearly to the same extent. In contrast height remains stable in Lewogoso, and even increases on average throughout the final four periods.

The underlying differences between the sedentary and pastoral samples are more clearly shown in Figures 5 and 6. These convert the continuous Z-score values for both weight-for-age and height-for-age into discrete measures of "wasting" (below -2 Z-scores for weight) and "stunting" (below -2 Z-scores for height) for pastoral Lewogoso and a

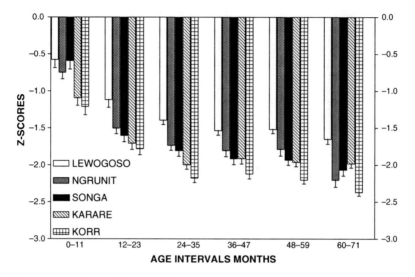

Figure 3. Weight-by-age Z-scores, all five communities, means and standard errors of the mean.

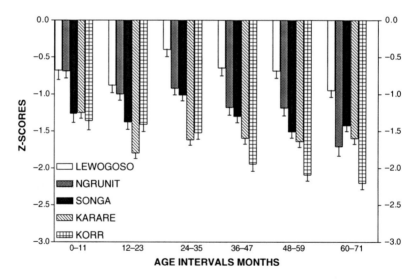

Figure 4. Height-by-age Z-scores, all five communities, means and standard errors of the mean.

pooled sample consisting of the four sedentary communities, omitting in both cases the final catch-all category. Presented in this fashion these data clearly show for lower rates of both wasting and stunting at all ages for children in pastoral Lewogoso campared to the pooled sedentary sample.

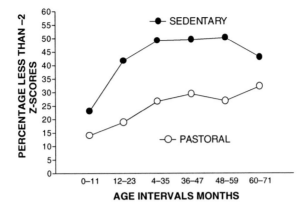

Figure 5. Measures of malnutrition for weight-by-age, pastoral versus sedentary samples, wasting defined as below −2 Z-scores.

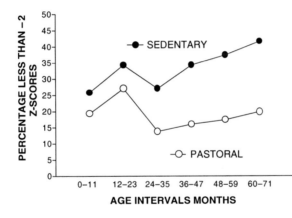

Figure 6. Measures of malnutrition for height-by-age, pastoral versus sedentary samples, stunting defined as below −2 Z-scores.

5.2. Accounting for the Differences: Diet, Morbidity, and Sample Composition

What factors account for the large, consistent growth differences between children from nomadic and sedentary communities? Our previous analyses revealed that inter-community differences arose primarily from dietary change, with milk remaining a central staple of the nomadic diet. For sedentary communities, children's milk intake decreased dramatically, replaced largely by grains. These results are understandable, given that the nomadic community of Lewogoso moves with their livestock, while livestock owned by sedentary village residents are often herded far away in order to take advantage of seasonally fluctuating, locally distributed water and vegetation. We previously found little or no difference in morbidity between communities. These findings suggested that it is the nutritional aspect of the nutrition-morbidity synergism that most affects Rendille/Ariaal children who settle.

Photo 5. Research assistant and research subject, AnnaMarie Aliayaro and son, Tony Living in Korr.

As shown in Figures 7 and 8, our previous cross-sectional results are replicated in the present longitudinal data. Pooling the sedentary groups again, Figure 7 shows daily average milk intake measured in cups for nomadic and sedentary children across the study period. It reveals nomadic Lewogoso children consuming up to three times the average reported cups of milk as the children from the four sedentary communities. These large community differences remain throughout the study period, even during the drought periods of 1995/6 when milk intake fell for Lewogoso.

Figure 8 again contrasts sedentary with nomadic samples, in this case revealing pooled data regarding average days ill with fever, diarrhea and colds over each two-month sampling interval. Here the contrast between sedentary and nomadic communities is not as dramatic as for milk consumption, and is harder to interpret. Only once in the seventeen measurements were the average sick days higher for Lewogoso than for the pooled sedentary sample. These results indicate that morbidity differences also make an important contribution to childhood growth patterns.

We further tested this assumption via Generalized Estimating Equations (Liang and Zeger, 1986) using the SAS® GENMOD program, as illustrated by Allison (1999: 184–188). We coded malnutrition as a dichotomous dependent variable (0 = malnourished, 1 = well nourished) for each child every time he/she was surveyed. As with previous analyses, measurements less than two negative standard deviations from the median of the reference population (<-2 S.D.) were considered as mild-to-moderate malnutrition (World Health Organization, 1986).

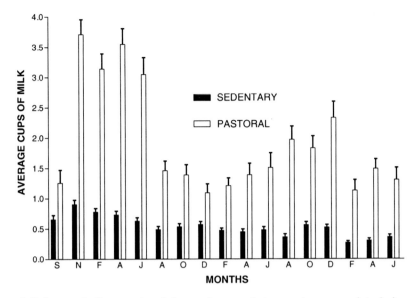

Figure 7. Daily cups of milk over study period, pastoral versus sedentary samples, means and standard errors of the means.

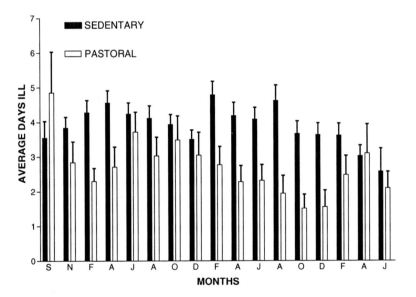

Figure 8. Days ill over study period, pastoral versus sedentary samples, means and standard errors of the means.

This approach also permitted us to examine other potentially important explanatory variables. Specifically, we considered family wealth, maternal parity and child gender composition differences in the two samples as potentially affecting growth patterns. Any of these could mean unequal access to food or child health care, the latter two in stemming from the Rendille custom of primogeniture (cf. Roth, 2000) and Rendille and Ariaal societies' patrilineal, patrilocal nature.

Table 2. Analysis of GEE Parameter Estimates, all Five Communities, Total Study Period.

Parameter		Estimate	Std Err	Z	Prob.
A. Weight-for-age					
INTERCEPT		−0.1456	0.1416	−1.03	0.3037
BREASTFED YES		−0.6562	0.1311	−5.01	<0.0001
ILLNESS		0.0272	0.0074	3.68	0.0002
MILK		−0.2307	0.0454	−5.08	<0.0001
WEALTH	POOR	0.5838	0.1518	3.85	0.0001
SEX	MALE	−0.2037	0.1499	−1.36	0.1743
B. Height-for-age					
INTERCEPT		−0.7954	0.1680	4.74	<0.0001
BREASTFED YES		0.0183	0.1570	0.12	0.9073
ILLNESS		0.0123	0.0069	1.80	0.0725
MILK		−0.1457	0.0492	−2.96	0.0030
WEALTH	POOR	0.5076	0.1438	2.77	0.0056
SEX	MALE	−0.3306	0.1803	−1.87	−0.0620

In Table 2 GEE analysis of the pooled five-community is presented for weight-by-age and height-by-age. For each run dichotomous and continuously distributed independent variables include the following.

1. **SEX**—this was converted into a dummy variable coded "0" for males and "1" for females.
2. **WEALTH**—interviewees were asked to rank their household, as well as their neighbors as either, "RICH", "SUFFICIENT" AND/OR "POOR". For the present studies differences in defining criteria of wealth for Korr versus Lewogoso (cash income versus animal ownership) led to dichotomizing this variable into "POOR" and "SUFFICIENT" strata.
3. **MILK**—a continuous variable recording cups of milk consumed in the past twenty-four hours.
4. **ILL**—a continuous variable denoting days ill with diarrhea, respiratory disease and/or fever in the past 60 days.
5. **BREASTFEEDING**—coded as a dichotomous variable, e.g., "breastfeeding—yes/ no" this variable serves two purposes. First it further represents the nutrition–infection synergism as breast milk contains both nutrients and maternal antibodies. Second it serves to partially model the age effect noted in the previous bivariate analyses of age and z-scores, where growth faltering was associated with increasing age in the sedentary communities. It was found by Fujita, Chapter 11, this volume, that breastfeeding patterns did not vary by community, thus eliminating this potential cultural variable.

Table 2 presents results for both height-and weight-for-age measurements for all five communities over the entire study period. Table 2A shows data for the weight-by-age analysis. With the exception of the SEX variable, all independent variables were highly ($p < 0.0001$) statistically significant. As expected, days ill (ILL) were positively associated with malnutrition ($Z = 3.68$, $p < 0.0002$) while milk consumed (MILK) was even more strongly negatively associated with malnutrition ($Z = -5.08$, $p < 0.0001$). Breastfeeding children were negatively associated with childhood malnutrition ($Z = -5.01, p < 0.0001$), illustrating the beneficial nutritional and anti-infection properties of mothers' milk.

Economic status (WEALTH) showed families from the poor stratum positively associated with malnutrition ($Z = 3.85$, $p < 0.0001$). Finally, while not statistically significant the SEX variable ($Z = -1.36$, $p = 0.1743$) shows a negative association between boys and malnutrition.

Many of these relationships hold when considering height-by-age, as shown in Table 2B. Thus MILK retains a strong negative association with malnutrition ($Z = -2.96$, $p = 0.003$), while poor households represented by the WEALTH variable still feature a highly significant positive association ($Z = 2.77$, $p = 0.0056$). While retaining the same algebraic signs as in the weight-by-age analysis, both days ill (ILL, $Z = 1.80$, $p = 0.0725$) and the variable denoting male children (SEX, $Z = -1.87$, $p = 0.0620$) are statistically non-significant, although both are close to the α 0.05 level.

Taken together, this analysis reveals the importance the nutrition-infection synergism plays in childhood malnutrition for Ariaal and Rendille children. Since this is clearly very important in the case of breastfeeding, we performed a subsequent GEE analysis based only on non-breastfeeding children. We further subdivided this analysis into times representing a normal rainfall year and following drought years, to see how independent variables fared under differing environmental conditions. In doing so we chose the first eight sampling times (September, 1994 until December, 1995) as representing the normal bimodal rainfall pattern characteristic of East Africa (see Figure 2). The remaining nine sampling times were either characterized by drought conditions (from February 1996 until February 1997), or by the excessively heavy rainfall (>300 mm) in April 1997.

Tables 3 and 4 present the results for these analyses. In Table 3, representing normal conditions, the most important independent variables are MILK and WEALTH. The former is strongly negatively related to malnutrition for both weight ($Z = -3.56$, $p < 0.0001$) and height ($Z = 2.52$, $p < 0.001$). Poverty, as coded for by the WEALTH variable, is positively associated with malnutrition for both weight ($Z = 4.15$, $p < 0.0001$) and height ($Z = 3.67$, $p < 0.0001$). Days ill and male children, the latter represented by the SEX variable, are not significantly associated with either measure of malnutrition.

Turning to Table 4 which presents GEE results during drought and heavy rainfall times, the nutrition-infection synergism is seen again, with MILK negatively associated with malnutrition, measured by both weight ($Z = -3.12$, $p = 0.0018$) and height ($Z = -2.75$, $p = 0.0059$). Milk is the only statistically significant variable in both measurements, with days ill

Table 3. Analysis of GEE Parameter Estimates, Normal Rainfall.

Parameter		Estimate	Std Err	Z	Prob.
A. Weight-for-age					
INTERCEPT		−0.2835	0.1806	−1.57	0.1165
ILLNESS		0.0138	0.0115	1.20	0.2321
MILK		−0.2100	0.0590	−3.56	0.0004
WEALTH	POOR	0.8533	0.2058	4.15	<0.0001
SEX	MALE	−0.3517	0.2033	−1.73	0.0837
B. Height-for-age					
INTERCEPT		−0.7995	0.2020	−3.96	<0.0001
ILLNESS		0.0067	0.0103	0.65	0.5134
MILK		−0.1522	0.0604	−2.52	0.0118
WEALTH	POOR	0.8532	0.2327	3.67	0.0002
SEX	MALE	−0.3543	0.2317	−1.53	−0.1262

Table 4. Analysis of GEE Parameter Estimates, Drought and Heavy Rainfall.

Parameter		Estimate	Std Err	Z	Prob.
A. Weight-for-age					
INTERCEPT		−0.0229	0.1968	−0.12	0.9073
ILLNESS		0.0405	0.0126	3.20	0.0014
MILK		−0.2996	0.0961	−3.12	0.0018
WEALTH	POOR	0.6348	0.2171	2.92	0.0035
SEX	MALE	−0.2437	0.2099	−1.16	0.2456
B. Height-for-age					
INTERCEPT		−0.7520	0.2321	−3.24	0.0012
ILLNESS		0.0200	0.0118	1.69	0.0907
MILK		−0.3811	0.1384	−2.75	0.0059
WEALTH	POOR	0.4822	0.2574	1.87	0.0610
SEX	MALE	−0.4406	0.2510	−1.76	−0.0792

(ILL) only significant in the analysis of weight-by-age ($Z = 3.20$, $p = 0.0014$). The WEALTH variable again shows children of poor families suffering malnutrition, significantly so for weight-for-age ($Z = 2.92, p = 0.0018$), and close to the α 0.05 level for height-for-age ($Z = 1.87, p = 0.0610$). While SEX is not statistically significant for either measure, as in all previous analyses it retains a negative algebraic sign, indicating that girls relative to boys are more susceptible to malnutrition.

What is clear in all analyses is the synergism between nutrition and infection, represented here by milk consumption and days ill. These findings support our previous hypotheses that sedentism among Rendille and Ariaal would be associated with changing dietary patterns featuring a reduction in milk consumption and increased morbidity due to the spread of density-dependent disease and/or susceptibility due to malnutrition. Unlike our previous findings (cf. Nathan et al., 1996) economic stratification plays an important role in child health in the present analysis, suggesting continuing economic differentiation over time in what were previously considered egalitarian populations. The policy implications of these results are considered in the next section.

6. SUMMARY AND DISCUSSION

This initial examination of longitudinal growth data collected over a three-year period revealed far poorer growth patterns in a sample of children from four sedentary Rendille and Ariaal communities, relative to same-aged children from the nomadic Ariaal community of Lewogoso. Analysis of dietary and morbidity patterns in both communities shed possible light on the underlying causes for the large differences. As in our previous cross-sectional studies (cf. Nathan et al., 1996; Fratkin et al., 1999) the nomadic group's diet featured significantly more milk consumption throughout the entire study period. This finding is hardly surprising since nomadic communities are always with their animals, while sedentary communities often are separated from their herds.

Unlike our previous studies, analysis of morbidity data revealed some significant differences between the sedentary and nomadic samples. Together these results support our initial hypothesis that child nutrition and morbidity would worsen in the transition to sedentism for formerly nomadic Rendille and Ariaal pastoralists. In addition to this specific example of the long-recognized nutrition-infection synergistic effect on child growth, we also found socio-economic variables, in the form of household wealth differentiation, exacerbating levels of childhood malnutrition. These consistent findings point to maladaptive biological consequences of sedentism for children in Rendille and Ariaal populations.

At the same time that we make the above observations, within the broader framework we have always recognized that sedentarization confers both benefits and constraints. In terms of policy implications the real question now is how to reconcile these negative biological findings with other possibly beneficial social consequences of Ariaal and Rendille sedentarization. Included among these benefits are increased access to public education, health facilities and larger markets, and increasing female involvement in all three (see Fratkin and Roth Chapter 2, this volume; Fratkin and Smith, Chapter 8, this volume). All these factors have the potential to beneficially influence childhood health. Yet at present they do not supercede the negative childhood health consequences of sedentarization for Ariaal and Rendille. Indeed, the fact that the nomadic Ariaal sample exhibits better growth patterns in bothwet and dry years argues strongly for the pastoral existence as a stronger and more flexible adaptation to the cyclical droughts and accompanying famines that characterize East Africa. The challenge for the future will be to develop policies that ensure child health under conditions of rapid socio-economic change represented by the transition from nomadic pastoralism to sedentism.

REFERENCES

Allison, P.D., 1995, *Logistic Regression Using the SAS® System: Theory and Application*. Cary, NC: SAS Institute.

Brainard, J.M., 1990, Nutritional status and morbidity on an irrigation project in Turkana District, Kenya. *American Journal of Human Biology* 2:153–163.

Campbell, B.C., P.W. Leslie, M.S. Little, J.M. Brainard, and M.A. DeLuca, 1999, Settled Turkana. In *Turkana Herders of the Dry Savanna: Ecology and Biobehavioral Response of Nomads to an Uncertain Environment*, edited by M.A. Little and P.W. Leslie, pp. 333–352. New York: Oxford University Press.

Chabasse, D., C. Roure, A.G. Rhaly, P. Rangque, and M. Quilici, 1985, Health of Nomads and Semi-Nomads of the Malian Gourma: An Epidemiological Approach. In *Population, Health and Nutrition in the Sahel*, edited by A. Hill, pp. 319–333. London: Routledge and Kegan-Paul.

Center for Disease Control, 1997, *Epi Info, Version 6*. Atlanta: United States Center for Disease Control.

Ensminger, J., 1992, *Making a Market: The Institutional Transformation of an African Society*. New York: Cambridge University Press.

Eveleth, P.B. and J. Tanner, 1990, *Worldwide Variation in Human Growth*. Cambridge: Cambridge University Press.

Fleuret, P. and A. Fleuret, 1991, Social Organization, Resource Management, and Child Nutrition in the Taita Hills, Kenya. *American Anthropologist* 93:91–114

Fratkin, E., 1997, Pastoralism: Governance and development issues. *Annual Review of Anthropology* 26:235–261.

Fratkin, E., 2004, *Ariaal Pastoralists of Northern Kenya*, second edition. Needham Heights, MA: Allyn and Bacon.

Fratkin, E. and E.A. Roth, 1990, Drought and economic differentiation among Ariaal Pastoralists of Kenya. *Human Ecology* 18:385–402.

Fratkin, E. and E.A. Roth, 1996, Who survives drought: Measuring winners and losers among the Ariaal Rendille pastoralists of Kenya. In *Case Studies in Human Ecology*, edited by D. Bates and S. Lees, pp. 159–174. Plenum Press: New York.

Fratkin, E. and K. Smith, 1995, Women's changing economic roles with pastoral sedentarization: Varying strategies in alternative Rendille communities. *Human Ecology* 23:433–454.

Fratkin, E., M.A. Nathan, and E.A. Roth, 1996, Sedentism and child health among Rendille pastoralists of Northern Kenya. *Social Sciences and Medicine* 43:503–515.

Fratkin, E., E.A. Roth, and M.A. Nathan, 1999a, Health consequences of pastoral sedentarization among Rendille of Northern Kenya. In *The Poor are Not Us: Poverty and Pastoralism*, edited by D.M. Anderson and V. Broch-Due, pp. 149–163. Oxford: James Currey Ltd.

Fratkin, E., E.A. Roth, and M.A. Nathan, 1999b, When nomads settle: commoditization, nutrition and child education among Rendille pastoralists. *Current Anthropology* 40(5):729–735.

Fujita, M., E.A. Roth, M.A. Nathan, E.M. Fratkin, 2004, Sedentism, Seasonality and Economic Status: A Multivariate Analysis of Maternal Dietary and Health Statuses Between Pastoral and Agricultural Ariaal and Rendille Communities in Northern Kenya. *American Journal of Physical Anthropology.*

Galvin, K., 1985, *Food procurement, diet, activities and nutrition of Ngisonyonka Turkana pastoralists in an ecological and social context.* Ph.D. Dissertation, State University of New York, Binghamton.

Galvin, K., 1992, Nutritional ecology of pastoralists in dry tropical Africa. *American Journal of Human Biology* 4:209–221.

Galvin, K.A. and M.A. Little, 1999, Dietary intake and nutritional status In *Turkana Herders of the Dry Savanna: Ecology and biobehavioral response of nomads to an uncertain environment,* edited by Michael A. Little and Paul W. Leslie, pp. 125–145. New York: Oxford University Press.

Hamill, P.T. Drizd, C. Johnson, R. Reed, A. Roche and W. Moore, 1979, Physical growth: National Center for Health Statistics percentiles. *American Journal of Clinical Nutrition* 32:607–629.

Hiernaux, J, 1964, Weight/height relationships during growth in Africans and Europeans. *Human Biology* 36:273–293.

Hill, A., 1985, Health and Nutrition in Mali. Population, Health and Nutrition in the Sahel: Issues in the Welfare of Selected West African Communities. London: Routledge and Kegan Paul.

Johnston, F.E. and M.A. Little, 2000, History of human biology in the United States of America. In, *Human Biology: An Evolutionary and Biocultural Perspective,* edited by S. Stinson, B. Bogin, R. Huss-Ashmore, and D. O'Rourke, pp. 27–46. New York: Wiley-Liss.

Liang, K. and S. Zeger, 1986, Longitudinal data analysis using generalized linear models. *Biometrika* 73:13–22.

Little, M., 1997, Adaptability of African pastoralists. In *Human Adaptability: Past Present and Future,* edited by Stanley Ulijaszek and Rebecca Huss-Ashmore, pp. 29–60. Oxford: Oxford University Press.

Little, M. and P. Leslie, 1999, *Turkana Herders of the Dry Savanna.* Oxford: Oxford University Press.

Little, M., S. Gray, and P. Leslie, 1993, Growth of nomadic and settled Turkana infants of north-west Kenya. *American Journal of Physical Anthropology* 92:335–344

Little, P.D., 1994, Maidens and Milk Markets: The Sociology of Dairy Marketing in Southern Somali. In *African Pastoralist Systems: An Integrated Approach,* edited by E. Fratkin, K. Galvin, and E.A. Roth, pp.165–184. Boulder: Lynne Rienner Publishers.

Martorell, R., 1989, Body size, adaptation and function. *Human Organization* 48(1):15–20.

McCabe, J.T., S. Perkin, C. Schofield, 1992, Can conservation and development be coupled among pastoral people? An examination of the Maasai of the Ngorongoro conservation area, Tanzania. *Human Organization* 51:353–366.

Nathan, M.A., E. Fratkin, and E.A. Roth, 1996, Sedentism and child health among Rendille pastoralists of northern Kenya. *Social Science and Medicine* 43: 503–515

Nestel, P., 1986, A society in transition: developmental and seasonal influences on the nutrition of Maasai women and children. *Food and Nutrition Bulletin* 8: 2–14.

Panter-Brick, C., 1998, Biological anthropology and child health: context, process and outcome. In *Biosocial Perspectives on Children.* edited by Catherine Panter-Brick. Cambridge: Cambridge University Press.

Pelletier, D., 1994, The potentiating effects of malnutrition on child mortality: epidemiological evidence and policy implications. *Nutrition Reviews* 52(12):409–415.

Roth, E., 1991, Education, tradition and household labor among Rendille pastoralists of northern Kenya. *Human Organization* 50:136–141.

Roth, E., 1996, Traditional pastoral strategies in a modern world: An example from Northern Kenya. *Human Organization* 55:219–224.

Roth, E., 2000, On pastoralist egalitarianism: Primogeniture and Rendille demography. *Current Anthropology* 41:269–271.

Scrimshaw, N., C. Taylor, and J. Gordon, 1968, Interaction of nutrition and infection. *World Health Monograph Series,* 57.

Sellen Daniel, W., 1996, Nutritional status of Sub-Saharan African pastoralists: A Review of the Literature. *Nomadic Peoples* 39:107–134

Shell-Duncan, B., 1995, Impact of seasonal variation in food availability and disease stress on the health status of nomadic Turkana children: A longitudinal analysis of morbidity, immunity, and nutritional status. *American Journal of Human Biology* 7:339–355.

Simondon K.B., E. Bénéfice, F. Simondon F., V. Delaunay, and A. Chahnazarian, 1993, Seasonal variation in nutritional status of adult and children in rural Senegal. In *Seasonality and human ecology,* edited by S.J. Ulijaszek and S.S. Strickland, pp. 166–183. Cambridge: Cambridge University Press.

Smith, K., 1999, The Farming Alternative: Changing Age and Gender Roles among sedentarized Rendille and Ariaal. *Nomadic Peoples* (NS) 3 (2): 131–146.

Statistical Analysis System, 1997, *SAS/STAT User's Manual*. Cary, NC: SAS Press.

Stinson, S., 2000, Growth variation: Biological and cultural factors. In *Human Biology: An Evolutionary and Biocultural Perspective*, edited by S. Stimson, B. Bogin, R. Huss-Ashmore, and D. O'Rourke, pp. 27–46. New York: Wiley-Liss.

Tanner, J.M., 1981, *A History of the Study of Human Growth*. Cambridge: Cambridge University Press.

Waters-Bayer, A., 1988, *Dairying by Settled Fulani Women in Central Nigeria: The Role of Women and Implications for Dairy Development*. Kiel: Wissenschaftsverlag Van Kiel.

World Health Organization, 1986, Use and interpretation of anthropometric indicators of nutritional status. *Bulletin of the World Health Organization* 64:929–941.

Zaal, F. and T. Dietz, 1999, Of Markets, Maize, and Milk: Pastoral Commoditization in Kenya. In *The Poor are not Us: Poverty and Pastoralism in Eastern Africa*, edited by D.M. Anderson and V. Broch-Due, pp. 163–198. Oxford: James Currey.

Chapter 10

Health and Morbidity among Rendille Pastoralist Children

Effects of Sedentarization

MARTHA A. NATHAN M.D., ERIC ABELLA ROTH, ELLIOT FRATKIN,
DAVID WISEMAN M.D. AND JOAN HARRIS R.N.

1. INTRODUCTION

Children throughout the developing world are at risk for undernutrition and infectious diseases. Poor nutrition not only hinders physical and cognitive development (Sigman et al., 1998; Walker et al., 1998), but deleterious immune effects invite viral and bacterial infection (Kossmann, 2000a; Kossmann, 2000b). Conversely, serious infection in an otherwise well-nourished child absorbs much-needed calories and often prevents intake of food, leading to undernutrition and wasting.

According to UNICEF (2001), malnutrition is associated with half of all deaths in under-fives worldwide. The first ranks of the global disease burden and source of child mortality for children in the developing world include diarrhea, respiratory infections and malaria, with measles declining precipitously in many areas due to effective immunization (WHO, 2003). Diarrhea, respiratory infection, and malaria are the three infections most influenced by environmental exposures (WHO, 1997). Thus, when evaluating the health of children in the developing world, it is important to assess the prevalence and impact of these three disease sets.

MARTHA A. NATHAN M. D. • Brightwood Health Center, Springfield MA 01107.
ERIC ABELLA ROTH • Department of Anthropology, University of Victoria, Victoria, British Columbia, Canada V8W 3P5.
ELLIOT FRATKIN • Department of Anthropology, Smith College, Northampton, Massachusetts, 01063.
DAVID WISEMAN M. D. AND JOAN HARRIS R. N. • Hornby Island, British Colombia, Canada.

Under-five mortality in Kenya is one of the world's highest at 122 per thousand. Per year 132,000 under-five deaths are reported (UNICEF, 2004). The major infectious killers of Kenyan children remain diarrhea, acute respiratory infection, and malaria, although HIV/AIDS is fast becoming a risk for childhood mortality (National Research Council, 1993; Omondi-Odhiambo, 1984; UNICEF, 2004).

Pastoralist Rendille children in Marsabit District, Kenya, survive in what would seem to outsiders to be a particularly bleak, child-hostile environment, growing up in an arid and isolated region where food, clean water, sanitation, health facilities, and schools are all in short supply, and infectious diseases are constant threats. What is known of the health problems of pastoralists is limited and underscores the high prevalence of infection. Hill's edited volume (1985) on health, nutrition, and demography in Mali assembled several non-controlled studies of farming, agro-pastoral, and pastoral groups. One study by Chabasse et al. (1985) reported that nomadic groups had higher rates of tuberculosis, brucellosis, syphilis, trachoma, and child mortality (children five and under) than settled agricultural populations. However, the latter suffered higher rates of bilharzia, and parasitic infections and more malaria and anemia, particularly among those groups living close to rivers.

The South Turkana Ecosystem Project (STEP) of the late 1980s researched ecology, health, nutrition, and fertility of nomadic Turkana of Kenya (Little and Leslie, 1999). Here researchers found that settled Turkana experienced reduced fertility, increased morbidity (particularly from malaria) and increased child mortality. Settled children under five showed more growth stunting than nomadic children, although settled children over five were heavier, which was attributed to greater role of carbohydrates in their diets, particularly for children receiving supplemental feeding in schools. Nomadic Turkana women, however, were taller, heavier, and had lower blood pressure than settled women (Brainard, 1990; Campbell et al., 1999; Galvin, 1992). Previously, Brainard (1986) found that nomadic Turkana suffered substantially higher infant mortality than settled Turkana agriculturalists, and Murray et al. (1980) noted increased iron-deficiency but overall decreased morbidity among pastoralist Turkana when compared with settled fish-eating Turkana.

2. RENDILLE SEDENTARIZATION PROJECT

The Rendille Sedentarization Project was undertaken to evaluate the health and nutrition of Rendille pastoralists and assess the impact of the many aspects of sedentarization, focusing particularly on women and children. When pastoralists settle in Marsabit District they change their relationship to towns, clinics, work, markets, schools, and water. Settling introduces Rendille children to exposure to disease in more densely populated settlements but may also bring increased sanitation and safer child-rearing practices through maternal education (see Roth and Ngugi, this volume, Chapter 13).

Long distances and lack of transportation limit access to preventive and curative services for traditional Rendille pastoralists. Average distances traveled by patients to Laisamis Hospital, the closest hospital for our nomadic study population, were 60 km according to a study by the Ministry of Planning and National Development. (Ministry of Planning and National Development, Kenya, 1994b). Moving near towns shortens that distance considerably allowing preventive—immunization—and early interventive services for acute serious infections.

Settling may also introduce environmental change. Rendille traditionally live in the dry lowlands of Marsabit District in small nomadic villages like Lewogoso. There and in

Photo 1. Ariaal mother and daughter in Lewogoso.

the larger famine relief settlements of Korr and Kargi the annual rainfall is 200 mm or less. Beginning in the 1970s, some lowland pastoralists moved up Marsabit Mountain to the wetter and cooler highlands, with annual rainfall 800 mm, to engage in farming and market activities (Ministry of Planning and National Development, Kenya, 1994a, see Fratkin, Adano, and Witsenburg, Smith chapters this volume). Two of our study villages, Songa and Karare, represent this move. Clean water for drinking and sanitation may thus be more readily available in these highland communities, but also the ecology of infectious agents causing respiratory diseases and malaria may be affected and therefore change the disease patterns in infants and young children. The *anopheles* mosquito needs for its life cycle calm pools of water more available on the mountain, but does not survive as well in cooler climates at higher altitudes. Open water does occur in lowland arid towns like Korr, particularly after rainfall and also near open-well sites, giving cyclic opportunity for *anopheles* reproduction and malaria epidemics. Cool, damp weather may also promote more exposure to wood smoke, a respiratory irritant inviting infection.

Settlement may also influence the effects of drought on food supply and disease. Any assessment of child morbidity in Northern Kenya must evaluate whether settlement makes children more or less susceptible to the deleterious effects of rainfall failure.

Two major questions our research attempts to address are: How do Rendille pastoralist children fare in regard to the dual threat of malnutrition and infection, and What is the effect of sedentarization? The present study is an attempt to evaluate the effect of settling— and particular the site and mode of settling—on disease patterns among Rendille, focusing particularly on children under six years of age. To this end we used diagnoses recorded for clinic visits in the areas of our study population to explain location influences and our bimonthly longitudinal sampling of communities (see Roth et al., this volume, Chapter 9; Fujita et al., this volume, Chapter 11). For further inquiry into differences, in 1995 a pediatrician and nurse performed physical exams and documented their findings.

3. MATERIALS AND METHODS

3.1. Materials

As described in previous chapters in this volume one of our data bases consists of thirty-five women from five different locations interviewed on a bi-monthly basis about the health and diet of their children aged five and under over a three year period from September 1994 to June 1997. Ages of children were determined by means of immunization records or an accepted events calendar. Among other questions, mothers were asked by field assistants to recall the number of days each child suffered from diarrhea, respiratory illness or colds, and fever over the previous thirty days.

3.1.1. Study Villages and Associated Clinics

The five villages, described in Chapter 2, are:

1. Lewogoso, an isolated and mobile arid lowland pastoralist settlement with an approximate population of 250 and no school or clinic;
2. Korr, a desert lowland famine-relief center, population approximately 2500, with a school and clinic/dispensary run by the Catholic Church;

Photo 2. Martha Nathan and AnnaMarie Aliyaro interview mothers in Pastoral Ariaal community of Lewogoso.

3. Ngrunit, population 1000, a small isolated town at the base of the Ndoto Mountains, mid-range in climate between highland and lowlands, with a primary school and clinic/dispensary run by the African Inland Church;

4. Songa, an irrigated agricultural community on Marsabit Mountain with a population of approximately 2500; and

5. Karare, a small town (like Korr) located on Marsabit Mountain, which, coupled with the nearby large agricultural scheme Nasikakwe, has a population also 2500. Both Karare/Nasikakwe and Songa have government-run clinic/dispensaries and primary schools.

The altitude of Korr and Lewogoso is 400 meters and rainfall averages 200 mm/year, with rains highly seasonal in the spring and fall. Ngrunit, though near the same altitude, is located at the base of the Ndoto Mountains and thus is served by a constant water source—the stream providing runoff from the mountains. Songa and Karare are located in the highlands of Marsabit Mountain. Songa is located in a forested area at approximately 1250 m and Karare/Nasikakwe on the drier western side of Marsabit Mountain at approximately 1400 meters. Both Karare and Songa are located within 20 km of the district capital of Marsabit town.

Laisamis hospital, the second largest in the district, and its outpatient clinic serve the lowlands and Ngrunit. Marsabit Hospital is the largest in the district and the only one with a permanent doctor.

Diagnosis records/month were obtained for 1994–1997 from the outpatient clinics at Marsabit and Laisamis Hospital and from clinics at Korr, Ngrunit and Karare. Marsabit Hospital serves the district capital Marsabit town whose patients are drawn mainly from the highland regions. Laisamis Hospital is located south of Marsabit town and serves the arid lowlands including Korr and Lewogoso. Clinics served adults and children. Medical

Photo 3. 'Well-baby' clinic at Karare dispensary.

rationale for visits was designated by local clinic personnel. Visits for "Diseases of the Respiratory System" were combined with "Pneumonia" to create the category "Respiratory Diseases". Disease categories were calculated as a percentage of total number of visits to that clinic per month or per year.

Korr and Laisamis were designated "dry lowland" clinics due to similar rainfall, altitude and access to water. Marsabit and Karare were designated "highland" clinics. Ngrunit was examined separately because of its unique access to water despite its low altitude and low rainfall. Similarly, in calculating child morbidity means, Songa and Karare were at times designated "highland" communities, Korr and Lewogoso "dry lowland" and Ngrunit was calculated separately.

3.1.2. Physical Examinations

In July 1995 pediatrician David Wiseman MD and nurse Joan Harris, both with previous experience in Marsabit District, examined available study children at all five villages and recorded their findings. Findings were expressed as percent of total children examined in that village.

3.1.3. Rainfall Data

We also made use of rainfall data, recorded for the district from records from Marsabit Station, to delineate possible environmental effects on health by site and over time. As can be seen from Figure 1, 1995 was a normal year, with a total rainfall of 855 mm, featuring the usual East African bimodal rainfall pattern with significant rains in the spring and the fall. However, 1996 was a drought year in which both season's rains failed and total rainfall only measured 320 mm.

3.2. Methods

The marked size disparity among the clinics led to the need to use percentages to indicate relative differences in incidence of diseases. Unfortunately this precluded standard

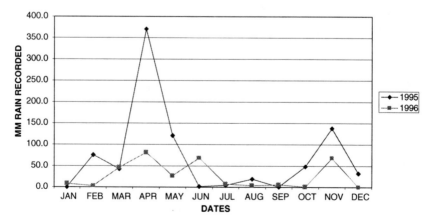

Figure 1. Rainfall recorded Marsabit station, 1995 and 1996.

statistical analysis, for example comparing different clinics and communities by contingency analysis. Because of this problem we turned again to the General Estimating Equation, designed for the analysis of longitudinal data, and described in a previous chapter (Roth et al., Chapter 9, this volume). In this case we used days ill from three specific diseases, diarrheal diseases, fevers and respiratory illness, generated by the longitudinal (1994–97) bimonthly interviews of Rendille and Ariaal mothers in the five study communities as the independent variables in our models. To search for the effects of environmental and temporal variation, we constructed three separate models using differing dichotomous categorical variables as dependent variables. These included: (1) a nomadic (Lewogoso) versus sedentary (Korr, Ngrunit, Songa and Karare) dicho-tomy, (2) a highland-lowland dichotomy (highland = Songa, Karare, lowland = Korr) and, (3) a model comparing data from 1995 to those from 1996.

4. RESULTS

4.1. Clinic Diagnosis Data

To evaluate the overall geographic and rainfall-based distribution of the main categories of infectious diseases of the area, we documented numbers of visits for respiratory diseases, malaria, and diarrhea as percentages of the total number of visits for each clinic for the years 1995 and 1996.

As shown in Table 1, Marsabit clinic was by far the busiest, with over 13,000 recorded visits per year. Karare was the smallest with approximately 1/10 the volume, around 1500 visits/year. Because of the disparity in volume, the Marsabit clinic numbers dominate evaluation of the highland clinic data.

As can be seen in Figure 2 and 3, diarrhea visit rates were relatively low compared to visits for respiratory and malarial illnesses. However, in both 1995 and 1996, the rates of diarrhea diagnoses in Karare as a percent of total clinic visits were approximately twice as high as for any of the other clinics.

For malaria and respiratory disease, however, there is a marked difference between the highland clinics and the dry lowlands. Visits for malaria in the dry lowlands, seen in Figure 4, were approximately twice as common per clinic volume than in the highlands for both the normal and dry years. Conversely, visits for respiratory disease in the highlands were more than twice as frequent/clinic volume as for the dry lowland clinics in both 1995 and 1996 as presented in Figure 5. Ngrunit shows a pattern consistent with a highland clinic with lower rates of malaria and higher rates of respiratory disease. Korr and Laisamis each had a higher percentage of visits for malaria than either Marsabit or Karare, and each dry lowland clinic had a lower percentage of visits for respiratory disease than the highland

Table 1. Number of Patient Visits/Clinic/Year.

	1995	1996	Total
Karare	1554	1437	2991
Korr	6714	5776	12490
Laisamis	2718	2025	4743
Marsabit	13474	13655	27129
Ngurunit	3530	2074	5604

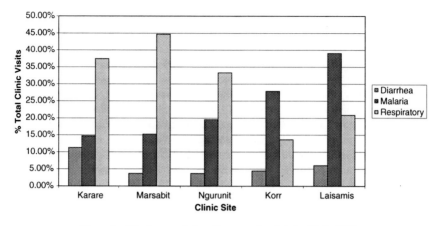

Figure 2. Diarrhea, respiratory diseases and malaria by Clinic, 1995.

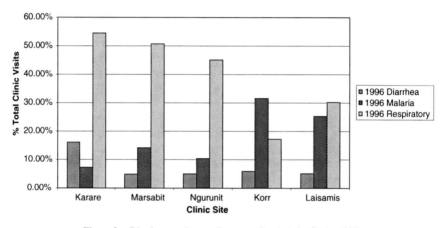

Figure 3. Diarrhea, respiratory diseases and malaria by Clinic, 1996.

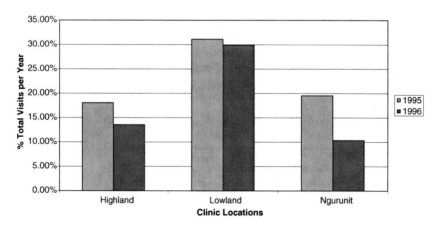

Figure 4. Malaria: Highland, dry lowland, Ngurunit by year.

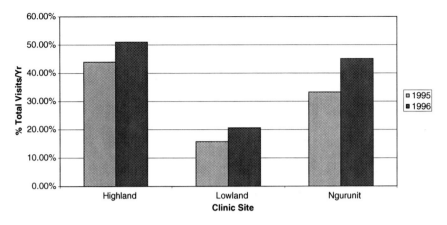

Figure 5. Respiratory disease: Highland, lowland, Ngrunit.

clinics in both years. Further, malaria decreased in incidence in all sites in the dry year, 1996 vs. 1995. This effect was least noticeable in the large Marsabit clinic, but still was evident. Respiratory disease visits unexpectedly increased in frequency in the drought year in all the clinics.

4.2. Child Morbidity Data

Mother's reports of child illness days per previous month for each of six interviews/year were combined for each village and the mean symptom-days/child/month was plotted for each village for 1995 and 1996, respectively in Figures 6 and 7. These showed markedly fewer days of diarrhea and colds/month/child for children in the nomadic community of Lewogoso than any of the four sedentary villages in both 1995 and 1996. Korr, the other lowland sample, had lower rates of diarrhea/child/month than did the highland towns of Songa and Karare, but had a higher incidence of cold days/child/month than any of the other towns in 1996 and the second highest of all the towns in the normal year 1995. Thus respiratory illness incidence for children did not seem to fit the highland: dry lowland dichotomy established by the clinic data where the highland clinics treated a higher ratio of respiratory diseases/clinic volume than did the dry lowlands clinics.

Fever days/child/month, which include illness from malaria but certainly not exclusively, were highest in Lewogoso in 1995 and lowest in Korr of all the communities. In Lewogoso in the normal rainfall year of 1995, there were more fever days/child/month than cold days/child/month, the only village and year in which this pattern occurred. Fever days/child/month appeared to decrease overall in the dry year, except in Korr, the lowland town, where they increased slightly as shown in Figure 8. Interestingly, cold days decreased in every village but Korr in the drought year as seen in Figure 9. Ngrunit's child morbidity pattern was very similar to Korr's: high rates of respiratory disease, low rates of diarrhea relative to the highland towns and low rates of fever relative to the highland towns and Lewogoso.

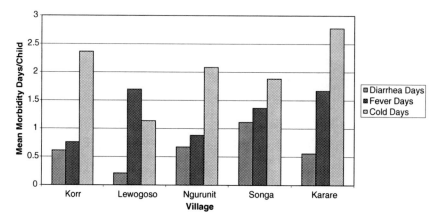

Figure 6. Morbidity days/child, 1995 by village.

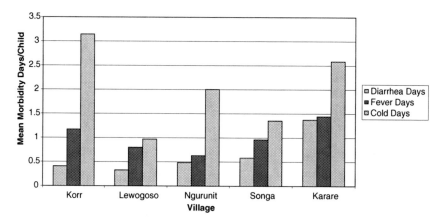

Figure 7. Morbidity days/child, 1996 by village.

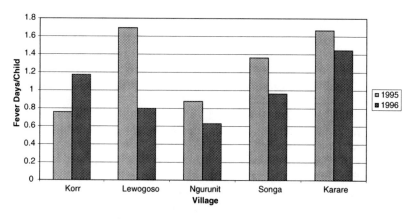

Figure 8. Fever days/child by year and village.

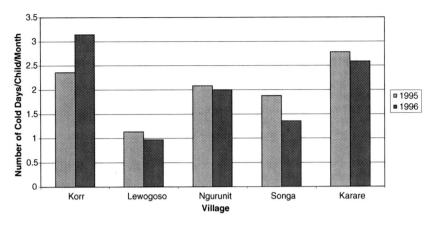

Figure 9. Cold days/child by year and village.

Table 2. Analysis of GEE Parameter Estimates, Sedentary
Communities versus Lewogoso.

Parameter	Estimate	Std Err	Z	Prob.
1995 ($n = 2186$)				
INTERCEPT	1.1550	0.1397	8.27	<0.0001
DIARRHEA	0.2839	0.0777	3.65	0.0003
COLDS	0.1739	0.0412	4.22	<0.0001
FEVER	−0.1046	0.0248	−4.23	<0.0001
1996 ($n = 1850$)				
INTERCEPT	1.1100	0.1490	7.45	<0.0001
DIARRHEA	0.1263	0.0554	2.28	0.0228
COLDS	0.1816	0.0362	5.01	<0.0001
FEVER	−0.0182	0.0382	−0.48	0.6340

Statistical analysis of the data using GEE as shown in Table 2 verifies that children surveyed in Lewogoso in both years suffered markedly lower numbers of days of colds/child/month in both 1995 and 1996 ($p < 0.0001$ for both years) than children in the other villages combined. Their mothers also reported significantly fewer numbers of days of diarrhea/child/month ($p < 0.0003$ for 1995, the normal year and <0.02 for 1996, the dry year). The number of days of fever/child/month was statistically higher for children in Lewogoso than for all those in the other villages in 1995. However, though it was higher in 1996, that difference did not achieve statistical significance ($p = 0.6$).

To determine if the morbidity for Lewogoso children was based on altitude and climate, we compared highland and lowland children, using Songa and Karare to represent the highlands and Korr the lowlands. Results, shown in Table 3, indicate a partial reversal of the clinic data, with Korr children suffering statistically significantly **less** fever than the highland children, particularly in the normal year, but also in the drought year ($p < .0001$ and $p < .04$ respectively). Moreover, the Korr lowland children in our study suffered significantly **more** days of respiratory illness in 1995, and more in 1996, though the difference in the drought year was not significant. Diarrhea days were significantly fewer among

Table 3. Analysis of GEE Parameter Estimates, Highland
(Songa and Karare) versus Lowland (Korr) Communities,
Probability = Lowlands.

Parameter	Estimate	Std Err	Z	Prob.
1995 (n = 1378)				
INTERCEPT	−0.7026	0.1525	−4.61	<0.0001
DIARRHEA	−0.0145	0.0365	−0.40	0.6914
COLDS	0.0243	0.0230	1.06	0.2909
FEVER	−0.1517	0.0295	−5.14	<0.0001
1996 (n = 1195)				
INTERCEPT	−0.8651	0.1625	−5.33	<0.0001
DIARRHEA	−0.1480	0.0421	−3.52	0.0004
COLDS	0.0961	0.0239	4.02	<0.0007
FEVER	−0.0581	0.0276	−2.10	0.0353

Korr children than among the highlands children in 1996, ($p = 0.0004$) and were fewer in 1995, but the difference was not statistically significant as seen in Table 3.

4.3. Physical Exams of Children

In the dry season following the spring rains of 1995, 38 study children were examined in Karare village, 34 in Korr, 36 in Songa, 37 in Lewogoso, and 35 in Ngrunit in the dry season following the spring rains of 1995. Of the three main forms of morbidity evaluated in the longitudinal data, only respiratory ailments could be assessed. As shown in Figure 10, at this particular time, Ngrunit children were least likely to have colds, followed by children from Lewogoso. With the fluidity of change in viral respiratory illness, the significance of these differences is useful in its support of the long-term morbidity data.

5. SUMMARY AND DISCUSSION

Our analysis of two-year data collection on the incidence of respiratory diseases, fevers, and diarrhea among settled and nomadic Rendille children in Marsabit district revealed that nomadic pastoralist children suffered significantly less morbidity from diarrhea and respiratory disease than did children from the settled towns. They endured statistically higher numbers of days of fever than did the settled children in the year of normal rainfall, but not in the dry year. Settled children from the other dry lowland town also suffered less diarrheal disease than did children of the highland town, but differed from the nomadic children in that their mothers reported more days of respiratory illness than did the mothers of the settled children in the highlands. Except for the children of Korr, who endured more days of respiratory illness in the drought year, children in all the villages suffered less malaria and respiratory disease in the drought year.

These findings are compared to clinic data that displayed a marked highland: dry lowland dichotomy. Respiratory illnesses constituted the plurality of diagnoses for highland clinics, whereas malaria diagnoses outweighed respiratory in dry lowland clinics for both years with the exception of Laisamis in the drought year. Clinic data showed that respiratory illness increased in the drought year, whereas fever days decreased in highlands, dry lowlands and Ngrunit.

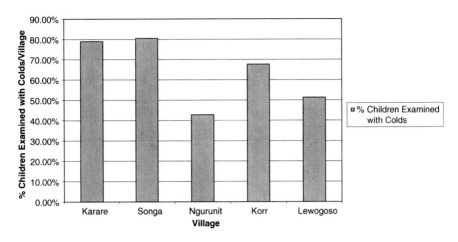

Figure 10. Children with upper respiratory infections by location, July 1995.

The statistically highly significantly lower rates of diarrhea and respiratory diseases for the nomadic children were unexpected. These families live long distances from clinics and lack clean water supply and access by mothers to education. The lower number of diarrhea days and cold days held in normal and drought years.

Decreased child morbidity for the nomadic children could be the result of a multitude of environmental determinants. Lewogoso is a dry lowland site and dry lowland clinics reported relatively fewer diagnosed respiratory illnesses for their total patient population than did the highlands. However, Korr children living in the same geographic environment had an incidence of colds significantly higher than the highlands. It is possible that the year 1996 was an anomaly in Korr, with one or more epidemics of respiratory illnesses. This is supported by the finding that only in Korr did respiratory illnesses increase in the drought year. However, even in 1995, Korr children still experienced more cold days than the highlands, indicating that Lewogoso's advantage for respiratory illness may not be based on geography.

Korr is a windy and dusty environment, more so than nearby Lewogoso or Ngrunit, and it is possible that chemical irritation or allergic reaction to the desert dust provides is the source of children's runny noses and coughs. However, that still does not resolve the conundrum: Korr clinic diagnoses of the general population revealed a typical dry lowland pattern of high incidence of malaria and low incidence of respiratory illness. It is possible, however, that (1) the dust and wind affect particularly children or (2) the increased respiratory illness derives from malnourishment among Korr children (see Nathan et al., this volume).

The highland: dry lowland dichotomy for respiratory illness and malaria was marked in the clinic data. Possible reasons for the relative increase in respiratory illness in the highlands are (1) cooler, somewhat wetter climate with increase in airborne molds and spores; (2) more frequent use of indoor wood and charcoal fires and (3) denser populations promoting greater exposure to illness.

The transmission of malaria is not based on human-to-human contact but the presence of the mosquito, which is promoted by standing water and inhibited by lower temperatures and higher altitudes. Clinic data support the greater influence of altitude and temperature on the epidemiology of malaria in Marsabit district.

Photo 4. Pastoral mother with first born child, Lewogoso.

Thus again our findings that the settled children of Korr suffered less fever than the highland children were somewhat anomalous. However, particularly in children, fever does not necessarily mean malaria: it could well represent viral or bacterial illness or tuberculosis. The preponderance of fever over respiratory illness characteristic of the lowland clinics was found, though, among the children of Lewogoso in the normal year, when Lewogoso children had a statistically higher number of fever days than all the settled children.

Child morbidity did not fully mirror the clinic findings on the effect of drought. Drought years bring great suffering but less fever and malaria, with the only exceptions the incidence of fever in Korr, and diarrhea in Karare each of which increased in 1996. However, in the general population clinic data, respiratory illnesses increased in the drought year, but for the children in all the villages but Korr, the number of cold days increased. Again, it is clear that adult disease patterns cannot necessarily be generalized to children.

It is of particular interest that nomadic Lewogoso children suffered the least number of diarrhea days in each year of any of the communities. The nomads have the most tenuous access to water—walking long distances to fetch it from Ngrunit in the dry season and relying on rain catchments in the rainy season. They also have no formal sanitation system, simply walking outside the village to relieve themselves.

All of the other settled communities have piped water, although the purity of that water is by no means assured. The community whose water, though accessible, is most apt to be contaminated is Karare, where cattle have access to the water source at Gof Bongole crater, and where, in the latter part of the drought year there appears to have been an epidemic of diarrhea affecting the children in our study and clinic patients. Thus simple access to water may not be key to health for Rendille children. Key is access to *clean* water, which may not be the case when nomads settle.

There were many missing data points for this study: in the 3-year bimonthly schedule, mothers sometimes failed to appear for survey interviews and exams. It is possible, then, that the improved morbidity profile for respiratory and diarrheal diseases for the nomadic children stems from a systematic sampling error, i.e., mothers move to town when their children become sick. However, the increased rate of fever illnesses for the Lewogoso children contradicts that theory.

The other unknown variable in this study is the prevalence of HIV. Diagnosis of HIV was not *recorded in clinic records* during the period of our study, but was acknowledged with increasing frequency (personal communication, Marsabit Hospital physician) in the district from 1995 to 1997. Its impact is simply unknown, but can be presumed to affect the settled communities more than the more isolated nomads. HIV infection can be represented by any of the forms of morbidity we studied: diarrhea, respiratory disease or fever.

However, the striking decrease in diarrheal and respiratory diseases for the nomadic children vs. settled children coupled with the previous findings of a relative decrease in malnutrition and stunting indicate an unexpected edge for health and growth of nomadic Rendille children.

As with previous studies, the policy implications of our findings are significant. Though pastoralism is not an option for all those in northern Kenya, the decrease in diarrheal and respiratory illness for pastoralist children is important for those policy-makers interested in decreasing child mortality and morbidity for African children, particularly those involved in settling of traditional nomads. According to our findings, the consequences of settling for the health and nutrition of pastoralist children may be negative ones.

REFERENCES

Brainard, J., 1986, Differential Mortality in Turkana: Agriculturalists and Pastoralists. *American Journal of Physical Anthropology* 70: 525.

Brainard, J., 1990, Nutritional status and morbidity on an irrigation project in Turkana District, Kenya. *American Journal of Human Biology* 2:153–163.

Campbell B.C., P.W. Leslie, M.A. Little, J.M. Brainard, M.A. DeLuca, 1999, Settled Turkana. In *Turkana Herders of the Dry Savanna: Ecology and Biobehavioral Response of Nomads to an Uncertain Environment*, edited by M.A. Liittle and P.W. Leslie, pp. 333–352. New York: Oxford University Press.

Chabasse, D., C. Roure, A.G. Rhaly, P. Rangque, and M. Quilici, 1985, The Health of Nomads and Semi-Nomads of the Malian Gourma: An Epidemiological Approach. In *Population, Health and Nutrition in the Sahel*, edited by A.G. Hill, pp. 319–333. London: Routledge and Kegan-Paul.

Galvin K. 1992, Nutritional ecology of pastoralists in dry tropical Africa. *American Journal of Human Biology* 4:209–221.

Kossmann, J., P. Nestel, M.G. Herrera, A. El Amin, W.W. Fawzi, 2000a "Undernutrition in relation to childhood infections: a prospective study in the Sudan." *European Journal of Clinical Nutrition* 54(6): 463–472.

Kossmann, J., P. Nestel, M.G. Herrera, A. El Amin, W.W. Fawzi, 2000b "Undernutrition and childhood infections: a prospective study of childhood infections in relation to growth in the Sudan." *Acta Paediatrica* 89(9): 1122–1128.

Little, M.A. and P.W. Leslie, 1999, *Turkana Herders of the Dry Savanna: Ecology and Biobehavioral Response of Nomads to an Uncertain Environment*. New York: Oxford University Press.

Ministry of Planning and National Development, Kenya, 1994a, Marsabit Development Programme. Health and Nutrition Survey. Nairobi: Government Printing Office.

Ministry of Planning and National Development, Kenya, 1994b, Marsabit District Development Plan, 1994–96. Nairobi: Government Printing Office.

Murray, M.J., A.B. Murray, and C.J. Murray, 1980, An Ecological Interdependence of Diet and Disease? A Study of Infection in One Tribe Consuming Two Different Diets. *American Journal of Clinical Nutrition* 33: 697.

Nathan, M.A., E.M. Fratkin, and E.A. Roth, 1996, Sedentism and Child Health Among Rendille Pastoralists of Northern Kenya. *Social Science and Medicine* 43(4): 503–515.

National Research Council, 1993, *Effects of Health Programs on Child Mortality in Sub-Saharan Africa*, pp. 74–107. Washington: National Academy Press.

Omondi-Odhiambo, A.M. Vorhoeve, and J.K. van Ginneken, 1984, "Age Specific Infant and Child Mortality and Cause of Death" In *Maternal and Child Health in Northern Kenya: An Epidemiological Study*, edited by K. van Ginneken, A.S. Muller, pp. 213, Sidney, Australia: Croom Helm.

Sigman, Marian, Michael P. Espinosa, and Shannon E. Whaley, 1998, "Mild Malnutrition and the Cognitive Development of Kenyan Schoolchildren." In *Nutrition, Health and Child Development*, edited by PAHO (Pan American Health Organization), pp. 91–103. Washington DC: Pan American Health Organization and the World Bank.

UNICEF, 2001, *Progress since the World Summit for Children: A Statistical Review*. New York: United Nations Children's Fund UNICEF.

UNICEF, 2004, "At a Glance: Kenya—The Big Picture," http://www.unicef.org/infobycountry/kenya.html.

Walker, Susan P., Christine A. Powell, and Sally M. Grantham-McGregor, 1998, Early Childhood Supplementation and Cognitive Development. In *Nutrition, Health and Child Development*, edited by PAHO (Pan American Health Organization), pp. 69–81. Washington DC: Pan American Health Organization and the World Bank.

WHO, 1997, *Health and Environment in Sustainable Development: Five Years after the Earth Summit*, WHO, Geneva: World Health Organization.

WHO, 2003, *State of the World's Vaccines and Immunization*. Geneva: World Health Organization.

Chapter 11

Sedentarization and Seasonality
Maternal Dietary and Health Consequences in Ariaal and Rendille Communities in Northern Kenya

MASAKO FUJITA, ERIC ABELLA ROTH, MARTHA A. NATHAN M.D., AND
ELLIOT FRATKIN

1. INTRODUCTION

In response to increasing population pressure, loss of grazing lands to private farms and ranches, political insecurity, and increased wealth stratification resulting from the commoditization of the pastoral economy, livestock-keeping populations of the world are increasingly shifting to more sedentary subsistence strategies such as agriculture, milk marketing, or wage-earning (Fratkin, 1997, 2001). Sedentarization triggers a variety of changes, simultaneously and complexly affecting maternal diet and health. Maternal health holds a key to the public health because women play a central role in the health of their children and families (Leslie, 1991). We are, however, only beginning to understand the consequences of sedentism for maternal health (Little et al., 1988, 1992; Shell-Duncan and Yung, 2004). Relocation, migration, and subsistence change for indigenous peoples may lead to nondirected or unplanned dietary changes with potential adverse effects on the population's health (for review see Kuhnlein and Receveur, 1996). Research on the nutritional consequence of sedentarization of pastoralists on children supports this negativity (Fratkin et al., 1999a, b; Nathan et al., 1996, Roth et al., this volume, Chapter 9; Shell-Duncan and

MASAKO FUJITA • Department of Anthropology, University of Washington, Seattle, Washington 98195-3100
ERIC ABELLA ROTH • Department of Anthropology, University of Victoria, Victoria, British Columbia, Canada V8W 3P5   MARTHA A. NATHAN M. D. • Brightwood Health Center, Springfield Massachusetts 01107   ELLIOT FRATKIN • Department of Anthropology, Smith College, Northampton, Massachusetts 01063.

Obiero, 2000). Some studies also indicate less favorable health status of settled women, relative to nomadic mothers' health status (for reviews of this argument see Nathan et al., 1996; Sellen, 1996). To date, however, most studies of sedentarization and maternal health are cross-sectional. This is limiting because pastoralists and former-pastoralists occupy ecological settings strongly characterized by high seasonality.

Because pastoral subsistence depends on the seasonal patterns of water and plant availability and corresponding geographic mobility of livestock and humans, seasonality is a crucial factor for understanding pastoralism (Dyson-Hudson and Dyson-Hudson, 1981; Sellen, 1996). Pastoralists are identified as vulnerable to seasonal dietary stresses (Little et al., 1983; Nestel, 1989) and to drought conditions (Ghai et al., 1979). Therefore a seasonal perspective is also crucial to fully understand the health consequences of sedentism. Seasonal fluctuations of vital resources that affect the seasonal pattern of nutritional stresses of pastoralists may be modified during sedentarization due to multiple socioecological factors, including improved access to reliable water sources, food storage facilities, cash, and markets. At the same time, seasonal work patterns are expected to change in the alternative production systems, such as agriculture, adding further complexity to the dietary and health seasonality.

Previous studies indicate that seasonal stresses or other aspects of dietary patterns that may negatively affect health are not limited to pastoralism, and the health issues of settled, formerly pastoral populations are serious, especially for women and children. Fulani, women in settled agricultural community show critical signs of nutritional seasonal stress associated with their agricultural activities (Hilderbrand, 1985). Among the Maasai, entry into a cash economy has not contributed to their dietary breadth for women and children (Nestel, 1986). Finally, for the Turkana, settled women's diets are deficient in protein and other nutrients, which may put them at high risk during pregnancy (Campbell et al., 1999).

Additionally, increased population density in sedentary community may increase the risk of infectious disease. In this regard settled Turkana populations are reported to seldom use pit latrines and are frequently exposed to contaminated drinking water (Brainard, 1990). Furthermore, high levels (self-reported) of malaria in settled Turkana irrigation schemes are a suspected contributor to fetal losses during the first trimester of pregnancy (DeLuca, 1996).

With specific reference to maternal health, sedentism accompanies a host of complex and varied changes in women's lives, for example in their work roles, access to food, cash, market, and healthcare (Fratkin and Smith, 1994, 1995; Hjort, 1990; Smith, 1997, 1998; Talle, 1988). These cultural and socioeconomic changes in turn may affect the health of individual mothers differently depending on their circumstances (i.e., socioeconomic, reproductive, etc.). Sedentarization is not a uniform process (Fratkin and Smith, 1994, 1995), and the health consequences of sedentism on women vary within and between communities (Shell-Duncan and Yung, 2004). Little is known how such complex consequences may manifest in a longer period of time reflecting seasonal variations. To more fully understand the health consequences of subsistence change it is necessary to take a longitudinal perspective and examine variability, although this is logistically more difficult than cross-sectional studies, given both the remoteness and high mobility of pastoralists (Sellen, 1996).

In this chapter, we examine the consequences of the shift from pastoralism to sedentary agriculture on maternal diet and health among the Ariaal and Rendille in northern Kenya. Using dietary, morbidity, and anthropometric data generated by bimonthly repeated surveys in the pastoral community of Lewogoso and the sedentary agricultural community of Songa covering one year period (1994–1995), we attempted to answer two related

Photo 1. Mothers and child in pastoral settlement of Lewogoso.

questions: (1) Does maternal diet and morbidity change with sedentism (2) and, if so, how do these changes affect maternal anthropometry? To address these questions, the two communities are compared in terms of the degree of maternal health disparities within each community and across the seasons. Further, the effects of lactation on maternal health within each community are also considered. The following section delineates the background of the study and possible pathways through which study variables may operate.

2. SEDENTARIZATION: IMPLICATIONS FOR DIET AND HEALTH

To investigate the biosocial concomitants of this process the Rendille Sedentarization Project collected dietary, morbidity, and anthropometric data over a three-year period (1994–7) for five Ariaal and Rendille communities. Here we present an analysis of maternal data collected for two highly diverse communities over a one-year period, September 1994–August 1995 (Fujita, 2000, 2001, 2002). These are the Ariaal and Rendille sedentary highland community of Songa, and the pastoral Ariaal sub-clan community named Lewogoso. We limited the present analysis to this period because it was characterized by the bimodal seasonal rainfall of East Africa, as shown in Figure 1, while the following years witnessed the failure of one or both or these rains, leading to drought conditions. The importance of rainfall in the East African savanna ecosystem for pastoralists can not be overstated, as the amount of livestock products (milk, blood) seasonally available to them depends upon browse or graze levels which are largely regulated by rainfall (cf. Coughenour et al., 1985). Our goal was to examine differences between a sedentary agricultural group and a pastoral group over a year characterized by this bimodal seasonal rainfall pattern, rather than incorporating drought years, in order to establish a baseline for seasonal change in our longitudinal data (Huss-Ashmore et al., 1988; Ulijaszkek and

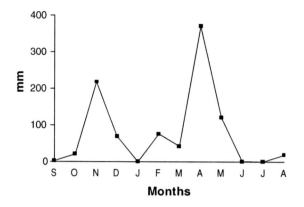

Figure 1. Rainfall: September 1994–August 1995.

Strickland, 1993). The present analysis was limited to a comparison of two of the Project's five communities to develop pilot methodologies for a subsequent analysis incorporating all five communities over the entire three-year period.

For the Ariaal and Rendille, sedentarization yielded a number of beneficial effects, including improved access to drinking water, education, health care, and a market economy (Fratkin, 1998; Roth, 1991; Smith, 1997, 1998). But the process of pastoral sedentarization also resulted in widening disparities in wealth distribution and access to food resources (Fratkin, 1997). Biological concomitants of sedentism for the Ariaal and Rendille include declining nutritional health of children, as evidenced by lower height- and weight-for-age when compared to samples from pastoral populations (Fratkin et al., 1999a, b; Nathan et al., 1996; Shell-Duncan and Obiero, 2000). First and foremost we think these socio-economic and biological differences reflect underlying changing dietary regimes. Pastoral diets generally are characterized as high in protein but low in calories, with marked seasonal variation in both protein and energy content (Galvin, 1985, 1992; Galvin and Little, 1999; Little et al., 1993; Nathan et al., 1996; Nestel, 1986; Shell-Duncan, 1995). The lean seasons for Ariaal and Rendille pastoralists come at the end of the two dry seasons (November–March and May–August) when livestock pasture becomes scarce, in turn limiting both drinking water and milk availability for human consumption (Fratkin, 1991). During dry periods, small stock are increasingly sold to purchase foods including grains (maize meal or *posho*) and other carbohydrates (sugar to mix with tea). Despite this seasonal stress, the milk-based, high-protein diet of pastoralists appears to contribute positively to their biosocial adaptation to a highly seasonal environment with limited resources for dietary energy (Galvin and Little, 1999). The positive ramifications of a pastoralist high-protein diet may be particularly significant for infants, pregnant women, and lactating mothers, because these groups not only are at risk from poor environments (Panter-Brick, 1998) but also have elevated nutrient requirements for dietary protein (Foster, 1992). Since protein is an indispensable nutrient for reproductively active pastoral women as well as for infants and growing children (Galvin and Little, 1999), the potential protein loss associated with agricultural sedentism may have a negative impact on maternal nutritional health. On the other hand, agriculture may buffer seasonality in food availability through the provision of food items that are less perishable and improved access to drinking and irrigation water.

Agricultural produce and sedentary life may facilitate more constant and improved access to cash economy, which may also reduce the seasonal stress associated with a milk-based pastoral diet.

With respect to economic differentiation, a shift in production systems may either improve or deteriorate individual nutritional status when socioeconomic status determines whether one benefits or suffers (Popkin et al., 1993; Shell-Duncan and Obiero, 2000; Shell-Duncan et al., 2001). An earlier study among the Kel Tamasheq in Central Mali (Wagenaar-Brouwer, 1985) demonstrated seasonal fluctuations of nutritional status of adult women differentiated by socioeconomic status. Similarly household economic status was an important factor in predicting seasonal weight fluctuation in a rural agricultural Ethiopian community (Ferro-Luzzi, 1990), while a study of mother-child pairs in Gamboula, a town in western Central African Republic (Andersson and Bergstrom, 1997) found socioeconomic status of mothers, mediated by long-term nutritional situations and living conditions, the most important factor for birth weight. Finally, in a year-long study of correlates of dietary intake of lactating mothers among nomadic pastoral Turkana, Gray (1994) found socioeconomic variables significant predictors of food consumption. For example, socioeconomic ranking of the herding camp was a greater contributor to seasonal variation in milk intake than was rainfall. Additionally, both composition and nutritional quality of maternal diets were altered by women's socioeconomic status; poor mothers consumed relatively higher calorie but poorer protein and micronutrients whereas wealthier mothers consumed lower calories but higher amounts of protein and micronutrients.

In a pastoral community, a family's health risk increases when their resource base is limited, for example, having few animals to trade for grain at times of need. These poorer families may first experience seasonal or drought-invoked nutritional stress. Nonetheless, pastoral families actively engage in reciprocity and exchange within the "moral economy" of redistribution from wealthier to poorer kin in pastoral communities, particularly with regard to milk, potentially minimizing the differences in dietary patterns by wealth status (Fratkin, 1998: 104; Grandin, 1988; Homewood, 1992; Talle, 1988). When pastoralists settle, however, wealth differences among families may increase, potentially weakening the reciprocal kinship-based distributive mechanism of seasonal food scarcity. Such wealth differentiation was evident among former Il Chamus pastoralists of northern Kenya (Little, 1985). Through settlement schemes in 1950's, former pastoralists who individually succeeded as businessmen and local retailers acquired livestock purely for investment purposes and hired local herders to look after their animals. This individualistic "success story" is a clear departure from the reciprocity-oriented and subsistence-based economy of the pastoral sector.

Today in sedentary agro-pastoral and agricultural communities of Ariaal and Rendille, certain families have a wider economic resource base, such as those engaged in the commercial livestock economy and those who take up cash-crop agriculture. This allows them not only to alleviate seasonal fluctuations in food availability but also to widen the variety of food in their diet. Typically, there are contrasting seasonal patterns of nutritional stress between agriculturists and pastoralists. Critical periods for agriculturists coincide with food shortages and high labor demands associated with farming and harvesting during pre-harvesting time (Simondon et al., 1993). Families with sufficient agricultural and/or pastoral resources may be able to even out the seasonal stresses associated with each subsistence mode. By contrast, poorer families who rely on their smaller pastoral or agricultural holdings for their subsistence and cash income are more likely to experience seasonal stresses distinct from wealthier families.

The number of poorer families in settled communities is larger than in pastoral ones, as it is the poor who are more likely to leave pastoral communities in search of alternative subsistence. The degree of dependency on the sale of farm or livestock products is also affected by household economic status. Among Ariaal and Rendille living near urban centers, women are active players who promote and sell surplus milk or vegetables to customers in town. Those who own sufficiently productive resources can successfully and regularly earn cash, whereas the poor without such resources cannot (Fratkin and Smith, 1994). By implication, wealthier women may be more successful in obtaining the stability of both food intake and nutritional quality. Shell-Duncan et al. (2001: 30) documented such cases of intra-community discrepancy among Rendille women in highland settlements where sedentism and development resulted "not only in overall improvements in dietary intake, but also increased disparity in economic status and risk of low dietary intake of milk, a key staple food."

Finally, as outlined previously for settled Turkana (cf. Campbell et al., 1999) sedentary agriculture may result in different morbidity patterns, with increased population density in settled communities giving rise to higher incidence and prevalence of infectious diseases, despite the availability of health facilities and immunization programs in sedentary communities (Nathan et al., 1996).

3. MATERIALS AND METHODS

To examine the above research questions and proposed pathways, this study compared longitudinal data from two Ariaal and Rendille communities. The first is Lewogoso, a subsistence-oriented pastoral community continuing their highly seasonal and mobile lifestyle (Fratkin, 1998). The second is Songa, an irrigation agricultural community of former pastoralists who grow a substantial amount of their produce for market sale, and spend their cash income to purchase the bulk of their diet (Smith, 1997).

3.1. Community 1: Pastoral Lewogoso

Lewogoso is a patrilineal and patrilocal Ariaal community of approximately 50 households (with an average of 5 people/household), located at the base of the Ndoto Mountains in Marsabit District. Lewogoso, like the majority of Ariaal communities, is largely self-sufficient with household herds averaging 12 camels, 20 cattle, and 50 small stock (Fratkin, 1998). Herders from Lewogoso move their animals in independent herding groups orbiting around semi-permanent locations. Boys herd milk camels close to the community so that they may return before nightfall each day, while non-milking camels are taken to distant herding camps (*fora*) for several weeks at a time. In contrast, cattle are herded by warriors and adolescent boys in more distant areas in highland valleys and rarely return to the community except during rainy periods when there is sufficient grazing. Because of this pattern milk from cattle may not be seasonally available to the pastoral community consisting of married men and women, and their small children. Small stock stay within the community, taken care of by women and girls unless it becomes too dry in which case they are grazed in distant camps (Fratkin, 1998).

Residents obtain the majority of their diet from their livestock, primarily in the form of milk, complemented by meat and blood. Due to this dependency, diet in Lewogoso is strongly affected by seasonal fluctuations of rainfall and consequent milk availability.

Photo 2. Tapping blood from goat in Lewogoso.

During the dry season milk becomes increasingly scarce, and blood from camels or meat from slaughtered small stock assume larger portions of the male diet, although these items generally do not contribute to women's diet. As the dietary contribution of milk diminishes during the dry season, people, especially women, consume increasing amounts of maizemeal, tea, and sugar, obtained by selling livestock, mostly cattle, and their skins (Fratkin, 1998).

3.2. Community 2: Sedentary Agricultural Songa

Songa, a permanent agricultural community near Marsabit Town, the district capital, exemplifies the cultural and socioeconomic changes associated with the sedentarization of Kenyan pastoralists. Located in the highland forests of Mt. Marsabit, Songa is a community of 2750 originally established by the African Inland Church in 1973 to provide an alternative means of subsistence for those Rendille who lost their livestock in the prolonged series of droughts in the late 1960s and 1970s (Smith, 1997, 1998, 1999). The drip irrigation system of Songa uses gravity fed pipes, enabling the population to raise a variety of agricultural crops without relying on high maintenance ditches. This reliable water source coupled with rich volcanic soil and stable market demand for agricultural produce from the district capital town of Marsabit, contribute to Songa's successful agricultural economy.

The main crops grown in Songa are *sukuma wiki* (kale), maize, and beans as well as tobacco and *mira'a* (*catha edulis*), a stimulant widely consumed by Muslim Somalis as well as Rendille and Boran of Marsabit Town (Smith, 1997). In theory, men control the production of cash crops such as maize, while women sell small quantities of subsistence garden surplus. In practice, however, men rarely harvest a large enough surplus of maize to sell in the market, while women are more successful in marketing food and stimulant crops (Smith, 1997). Surplus produce from the women's subsistence gardens may be small in

Photo 3. Women of Songa.

quantity at any one time but available all year around, in effect greatly contributing to the household's daily cash income. Everyday, local sellers in the town of Marsabit purchase from Songa most local crops including *sukuma wiki*, tomatoes, oranges, papayas, mangos, bananas, tobacco and *mira'a* (Smith, 1997).

3.3. Data

Data used in the analysis come from a larger data set collected over a three-year period (1994–1997) which compared five Ariaal and Rendille communities of varying degrees of sedentism and economic bases for health, nutritional, and socioeconomic information (Fratkin et al., 1999a, b; Fratkin and Roth, this volume). The Rendille Sedentarization Project surveyed approximately 40 women and their children in each community (total 202 women and 488 children) every two months for the three-year period. Surveys repeatedly measured the same subjects: consisting of forty mothers from pastoral Lewogoso and thirty-eight mothers from agricultural Songa. Survey questions included fertility histories, 24-hour dietary recall, 30-day morbidity recall, and anthropometric measurements. Women were selected on the basis of having a child under six years old. Detailed information on data collection for this original study is given in Nathan et al. (1996, see also Roth et al., Chapter 9 and Nathan et al., Chapter 10, this volume). In the following sections, we explain the data and methodologies for maternal data used in this analysis.

Ages were determined by referral to women's immunization records when possible, and when those were unavailable by reference to a historic events calendar developed and used in previous Rendille studies in consultation with Rendille field assistants. Ages of five mothers from Lewogoso and seven mothers from Songa were unavailable. Maternal ages were non-significantly different ($t = 0.4429$, $p = 0.6591$, mean Lewogoso = 30.1, mean Songa = 30.8) between communities. We used the mean age per sample to assign ages to those women we could not age otherwise. The number of pregnant and lactating mothers changed over the course of the year in both communities, but was roughly equivalent between communities. Breastfeeding patterns were similar for both communities, as was age of weaning, with no children over 24 months still breastfed and only one child under 18 months of age fully weaned.

To reconstruct maternal dietary patterns a detailed 24-hour dietary recall was performed. Each mother was asked to name foods they consumed the previous day. These were recorded separately for morning, afternoon, and evening meals, and counted servings of milk, starch (including cooked maize-meal, termed *posho*, porridge made with milk, or whole maize, rice, or wheat-based *chapati* bread), fat, tea, sugar, fruit (e.g., mango, papaya, bananas) or green vegetables, including local kale (in Kiswahili, *sukuma wiki*). Frequency of servings was reported rather than actual amounts consumed (e.g., calories or volume) which were not possible to observe or otherwise measure. An important exception was estimation of the amount of milk consumed based on standard metal cups widely used in the area, such that "one small cup" was listed as one cup, "one large cup" as 1.5 cups, and milk served with tea or porridge estimated at 0.25 cups. This scheme introduced standardization to the most important Rendille food source.

Famine relief was available to all communities to varying degrees. Maize and beans constituted the bulk of this food aid, distributed by local missions. As this food was available to all study communities—less to the nomadic community because of distance to distribution points—"poor" did not necessarily reflect lack of access to food.

Mothers were weighed using a SECA digital scale. Maternal triceps skinfold thickness (TSF) was measured with a Holtain caliper and mid-arm circumferences (MAC) were

Photo 4. Measuring mother's height, Lewogoso.

obtained via a Roche disposable tape. Eight Ariaal and Rendille assistants were hired for the three year period, working in pairs to conduct the surveys in the same community over the course of the study (one pair surveyed two communities, the others one community each). Initial monitoring of local assistants by the project manager (Fratkin) was repeated at the beginning of each year, and supervised at each survey episode by team supervisor. While we did not undertake formal tests for inter-observer error, for some measurements, exemplified by TSF, observers took three readings and then used the average.

Maternal morbidity data were gathered by asking mothers: (1) the number of days each of them was ill in the past one month and, (2) the category of disease, e.g., diarrhea, fever, and/or respiratory infections ("colds" or "cough"). These three types of illnesses are well known, and mothers showed no problem recalling how many days in a month they were ill.

Families were assigned to one of the three economic strata—poor, middle, and rich—based upon the composite measure of household wealth, livestock holding, wage income, remittances, and household expenditures. For pastoralists, wealth was determined according to the number of animals they owned (indicating both levels of food and access to income) and sold, where pastoral households were classified as "poor" if they owned less than 4.5 Tropical Livestock Units (TLUs) per capita, "middle" if 4.5–7.0 TLU/person, and "rich" if they owned more than 7.0 TLU (1 TLU = one 250 kg cow, 0.8 camel, or 10 goats/sheep, the minimum per capita level necessary for subsistence off livestock products in an arid environment). The agricultural community of Songa was stratified based on monthly cash income (poor <US$10, middle as $10–50, and rich if above $50). The criteria for wealth stratification were confirmed by using "indigenous wealth-ranking" methodology of Grandin (1983), where individuals are ranked by a consensus of village men and women including native enumerators.

3.4. Analytical Methods

Bivariate descriptive analysis of diet first delineated basic dietary patterns in the two communities. Next, to compare seasonal dietary and morbidity patterns for women from both the pastoral and agricultural communities, as well as wealth differentials within each community, a repeated measures multivariate analysis of variance (RM-MANOVA) was conducted. This determined interactions between the communities and economic status over time and estimated the contribution of independent variables—community affiliation, economic status, and seasonality—on each dependent variable. The latter included six major food items/groups (cups of milk and serving-frequencies of starch, sugar, fat, beans, and greens) for dietary analysis and days spent ill (diarrhea, fever, and cold) for morbidity analysis. The economic composition of the samples showed a binary distribution in both communities. However, there was a crucial difference between the communities as shown in Figure 2. About two thirds of agricultural Songa mothers were in the poor stratum, one third was in the middle stratum and there were no rich households. In contrast, over two thirds of Lewogoso mothers were in the middle stratum, and one third was in the rich stratum. There were only three mothers from the poor stratum in pastoral Lewogoso. The economic status of mothers from the two communities may not be comparable because of qualitative differences in the respective economic base. For pastoralists "wealth" consists of livestock, but for the sedentary farmers it consists of land or cash. Still, the classification of mothers by economic status within each community was considered not only valid but also important because such a factor could potentially influence maternal diet and health. RM-MANOVA allows comparison of the two communities in terms of economic differences that may reveal intra-community differences in diet and morbidity over time.

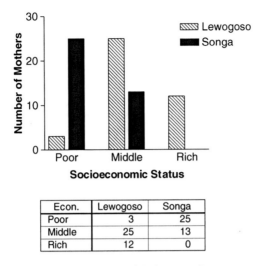

Econ.	Lewogoso	Songa
Poor	3	25
Middle	25	13
Rich	12	0

Figure 2. Sample composition by economic status.

In multivariate repeated analysis, statistical inquiries are two-fold: First, comparison of treatment groups (economic strata) and secondly, comparison within the repeated measures. In statistical terms, the former aspects are referred to as between-subjects effects while the latter are termed within-subject effects (Khattree and Naik, 1999). The hypothesis tested in this analysis also consist of two aspects: (1) whether or not sedentary agriculture increases differences in diet and morbidity, attributable to economic strata and (2) whether or not such differences depend upon seasonality (i.e., whether or not the seasonal pattern in between-economic strata differences within one community is different from the pattern in the other community). The former aspect can be tested in the between-subjects (between-economic strata) effects. The latter can be tested in the within-subject by between-subjects interaction effects, or more plainly, the interaction effects of time and economic difference (hereinafter TIME*ECON effects or interaction effects).

For this analysis, the poor stratum of pastoral Lewogoso was excluded because both communities overwhelmingly featured a binary distribution of economic strata. In addition, the poor stratum of Lewogoso consisted of only three mothers, and statistical results were considered likely to be artifacts of its small sample size. Finally, for statistical comparison, it is desirable to have an equal number of economic strata in the two communities. By eliminating the poor stratum of pastoral Lewogoso, it became possible to compare the two communities, each of which reflected the respective binary distribution of economic strata. This methodology could not account for the differential distribution of wealth in the two communities mentioned above. Nonetheless, it was capable of comparing the two communities in terms of the within-community dietary and morbidity differences attributable to mothers' wealth status differences.

Missing values were imputed as follows. For bounded missing values, occurring with both preceding and subsequent values available, the mean value of these two adjacent values was used. For multivariately missing data, that is, when the missing value had only preceding or last value available, that last value was carried forward. For initial values, the subsequent value was replicated (Solas, 1997). A total of 294 out of 6084 values (4.8%) were imputed in this manner.

Bivariate descriptive analysis of diet and repeated measures MANOVA for dietary and morbidity analyses were conducted using the SAS® System Version 6.12 (1986–1996). Graphs and plots were constructed with the GraphPad Prism Version 2.0 (1994–1995).

Next, to assess the biological ramifications of changing dietary patterns, we analyzed maternal anthropometric measurements via Generalized Estimating Equations (GEE) using the SAS® GENMOD program. GEE was developed by Liang and Zeger (1986) as a regression modeling approach to deal with correlated data, exemplified here by our longitudinal repeated measures methodology. These repeated measures from any one individual or cluster of individuals are correlated with each other and are therefore no longer independent. Generalized Estimating Equations can estimate the correlation between a single individual or cluster's response and provide a correct estimate of each effect's variance. GEE can utilize both continuous and discrete variables, allowing us to include continuous traits (e.g., age, milk intake, days ill) and discrete traits (e.g., pregnant and breastfeeding status) in the same statistical model. In particular, the ability to analyze categorical variables allowed us to include pregnant and/or breastfeeding mothers to our analysis. Otherwise we would be forced to omit these women from analysis because pregnancy and maternal-related bodily changes (cf. Little et al., 1992; Institute of Medicine, 1990, 1996) constitute confounding variables. Such omissions would be particularly debilitating to our sample, which was selected on the basis of having young children. In Rendille society women have on average slightly over six children during the course of their reproductive career (Roth, 1994, 1999), and fertile women spend a considerable amount of time in their reproductive period either pregnant or breastfeeding. This is exemplified in Figure 3, which presents the frequency of women breastfeeding, pregnant or neither for our six sample periods. As can be seen from this figure, excluding breastfeeding and pregnant women in our one-year sample results in different numbers of women per study periods and severely reduces each community's sample size per sampling time. Because of this, we first used GEE analysis for the overall sample and include age, pregnancy and breastfeeding as biologically and reproductively based independent variables which we then compare with diet, seasonality, morbidity, and

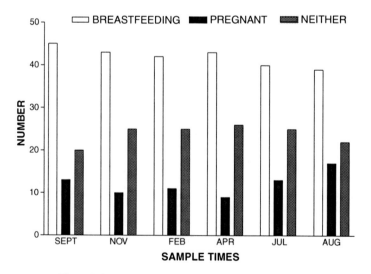

Figure 3. Sample composition by mother's reproductive status.

socio-economic variables. A separate run, based on only breastfeeding women, was then completed, using only the last four variables. The number of pregnant mothers was too small for analysis.

Age was coded as a continuous variable, while pregnancy and breastfeeding were coded as categorical, "Yes/No" variables. Continuous variables denoting average cups of milk and average reported days ill represented dietary and morbidity patterns. Economic strata were dichotomized as "poor" and "middle" for Songa and we combined the few poor families in Lewogoso with those from the "middle" stratum and contrasted this group with the rich households. To account for seasonality, we separated measurements taken during rainy seasons from those collected during dry periods.

GEE analyses were conducted using a battery of derived anthropometric indices as the dependent variable. These included maternal weight, uncorrected arm muscle area (AMA), and arm fat area (AFA). AMA and AFA are derived from measurements of triceps skinfold thickness (TSF) and mid-arm circumference (MAC). AMA was determined from the formula: $\text{AMA (cm}^2) = [\text{MAC} - (\text{TSF} * \pi)]^2/4 * \pi$. AMA represents the volume of the arm uncorrected for (or inclusive of) bone contribution but exclusive of the contribution of sub-cutaneous fat (Callaway et al, 1988; Gurney and Jelliffe, 1973; Howell, 1988). AFA represents the subcutaneous fat contribution, determined as the difference between total upper arm area (TAA) and uncorrected AMA (AFA = TAA − AMA), where $\text{TAA (cm}^2) = (\text{MAC}^2)/4 * \pi$ (Howell, 1988). Finally, body weight reflects muscles, body fat, skeletal supports, and body fluids. AMA and AFA are anthropometric indicators of body composition. Along with body weight, body composition is influenced by one's nutritional health status.

4. RESULTS

4.1. Dietary Comparison by Community

Figures 4a and 4b show food servings for pastoral Lewogoso and agricultural Songa respectively. Diet in Lewogoso consisted predominantly of milk. Mothers' diet showed large seasonal fluctuations in food compositions over the course of the six surveys largely depending upon the availability of milk. Mothers' diet was not diversified when milk was abundant. September 1994 was the time with the highest food servings, but it was not the best time due to the extreme scarcity of milk and consequently unbalanced nutrition. November 1994 and June 1995 were better times with abundant milk as well as generous total food servings (mean = 5.5 servings). The hungry month was August 1995 when milk was scarce and the total food servings recorded were the lowest of the year. Overall, mothers in Lewogoso appear to consume a lot of protein from milk, seasonally supplemented by a host of low-protein, high-calorie foods such as maize, sugar, and fat.

The dietary features of the two communities are summarized in Table 1. Mothers in Songa mostly consumed high-calorie and low-protein foods: maizemeal, sugar, and fat. Throughout the year, these three items occupied the overwhelming majority of mothers' diet. Contributions from milk, beans, and greens were persistently minor. Dietary composition in Songa remained stable, unlike Lewogoso. In this regard, dietary seasonality is considerably reduced in Songa. September 1994 was the time of hunger with the least total servings, while June 1995 was the best time with the most abundant food servings. Fluctuation of dietary intake, as measured by servings was equivalent to that in Lewogoso. Overall mothers in Songa apparently consumed larger amounts of carbohydrates and fat but

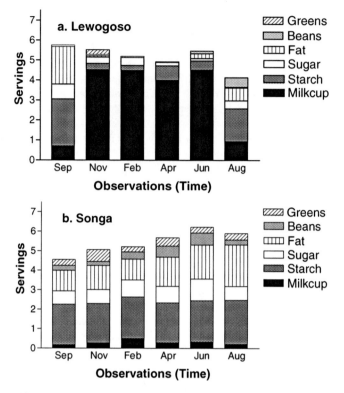

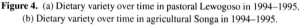

Figure 4. (a) Dietary variety over time in pastoral Lewogoso in 1994–1995.
(b) Dietary variety over time in agricultural Songa in 1994–1995.

Table 1. Dietary Features by Community.

Dietary features	Lewogoso	Songa
Best time	November 1994/June 1995	June 1995
Lean time[1]	September 1994/August 1995	September 1994
Dominant Food Items	Milk	Maize, fat, sugar
Macronutritional Characteristics	High milk-protein and low calorie intake	Lower milk-protein and high calorie intake
Seasonality	Highly seasonal milk intake and dietary variety	Constant intake of maize, sugar, and fat

[1] Lean time defined in terms of milk scarcity.

greatly reduced milk protein, relative to Lewogoso. Small amounts of milk consumed thus made an important protein contribution. Another potentially important difference, discussed in a later section, is that mothers in Songa regularly consumed beans.

4.2. Repeated Measures MANOVA Results for Diet

Table 2a shows results for diet in pastoral Lewogoso. Dependent variables included six major food items/groups (cups of milk and serving-frequencies of starch, sugar, fat,

Table 2. Summary Results for General Linear Models Procedure Repeated Measures Analysis of Variance. Tests of Hypotheses for Between-Subjects (Between-Economic Strata) Effects for Dietary Intake.

Food	Sources	DF[1]	Type III sum of squares	Mean-square	F value	Probability > F
a. Pastoral Lewogoso						
Milk	Economic strata	1	20.8057	20.80560	2.59	0.1162
	Error	35	280.7156	8.02044		
Starch	Econ	1	0.0038	0.00384	0.00	0.9442
	Error	35	27.1178	0.77479		
Sugar	Econ	1	0.0937	0.09370	0.24	0.6280
	Error	35	13.7261	0.39217		
Fat	Econ	1	0.8721	0.87208	1.62	0.2113
	Error	35	18.8261	0.53788		
Beans	Econ	1	0.4865	0.48648	3.41	0.0735
	Error	35	5.0000	0.14285		
Greens	Econ	1	0.0022	0.00216	0.47	0.4962
	Error	35	0.1600	0.00457		
b. Agricultural Songa						
Milk	Econ	1	0.0662	0.0662	0.35	0.5576
	Error	36	6.8025	0.1889		
Starch	Econ	1	0.0863	0.0863	0.23	0.6314
	Error	36	13.282	0.3689		
Sugar	Econ	1	0.9620	0.9620	2.39	0.1310
	Error	36	14.4984	0.4027		
Fat	Econ	1	2.1056	2.1056	3.52	0.0686
	Error	36	21.5128	0.5975		
Beans	Econ	1	8.7452	8.7452	24.56	0.0001
	Error	36	12.8205	0.3561		
Greens	Econ	1	8.6802	8.6802	15.35	0.0004
	Error	36	20.3548	0.5654		

[1] Degree of freedom reflects number of mothers.

beans, and greens) reported in 24-hour dietary recall interviews. The total lack of significant results indicates that economic differences within Lewogoso did not significantly influence the intake levels of any food items.

Table 2b shows the results for agricultural Songa. In contrast to Lewogoso, diet in Songa was influenced by economic status. Specifically, two food items—beans and greens—were highly significantly ($p < 0.001$) differentiated by mothers' economic status. Figures 5a and 5b show profile plots for beans and greens intake by economic status in Songa, respectively. These clearly reveal significant differentiation between economic strata. On average, mothers in the middle stratum had access to reasonable servings of both beans (mean 0.7 and range 0.3–1.0 servings/day) and greens (mean 0.7 and range 0.5–1.4 servings/day). Poor mothers, however, consumed less than half of these servings (mean 0.3 and range 0.1–0.4 servings of beans; mean 0.3 and range 0.1–0.6 servings of greens) all year round. Additionally, poor mothers had especially low access to both beans and greens toward the end of each dry season (February and August, respectively).

Table 3a shows results from the test of the hypothesis for TIME*ECON interaction effects for pastoral Lewogoso. In Lewogoso, both economic difference and seasonality ($p < 0.05$) simultaneously influenced only starch intake. Figure 6 shows time profile plots

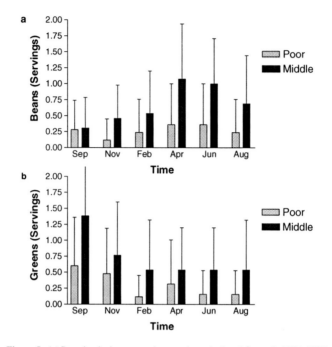

Figure 5. (a) Bean intake by economic status in agricultural Songa in 1994–1995.
(b) Greens intake by economic status in agricultural Songa in 1994–1995.

Table 3. Summary Results for General Linear Models Procedure Repeated
Measures Analysis of Variance for Dietary Intake.[1]

Food	Hypothesis tested	Value	F	Number DF	Dependent variables DF	Probability
a) Pastoral Lewogoso						
Milk	Time*Economic Strata	0.9601	0.2565	5	31	0.9327
Starch	Time*Economic Strata	0.6986	2.6740	5	31	0.0403
Sugar	Time*Economic Strata	0.7625	1.9304	5	31	0.1176
Fat	Time*Economic Strata	0.8514	1.0814	5	31	0.3899
Beans	Time*Economic Strata	0.8580	1.8196	3	33	0.1628
Greens	Time*Economic Strata	0.9866	0.4730	1	35	0.4962
b) Agricultural Songa						
Milk	Time*Economic Strata	0.9618	0.2541	5	32	0.9346
Starch	Time*Economic Strata	0.9192	0.5623	5	32	0.7280
Sugar	Time*Economic Strata	0.8917	0.7769	5	32	0.5737
Fat	Time*Economic Strata	0.9514	0.3264	5	32	0.8933
Beans	Time*Economic Strata	0.7811	1.7925	5	32	0.1426
Greens	Time*Economic Strata	0.8959	0.7430	5	32	0.5971

[1] Dependent variables: Food item from survey time 1–6.

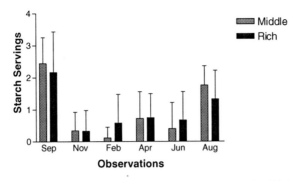

Figure 6. Starch intake by economic status in pastoral Lewogoso in 1994–1995.

for Lewogoso's starch consumption by economic status. Differences here varied noticeably but non-systematically, depending upon the time of the year. Table 3b shows results for agricultural Songa. No statistically significant TIME*ECON effects on any of the food items were found.

4.3. Repeated Measures MANOVA Results for Morbidity

RM-MANOVA on morbidity by economic status produced no statistically significant results (results not shown). This indicates that sedentarization did not translate into different morbidity patterns within communities.

4.4. GEE Results for Anthropometric Indices

Tables 4a and 4b present parameter estimates derived from the biological, reproductive, dietary, morbidity, seasonal, and socio-economic independent variables of the GEE analyses of weight, AMA, and AFA for pastoral Lewogoso (n. of observations = 238) and agricultural Songa (n. of observations = 227). Among the biological and reproductive variables, age and pregnancy were significantly ($p < 0.05$) positively associated with weight in Lewogoso. In Songa, age was also significant ($p < 0.05$) and positively associated with AMA. Among non-biological/reproductive variables, socioeconomic status was significant ($p < 0.05$) and positively associated with AFA in Lewogoso. Milk consumption was a significant positive predictor of AMA in Songa.

Table 5 summarizes results from the GEE analysis of anthropometry with the subsample of breastfeeding mothers in Songa (n. of observations = 117). This examined the significance of non-biological independent variables including milk, days ill, season, and economic differentiation, with once again weight, AMA, and AFA as dependent variables. For Songa, milk was significant ($p < 0.05$) and positively associated with both weight and AMA of breastfeeding women. AMA was also significantly negatively associated with rainfall seasonality. For Lewogoso (n. of observations = 133), no significant variables were found for any of the dependent variables (results not shown).

5. SUMMARY AND DISCUSSION

This study examined longitudinal dietary, morbidity, and anthropometric data for Ariaal and Rendille mothers from two communities, one engaged in pastoral livestock

Table 4. Results of GEE Analysis for Anthropometric Indices, Total Samples.[1]

Independent variables	Dependent variable = weight	Dependent variable = AMA	Dependent variable = AFA
a. Pastoral Lewogoso			
Age	0.2741*	0.2110	0.0899
	(0.1064)	(0.2418)	(0.0802)
Pregnancy	2.8091*	0.8120	−1.5276
	(1.3523)	(5.5117)	(1.2143)
Breastfeeding	1.9856	0.25395	0.8423
	(1.2418)	(3.3816)	(1.2406)
Milk	0.0901	1.3447	0.2141
	(0.1208)	(0.8627)	(0.1606)
Economic strata	−0.7553	2.1258	1.8806*
Rich vs. Middle and	(1.5801)	(4.2551)	(1.2374)
Poor			
Days Ill	0.1706	−0.9106	−0.2520
	(0.1667)	(0.8341)	(0.1288)
Season (Wet)	−0.0706	6.4996	0.8748
	(0.4695)	(4.2235)	(0.7533)
b. Agricultural Songa			
Age	0.1588	0.2451*	0.1955
	(0.122)	(0.0979)	(0.1169)
Pregnancy	2.2795	−1.9372	−2.3160
	(1.6659)	(1.8806)	(1.4954)
Breastfeeding	−2.3725	0.0697	−1.0083
	(1.2799)	(1.4620)	(1.3939)
Milk	1.5696	1.8056*	−0.2000
	(0.8791)	(0.8997)	(0.6899)
Economic strata	0.1805	−1.2297	−0.4004
Middle versus Poor	(1.4873)	(1.4534)	(1.3355)
Days Ill	−0.0698	0.0143	−0.1160
	(0.1463)	(0.2914)	(0.1035)
Season (Wet)	−0.2666	−0.7560	−0.0051
	(0.3640)	(0.5302)	(0.4005)

* = $p < 0.05$.
[1] Standard errors in parentheses.

Table 5. Results of GEE Analysis for Anthropometric Indices,
Breastfeeding Sample from Songa.[1]

Independent variables	Dependent variable = weight	Dependent variable = AMA	Dependent variable = AFA
Milk	1.920*	1.9069*	−0.0131
	(0.8612)	(0.9706)	(0.6681)
Economic strata	0.2170	−1.2243	−1.7938
Middle vs. Poor	(1.5817)	(1.4916)	(0.6475)
Days Ill	0.1314	0.0607	−0.2741
	(0.2123)	(0.3578)	(0.1831)
Season (Wet)	0.3780	−3.4347*	−0.1422
	(0.5465)	(0.6428)	(0.6475)

* = $p < 0.05$.
[1] Number of Observations = 117. Standard errors in parentheses.

production (Lewogoso) and the other practicing sedentary agriculture (Songa). We examined (1) whether and how maternal diet may change across seasons with agricultural sedentism, and (2) how it may affect the variability in maternal anthropometry through both within- and between-community comparisons of variability across seasons.

For the first question of whether diet changes with sedentarization, the answer was yes. Bivariate descriptive analysis indicated that Songa's sedentary agricultural diet was characterized as starch-based rather than milk-based, with a reduced seasonality in terms of dietary composition or variety. RM-MANOVA found no seasonal effects on diet in the sedentary agrarian sample, but revealed a significant seasonal effect in the pastoral group in terms of starch intake. This indicates that in agricultural Songa the effects of economic status on dietary intake did not depend upon the season, reflecting a generally reduced dietary seasonality. In contrast, the RM-MANOVA results for pastoral Lewogoso reflected the highly seasonal nature of pastoral mothers' diet in general, as well as the presence of seasonal differentiation of dietary patterns (starch consumption) attributable to economic status. This methodology further revealed significantly reduced bean and greens intake all year round for poor mothers in agricultural Songa relative to wealthier mothers in the same community. Poor mothers had especially limited access to beans and greens toward the end of each dry season (February and August, respectively). However, no seasonal or economic effects were found for morbidity patterns in either community.

For the second question of how maternal diet may affect within- and between-community variability in anthropometry, we used GEE procedures to test the effects of dietary, socio-economic, environmental and morbidity patterns associated with the transition to sedentary agriculture upon a battery of anthropometric indices. Along with the above factors, our models also included biological and reproductive factors—age, pregnancy, and breastfeeding—which have the potential to alter maternal body composition. The effects of agricultural diet on maternal anthropometry depended on both the household wealth status and reproductive status of mothers. Specifically, GEE analysis showed that age was the only significant biological predictor of maternal anthropometry in both communities (weight in Lewogoso and AMA in Songa). For reproductive variables, pregnancy was positively associated with weight in Lewogoso, but was insignificant in Songa. Breastfeeding was insignificant in both communities.

Among the array of non-biological/reproductive factors, socioeconomic status was a positive predictor of AFA in Lewogoso, while milk was a significant positive predictor of AMA in Songa. The former indicates that mid-arm fat mass was differentiated by socioeconomic status in Lewogoso. This was unexpected given the long-established milk sharing norm among pastoralists (Grandin, 1998; Homewood, 1992; Talle, 1988), and it might potentially indicate that non-milk foods are less readily shared across economic strata among pastoral Ariaal and Rendille, as seen in the seasonally differentiated starch intake patterns (see Table 3a). However, since this economic disparity in starch consumption was only seasonal and non-systematic (see Figure 6), there might be other factors also mediating the association between economic status and AFA of mothers in Lewogoso. We can speculate that one such potential factor may be differential energy expenditure patterns attributable to economic status, in which wealthier mothers have less strenuous workload although we cannot test this in this study.

The latter observation that in Songa higher consumption of milk was a predictor for larger AMA—a crude indicator of body protein—reveals the relative importance of milk nutrients, particularly milk protein (and probably micronutrients) in the agricultural community. Furthermore, additional GEE runs completed including only breastfeeding mothers in

Songa found milk a significant positive predictor for AMA and weight. This reflects an increased importance of milk for anthropometric status of lactating mothers relative to non-lactating mothers.

In addition, we found rainy season significantly inversely associated with AMA of lactating mothers in Songa. This is noteworthy. On the one hand, this is a reflection of breastfeeding mothers' heightened vulnerability to environmental change, relative to non-breastfeeding mothers in the same community. On the other hand, this indicates that reduced dietary seasonality in Songa did not provide a successful buffer for lactating mothers. It is curious that lactating mothers in Lewogoso did not show similar vulnerability despite more marked seasonal volatility in their diet. Overall, these results suggest that maternal diet has complexly changed in agricultural sedentsim, featuring the cereal staple, wider dietary variety, and lesser degrees of seasonal fluctuations, while the effects of such changes on maternal anthropometry depended on both household wealth status and reproductive status of mothers.

However, several methodological limitations on this study must be considered before discussing the ramifications of these results. First, the validity of 24-hour dietary recall data is difficult to determine, and may be limited due to selective recall and daily variation (Witschi, 1990). However, we believe that this study benefited from the strengths of 24-hour recall methodology. For example, the essentially open-ended nature of dietary recall methodology is useful for assessing mean nutrient intakes among groups with different dietary characteristics, and multiple repeated recall sessions are believed to improve rather than deteriorate the accuracy of estimates (Witschi, 1990).

The absence of maternal energy expenditure variables is another limitation. Variations in work and breastfeeding patterns for lactating mothers, for example, may make unmeasured contributions to the observed between- and within-community differences in some anthropometric indices. While mothers in both communities work equally strenuously, many activities have diverged (Fratkin and Smith, 1995; Smith, 1997). For example, mothers in agricultural Songa walk some hours to the capital town to sell garden produce. While pastoral mothers walk long distances to manually collect drinking water and firewood for the household, they also spend substantial time in a sitting position for manufacturing tasks at home, moving just arms and hands. Such activity differences may complexly affect maternal anthropometric values.

Similarly, breastfeeding patterns may shift with new subsistence activities as well as with increased access to cash and store-bought commodities. Such changes may affect the patterns of maternal nutrient expenditure even though our initial survey detected no observable differences in breastfeeding and weaning patterns. In pastoral communities, mothers introduce infants to non-breastmilk liquids and foods within the first months or weeks after birth to supplement or complement breastmilk (Gray, 1996; Sellen, 2002). Herds' milk may be used for such purposes, but in agricultural communities, sugar water, stewed maize, or baby formula may be more readily and constantly available, being less prone to the rainfall seasonality. This availability of non-breastmilk foods may facilitate earlier introduction of non-breastmilk foods or less frequent breastfeeding in an agricultural community (cf. Levine, 2000; Sellen, 1998, 2002). Further, a long hike to the town may physically separate mother and infant for hours, thus leading to infrequent breastfeeding. Regardless of economic bases, weaning patterns may vary depending on the household food supply (cf. Sellen, 2001). Therefore, our interpretation of anthropometric data regarding breastfeeding subsamples may be partially attributable to the above extraneous factors.

Our anthropometric analysis of breastfeeding mothers did not account for the changes in fat distribution associated with age and/or the stages of lactation (Institute of Medicine, 1991). Consequently our inferences regarding lactating mothers are approximate, and we cannot discuss variability within breastfeeding subsamples attributable to maternal age and stage of lactation.

With these limitations in mind, our results still have important implications in the debate about the pros and cons of maternal-child health among sub-Saharan African pastoralists (cf. Sellen, 1996; Little, 1997). The first part of this analysis documented dramatic dietary change for a sedentary agricultural Ariaal and Rendille community, relative to their pastoral counterparts, including dietary differentials attributable to socioeconomic status. Concentrating on breastfeeding women, one of the sub-groups most at risk during times of socio-economic change, the second part of our analysis points to negative consequences for reduced milk consumption for two of our anthropometric indices, weight and AMA. These results suggest that we are starting to see negative biological concomitants of sedentarization for maternal health similar to earlier analyses of child health (cf. Nathan et al., 1996; Fratkin et al., 1999a, b; Roth et al., this volume, Chapter 9; Shell-Duncan and Obiero, 2000).

Nutritional intervention programs would benefit from understanding the within- and between-community variability for identifying and targeting the risk populations and sub-populations and their vulnerable seasons. This study showed that the most striking dietary change for sedentary mothers was the reduction of milk intake, taken over by predominantly plant foods. This has potentially negative consequences of protein and micronutrient deficiencies because milk is a major source of protein, as well as micronutrients such as vitamin A (Nestel, 1986; Kagawa, 2000; McLaren and Frigg, 2001). Poor and lactating mothers were the hardest hit by the reduction of milk intake. Both groups may be at risk of protein deficiency because of their limited access to beans. Breastfeeding mothers were also found to benefit significantly by additional milk and/or bean intake, regardless of their wealth status. The potential reduction of micronutrients such as Vitamin A associated with the reduction of milk intake further puts these two subgroups of mothers at risk of micronutrient deficiency because of their high requirements for micronutrients to sustain their productive and reproductive function and health (Black, 2003; Foster, 1992; McLaren and Frigg, 2001).

Protein and micronutrient deficiencies incur serious health and functional costs such as impaired function and immunity, and increased infectious morbidity and mortality (Black, 2003; Keene, 2001; Dallman, 1987; McLaren and Frigg, 2001; Scrimshaw, 1984). For sedentary communities, possible remedies include increasing animal holdings to obtain more milk or strengthening existing ties to the pastoral sectors of their society to secure better access to milk. However, these suggestions may be unfeasible for an agricultural community like Songa. The former would fail because of the limited number of animals people can keep in the farming community without destruction to their farm and threats to livestock from ticks and other insects in the highlands. The latter is a possibility, but is not likely to lead to an increase of daily supply of milk given the long distance between animal camps and the community.

A more realistic solution for protein nutrient may be to improve the production and distribution of beans. Mixing maize and beans yields balanced amino acids roughly equivalent to animal protein (Foster, 1992). Taking advantage of this process of protein complementarity, beans can serve as a good alternative protein source that Songans can incorporate in their regular diet. This requires an increased production of beans for domestic use rather than for market sale so that local dishes such as *githeri*, made of maize and beans, become more readily available.

While our analysis does not provide estimates of micronutrients, the meager contribution of animal foods in agricultural Songa alone allows us to infer that mothers are essentially dependent on vegetables and fruits for their micronutrient intake. The potential problem of micronutrient inadequacy is complicated by dietary characteristics such as the source of micronutrients and/or the presence of enhancers that help improve micronutrients' bioavailability. For example, both iron and Vitamin A from animal source are more readily bioavailable nutritionally than those from plant sources (McLaren and Frigg, 2001; Davidsson, 2003). Nonetheless, when green leafy vegetables are consumed with fat, as seen in agricultural Songa, the vegetable-source Vitamin A improves its bioavailability (McLaren and Frigg, 2001). Similarly, complementary foods such as ascorbic acid may enhance the absorption of non-heme iron (Davidsson, 2003).

Nutritional consequences of agricultural sedentism are therefore more complex than what we can infer from dietary and anthropometric data alone. On the one hand, maternal diets based predominantly on plant foods in a rural agricultural community may be inherently disadvantaged micronutritionally relative to pastoral diets with high animal food contents. On the other hand, dietary stability and breadth along with large fat intake may partially or fully compensate such a disadvantage. It is important to monitor for clinical signs and biochemical markers for sub-clinical nutritional deficiencies, for example using hemoglobin concentration for iron (Shell-Duncan and McDade, 2004) and retinol or retinol-binding proteins for Vitamin A statuses, to further investigate nutritional consequences.

Sedentary agriculture provides new opportunities for pastoralists in East Africa, particularly for poor families with few animals. Recent research among Maasai of the Ngorongoro Conservation Area of Tanzania shows that maize cultivation is now an essential component of pastoral Maasai diet (McCabe, 2003). Yet, maize based diets are potentially problematic unless maize successfully supplements calories while coexisting with a steady supply of protein and micronutrients (Whitney et al., 1990). Beans and greens are crucial to these settled groups for protein and micronutrients, and their availability is directly affected by household wealth in farm size and productivity as well as season. The relative nutritional importance of these foods also depends on maternal reproductive status. Community policy makers would benefit from paying attention to the within-community disparities in key food items attributable to wealth status.

This study demonstrated differences in food intake within and between pastoral and agricultural communities of the same society (Rendille and closely related Ariaal), and showed the importance of longitudinal research in understanding the consequences of sedentarization for pastoralists of northern Kenya living in highly seasonal environments. Longitudinal data reflecting resource variability across seasons helped understand the seasonal dimensions of maternal dietary and nutritional health consequences in such a way that a cross-sectional study can not address. Using the results as baseline data, we hope to extend our analysis to the remaining two drought years to more realistically understand the maternal dietary and health consequences of sedentary agriculture. Drought is recurring in northern Kenya and is an important factor that imposes unavoidable force on the diet and health of inhabitants in the region. This may also allow us a large enough sample of pregnant women to analyze the effects of sedentarization upon this vital sub-group.

ACKNOWLEDGEMENTS. We are grateful to the Office of the President, Republic of Kenya, for permission to conduct fieldwork in Marsabit District, Kenya between 1994–1999. We also extend our appreciation to residents of Marsabit District who assisted our project, and especially to our field assistants Anna Marie Aliyaro, Larian Aliyaro, Korea Leala,

Daniel Lemoille, and Kevin Smith. We also thank Bettina Shell-Duncan for her comments and suggestions for drafts. This research was supported with grants from the National Science Foundation (SBR-9400145 and SBR-9696088), the Social Sciences and Humanities Research Council of Canada, the Mellon Foundation, the University of Victoria (Canada), Geisinger Medical Foundation, and Smith College, Northampton Massachusetts.

REFERENCES

Andersson, R. and Bergstrom, S., 1997, Maternal nutrition and socio-economic status as determinants of birthweight in chronically malnourished African women. *Tropical Medicine and International Health* 2:1080–1087.

Black, R., 2003, Micronutrient deficiency—an underlying cause of morbidity and mortality. *Bulletin of the World Health Organization* 81:79.

Brainard, J., 1990, Nutritional status and morbidity on an irrigation project in Turkana District, Kenya. *American Journal of Human Biology* 2:153–163.

Callaway, C.W., Chumlea, W.C., Bouchard, C., Himes, J.H., Lohman, T.G., Martin, A.D., Mitchell, C.D., Mueller, W.H., Roche, A.F., and Seefeldt, V.D., 1988, Circumferences. In *Anthropometric standardization reference manual*, edited by T.G. Lohman, A.F. Roche, and R. Martorell, pp. 39–54. Human Kinetics Books, Champaign.

Campbell, B.C., Leslie, P.W., Little, M.A., Brainard, J.M., and DeLuca, M.A., 1999, Settled Turkana. In *Turkana herders of the dry savanna: Ecology and biobehavioral response of nomads to an uncertain environment*, edited by M.A. Little and P.A. Leslie, pp. 333–352. Oxford University Press, New York.

Coughenour, M., Ellis, J., Swift, D., Coppock, D., Galvin, K., McCabe, J., and Hart, T., 1985, Energy extraction and use in a nomadic pastoral ecosystem. *Science* 230:619–625.

Dallman, P.R., 1987, Iron deficiency and the immune response. *The American Journal of Clinical Nutrition* 46:329–334.

Davidsson, L., 2003, Approaches to improve iron bioavailability from complementary foods. *The Journal of Nutrition* 133 (5 Suppl 1):1560S–1562S.

DeLuca, M.A., 1996, Survival analysis of intrauterine mortality in a settled Turkana population. *American Journal of Physical Anthropology* (supp). 22:96–97.

Dyson-Hudson, R. and Dyson-Hudson, N., 1981, Nomadic pastoralism. *Annual Review of Anthropology* 9:15–61.

Ferro-Luzzi, A., 1990, Social and public health issues in adaptation to low energy intakes. *The American Journal of Clinical Nutrition* 51:309–315.

Foster, P., 1992, *The World Food Problem*. Lynne Rienner Publisher, Boulder.

Fratkin, E., 1991, *Surviving Drought & Development: Ariaal Pastoralists of Northern Kenya*. Westview Press, Boulder.

Fratkin, E., 1997, Pastoralism: Governance and development issues. *Annual Review of Anthropology* 26:235–261

Fratkin, E., 1998, *Ariaal Pastoralists of Kenya: Surviving Drought and Development in Africa's Arid Lands*. Allyn & Bacon, Boston.

Fratkin, E., 2001, East African pastoralism in transition: Maasai, Boran, and Rendille cases. *African Studies Review* 44 (3):1–25.

Fratkin, E., Nathan, M.A., and Roth, E.A., 1999a, Health consequences of pastoral sedentarization among Rendille of northern Kenya. In *The Poor are not us: Poverty & Pastoralism in Eastern Africa*, edited by D.M. Anderson and V. Broch-Due, pp.149–162. Ohio University Press, Athens.

Fratkin, E., Roth, E.A., and Nathan, M.A., 1999b, When nomads settle: The effects of commoditization, nutritional change, and formal education on Ariaal and Rendille pastoralists. *Current Anthropology* 40: 720–735.

Fratkin, E., Smith, K., 1994, Labor, livestock and land: The organization of pastoral production. In *African Pastoralist Systems: An Integrated Approach*, edited by E. Fratkin, K.A. Galvin, and E.A. Roth, pp. 91–112. Lynne Rienner Publishers, Boulder.

Fratkin, E., Smith, K., 1995, Women's changing economic roles with pastoral sedentarization: Varying strategies in alternate Rendille communities. *Human Ecology* 23:433–454.

Fujita, M., 2000, *Sedentarization, Seasonality and Economic Differentiation: A Preliminary Analysis of Maternal Diet and Health in Ariaal-Rendille Communities in Northern Kenya*. Unpublished paper presented at the 99th Annual Meeting of American Anthropological Association.

Fujita, M., 2001, *New diet and new body: Negative consequences of dietary change in a young agricultural community in northern Kenya.* Unpublished paper presented at the Annual Meeting of the Canadian Anthropology Society.

Fujita, M., 2002, *Sedentarization, seasonality, and economic differentiation: maternal diet and health in Ariaal-Rendille communities in northern Kenya.* Master's Thesis, University of Victoria, Victoria, Canada.

Galvin, K., 1985, *Food procurement, diet, activities and nutrition of Ngisonyonka Turkana pastoralists in an ecological and social context.* Ph.D. dissertation, State University of New York, Binghampton.

Galvin, K., 1992, Nutritional ecology of pastoralists in dry tropical Africa. *American Journal of Human Biology* 4:209–221.

Galvin, K. and Little, M.A., 1999, Dietary intake and nutritional status. In *Turkana Herders of the Dry Savanna: Ecology and Biobehavioral Response of Nomads to an Uncertain Environment*, edited by M.A. Little and P.A. Leslie, pp. 125–145. Oxford University Press, New York.

Ghai, D., Godfrey, M., and Lisk, F., 1979, *Planning for Basic Needs in Kenya.* International Labour Organization, United Nations, Geneva.

Grandin, B.,1983, The importance of wealth effects on pastoral production: A rapid method for wealth ranking. In *Pastoral Systems Research in sub-Saharan Africa*, edited by International Livestock Centre for Africa, pp. 237–256. International Livestock Centre for Africa, Nairobi.

GraphPad Prism Version 2.0., 1994–1995, GraphPad Software, Inc. San Diego, CA. U.S.A.

Gray, SJ., 1994, Correlates of dietary intake of lactating women in south Turkana. *American Journal of Human Biology* 6:369–383.

Gray, SJ., 1996, Ecology of weaning among nomadic Turkana pastoralists of Kenya: Maternal thinking, maternal behavior, and human adaptive strategies. *Human Biology: An International Record of Research* 68:437–465.

Gurney, J.M. and Jelliffe, D.B., 1973, Arm anthropometry in nutritional assessment: nomogram for rapid calculation of muscle circumference and cross-sectional muscle and fat areas. *American Journal of Clinical Nutrition* 26:912–915.

Hilderbrand, K., 1985, Assessing the components of seasonal stress amongst Fulani of the Seno-Mango, central Mali. In *Population, Health and Nutrition in the Sahel: Issues in the Welfare of Selected West African communities*, edited by A.G. Hill, pp. 254–288. KPI, London.

Hjort, A., 1990, Town-based pastoralism in eastern Africa. In *Small town Africa: Studies in Rural-urban Interaction*, edited by J. Baker, pp. 143–160. Scandinavian Institute of African Studies, Uppsala.

Homewood, K.M., 1992, Development and the ecology of Maasai pastoralist food and nutrition. *Ecology of Food and Nutrition* 29:61–80.

Howell, W.H., 1998, Anthropometry and body composition analysis. In *Contemporary Nutrition Support Practice: A Clinical Guide*, edited by L.E. Matarese and M.M. Gottschlich, pp. 33–46. W.B. Saunders Company, Philladelphia.

Huss-Ashmore, RA., 1993, Agriculture, modernization and seasonality. In *Seasonality and Human Ecology*, edited by S.J. Ulijaszek and S.S. Strickland, pp. 202–219. Cambridge University Press, Cambridge.

Huss-Ashmore, R.A., Curry, J., and Hitchcock, R.K., 1988, *Coping with seasonal constraints.* The University Museum, University of Pennsylvania, Philadelphia.

Institute of Medicine, 1990, Body composition changes during pregnancy. In *Nutrition During Pregnancy*, edited by Subcommittee on Nutritional Status and Weight Gain During Pregnancy, pp. 121–136. National Academy Press, Washington, D.C.

Institute of Medicine, 1991, Summary, conclusions, and recommendations. In *Nutrition During Lactation*, edited by Subcommittee on Lactation Committee on Nutritional Status During Pregnancy and Lactation, Food and Nutrition Board, pp. 4–5. National Academy Press, Washington, D.C.

Institute of Medicine, 1996, *WIC nutrition risk criteria: A scientific assessment.* Committee on Scientific Evaluation of WIC Nutrition Risk Criteria. National Academy Press, Washington, D.C.

Kagawa, Y., 2000, *Shokuhin Seibunhyo (Standard Tables of Food Composition in Japan.) 2000.* Joshi Eiyo-Daigaku Shuppanbu, Tokyo.

Keene, A.M., 2001, Diagnostic assessment. In *Medical-surgical Nursing: Clinical Management for Positive Outcomes*, edited by J.M. Black, J.H. Hawks, and A.M. Keene, p.214. W.B. Saunders, Philadelphia.

Khattree, R. and Naik, D.N., 1999, *Applied multivariate statistics with SAS software*, 2nd Edition. SAS Institute Inc., Cary.

Kuhnlein, H.V. and Receveur, O., 1996, Dietary change and traditional food systems of indigenous peoples. *Annual Review of Nutrition* 16:417–442.

Leslie, J., 1991, Women's nutrition: the key to improving family health in developing countries? *Health Policy and Planning* 6(1):1–19.

Levine, N.E., 2000, Women's work and infant feeding; a case from rural Nepal. In *Nutritional Anthropology: Biocultural Perspectives on Food and Nutrition*, edited by A.H. Goodman, D. Dufour, and G. Pelto, pp. 320–332. McGraw-Hill. XXXX.

Liang, K.Y. and Zeger, S., 1986, Longitudinal data analysis using generalized linear models. *Biometrika* 73:13–22.

Little, M.A., 1997, Adaptability of African pastoralists. In *Human Adaptability: Past, Present and Future*, edited by J. Ulijaszek and R, Huss-Ashmore, pp. 29–61. Oxford University Press, Oxford.

Little, M.A., Galvin, K., and Leslie, P., 1988, Health and energy requirements of nomadic Turkana pastoralists. In *Coping with Uncertainty in Food Supply*, edited by I. de Garine and G.A. Harrison, pp. 288–315. Oxford University Press, Oxford.

Little, M.A., Gray, A., and Leslie, P., 1993, Growth of nomadic and settled Turkana infants of north-west Kenya. *American Journal of Physical Anthropology* 92:335–344.

Little, M.A., Leslie, P.W., and Campbell, K.L., 1992, Energy reserves and parity of nomadic and settled Turkana women. *American Journal of Human Biology* 4:729–738.

Little, M.A., Galvin, K., and Mugambi, M., 1983, Cross-sectional growth of nomadic Turkana pastoralists. *Human Biology: An International Record of Research* 55:811–830.

Little, P.D., 1985, Absentee herd owners and part-time pastoralists: The political economy of resource use in northern Kenya. *Human Ecology* 13:131–151.

McCabe, J.T., 2003, Sustainability and livelihood diversification among the Maasai of Northern Tanzania. *Human Organization* 62(2):100.

McLaren, D.S. and Frigg, M., 2001, *Sight and life manual on Vitamin A deficiency disorders (VADD)*. Task Force SIGHT and LIFE, Basel.

Nathan, M.A., Fratkin, E.M., and Roth, E.A., 1996, Sedentism and child health among Rendille pastoralists of northern Kenya. *Social Science and Medicine* 43:503–515.

Nestel, P., 1986, A society in transition: Developmental and seasonal influences on the nutrition of Maasai women and children. *Food and Nutrition Bulletin* 8:2–18.

Nestel, P., 1989, Food intake and growth in the Maasai. *Ecology of Food and Nutrition* 23:17–30.

Panter-Brick, C., 1997, Biological anthropology and child health. In *Biosocial perspectives on children*, edited by C. Panter-Brick, pp. 66–101. Cambridge University Press, Cambridge.

Popkin, B.M., Keyou, G., Fengying, Z., Guo, X., Ma, H., and Zohoori, N., 1993, The nutrition transition in China: A cross-sectional analysis. *European Journal of Clinical Nutrition* 47:333–346.

Roth, E.A., 1991, Education, tradition, and household labor among Rendille pastoralists of N. Kenya. *Human Organization* 50:136–141.

Roth, E.A., 1994, Reexamination of Rendille population regulation. *American Anthropologist* 95:597–611.

Roth, E.A., 1999, Proximate and distal demographic variables among Rendille pastoralists. *Humam Ecology* 27:517–536.

SAS System for Windows Version 6.12., 1989–1996, SAS Institute Inc., Cary, NC.

Scrimshaw, N.S., 1984, Functional consequences of iron deficiency in human populations. *Journal of Nutritional Science and Vitaminology* 30: 47–63.

Sellen, D.W., 1996, Nutritional status of Sub-Saharan African pastoralists: A review of the literature. *Nomadic Peoples* 39:107–134.

Sellen, D.W., 1998, Infant and young child feeding practices among African pastoralists: the Datoga of Tanzania. *Journal of Biosocial Science* 30:481–499.

Sellen, D.W., 2000, Seasonal ecology and nutritional status of women and children in a Tanzanian pastoral community. *American Journal of Human Biology* 12:758–781.

Sellen, D.W., 2001, Weaning, complementary feeding and maternal decision making in a rural east African pastoral population (Datoga). *Journal of Human Lactation* 17:233–244.

Sellen, D.W., 2002, Anthropological approaches to understanding variation in breastfeeding and application to promotion of "baby friendly" communities. *Nutritional Anthropology* 25:9–29.

Shell-Duncan, B., 1995, Impact of seasonal variation in food availability and disease stress on the health status of nomadic Turkana children: A longitudinal analysis of morbidity, immunity, and nutritional status. *American Journal of Human Biology* 7:339–355.

Shell-Duncan, B. and T. McDade, 2004. Use of combined measures from Capillary Blood to Assess Iron Deficiency in Rural Kenyan Children. *The Journal of Nutrition* 134:384–387.

Shell-Duncan, B., Obiero, W.O., 2000, Child nutrition in the transition from nomadic pastoralism to settled lifestyles: Individual, household, and community-level factors. *American Journal of Physical Anthropology* 113:183–200.

Shell-Duncan, B. and Yung, S.A., 2004, The maternal depletion transition in northern Kenya: the effects of settlement, development and disparity. *Social Science and Medicine*. 58(12): 2485–2498.

Shell-Duncan, B., Muruli, L., and Obiero, W.O., 2001, *The demographic and health consequences of sedentariza-tion among the Rendille of northern Kenya*. Unpublished report to the Government of Kenya, Office of the President.

Simondon, K.B., Bénéfice, E., Simondon, F., Delaunay, V., and Chahnazarian, A., 1993, Seasonal variation in nutritional status of adult and children in rural Senegal. In *Seasonality and Human Ecology*, edited by S.J. Ulijaszek and S.S. Strickland, pp. 166–183. Cambridge University Press, Cambridge.

Smith, K., 1997, *From livestock to land: The effects of agricultural sedentarization on pastoral Rendille and Ariaal of northern Kenya*. Ph.D. Dissertation, The Pennsylvania State University.

Smith, K., 1998, Farming, marketing, and changes in the authority of elders among pastoral Rendille and Ariaal. *Journal of Cross-Cultural Gerontology* 13:309–332.

Smith, K., 1999, Economic transformation and changing work roles among pastoral Rendille and Ariaal of northern Kenya. *Research in Economic Anthropology* 20:135–161.

Sobania, N., 1988, Pastoralist migration and colonial policy: a case study from northern Kenya. In *The ecology of Survival: Case Studies from Northeast African History*, edited by D.H. Johnson and D.M. Anderson, pp. 219–240. Lester Crook Academic Publishing, London.

SOLAS, 1997, *SOLAS for missing data analysis 1.0: The solution for missing values in your data*. Statistical Solutions Ltd., Cork.

Talle, A., 1988, *Women at a loss*. Stockholm studies in social anthropology No. 19. University of Stockholm, Stockholm.

Ulijaszek, S.J. and Strickland, S.S., 1993, *Seasonality and human ecology*. Society for the study of human biology symposium 35. Cambridge University Press, Cambridge.

Wagenaar-Brouwer, M., 1985, Preliminary findings on the diet and nutritional status of some Tamasheq and Fulani groups in the Niger delta of Central Mali. In *Population, Health and Nutrition in the Sahel: Issues in the Welfare of Selected West African Communities*, edited by A.G. Hill, pp. 226–253. KPI, London.

Whitney, E.N., Hamilton, E.M.N., and Rolfes, R.S., 1990, *Understanding Nutrition*. 5th Edition. West Publishing Co., St. Paul.

Witschi, J.C., 1990, Short-term dietary recall and recording methods. In *Nutritional Epidemiology*, edited by W. Willet, pp. 52–68. Oxford University Press, Oxford.

Chapter 12

Development, Modernization, and Medicalization

Influences on the Changing Nature of Female "Circumcision" in Rendille Society

BETTINA SHELL-DUNCAN, WALTER OBUNGU OBIERO, AND
LEUNITA AUKO MURULI

1. INTRODUCTION

Female "circumcision" is a practice involving the partial or complete removal of the external female genitalia. Although precise figures are difficult to obtain, a recent study
has estimated that 132 million women worldwide have experienced some form of genital
cutting (Toubia and Izette, 1998). Although this custom has been practiced for thousands of
years in some parts of Africa, only recently has it obtained enormous international attention, and become a topic having little parallel in its ability to arouse an emotional response.

The term "female circumcision" is a euphemistic description for what is really a
variety of procedures for altering female genitalia, and can be categorized into 4 main
types. The least extensive type, and only one that can be construed as analogous to male
circumcision, is often referred to as *sunna* (Arabic for "tradition" or "duty"), and involves
the removal of the clitoral prepuceot hood. In the majority of cases categorized as sunna,
the clitoral prepuce is removed with all or part of the clitoris as well. Therefore, in the medical literature, this form is often referred to as clitoridectomy (Toubia, 1994), and the World
Health Organization (WHO) classifies this form a Type I. A more extensive version of this
procedure entails clitoridectomy accompanied by the removal of the labia minora as well.

BETTINA SHELL-DUNCAN • Department of Anthropology, University of Washington, Seattle, Washington
98195   WALTER OBUNGU OBIERO • Family Health International, Arlington, Virginia 22201
LEUNITA AUKO MURULI • Institute of African Studies, University of Nairobi, Nairobi, Kenya.

This procedure is referred to as *excision*, or Type II by the WHO. The most radical form of female circumcision is known as Pharonic circumcision or *infibulation*. This practice involves complete removal of the clitoris, labia minora and labia majora, and stitching together the two cut sides so as to cover the urethra and vaginal opening, leaving only a minimal opening for the passage of urine and menstrual blood. Commonly a small stick is inserted to maintain the opening, and the legs of the girl are bound together to promote healing. In areas where some degree of medicalization has taken place, antibiotics and anesthesia may be used, and the opening may be stitched using catgut or silk suture rather than thorns. In the Sudan there is a variation known at *matwasat*, or "intermediate circumcision," which involves the same amount of cutting, but stitching together only the anterior two-thirds of the outer labia, leaving a larger posterior opening (Toubia, 1993). Both *matwasat* and infibulation are classified by the WHO at Type III.

Apart from these main types of genital cutting, some lesser-known variations have been reported (collectively referred to as Type IV by the WHO). These include *hymenectomy*; *zur-zur cuts* of the cervix, which are intended to remedy obstructed labor; *gishiri cuts*, which involve the cutting of the vaginal wall; and "symbolic circumcision", which involved nicking the clitoris with a sharp instrument to cause bleeding but no permanent alteration of the external genitalia (Shell-Duncan and Hernlund, 2000).

The general term "female circumcision" is often used to refer collectively to these procedures. With the spread of feminist consciousness and the development of an international women's health movement, objection to this term has been voiced, and the term female genital mutilation (FGM) has been offered as a more accurate name. The use of the word "mutilation" has, however, been criticized by African women's groups because it is thought to imply excessive judgment by outsiders and insensitivity toward individuals who have undergone the procedure (Eliah, 1996). The term "female genital cutting" has been proposed as a more neutral, less value-laden term. I agree that the term "mutilation" denotes condemnation and will use the terms female genital cutting, the euphemism female "circumcision" (with quotations to acknowledge the imprecision of this term), or the more precise descriptive term for each procedure: clitoridectomy, excision, and infibulation.

Some of the most contentious debates surrounding the practice of female "circumcision" emanate from the health risks associated with genital cutting procedures. The medical consequences of female genital cutting are central in two prominent—yet contradictory—arguments. On one hand, by emphasizing that female genital cutting exposes women to unnecessary, and often severe, health risks, a medical argument forms the foundation of most anticircumcision campaigns. On the other hand, any efforts to minimize the health risks through medical interventions are strongly opposed by anticircumcision advocates, based on the belief that medicalization counteracts efforts to eliminate the practice.

In the anticircumcision literature, fragments of information from different types of genital cutting, performed under widely varying conditions, are melded and repeated as the "medical sequelae."[1] As variations in the practice (degree of cutting, training of the circumcisor, sanitary condition, degree of medical support) are obliterated, presented is a seemingly objective, scientific discussion of the medical "facts" of a single practice—"genital mutilation." This discussion is often divided into three categories: short-term, long-term, and obstetrical consequences.

Short-term complications include hemorrhage, severe pain, local and systemic infection, shock from blood loss and, potentially, death. Infection is associated with delayed healing and the formation of keloid scars. In addition, pain and fear following the procedure can lead to acute urinary retention. *Long-term complications* are said to be associated more often with infibulation than with excision or clitoridectomy (Toubia, 1993), although this has been poorly researched. Possible long-term complications include genito-urinary problems, such as difficulties with menstruation and urination, which result from a near-complete sealing off of the vagina and urethra. Untreated lower urinary tract infections can ascend to the bladder and kidneys, potentially resulting in renal failure, septicemia, and death. Chronic pelvic infections can cause back pain, dysmenorrhoea (painful menstruation), and infertility. Another frequently mentioned complication is the formation of dermoid cysts, resulting from embedding epithelial cells and sebaceous glands in the stitched area. Additionally, if the clitoral nerve is trapped in a stitch or in the scar tissue, a painful neuroma (tumor of neural tissue) can develop. All forms of female genital cutting are alleged to be potentially associated with diminished sexual pleasure and, in certain cases, inability to experience a clitoral orgasm. Infibulated women may experience painful intercourse, and often have to be cut open for penetration to occur at all. *Obstetrical complications* are most often reported in association with infibulation. These include obstructed labor, excessive bleeding from tearing and de-infibulation during childbirth. Obstructed labor may lead to the formation of vesico-vaginal and recto-vaginal fistulae (opening between the vagina and the urethra or rectum, allowing for urine or feces to pass through the vagina). Some researchers have suggested that increased obstetrical risk exists for excised women as well (e.g., Epelboin and Epelboin, 1981). Scar tissue may contribute to obstructed labor since fibrous vulvar tissue fails to dilate during contractions, and hemorrhage may result from tearing through scar tissue. The occurrence of obstetrical complications in excised women has, however, never been systematically evaluated.

Further research is needed to understand the full range of health consequences of various forms of female circumcision for many reasons. One pressing reason is that health risks resulting from medically unnecessary genital cutting procedures are the cornerstone of arguments for opposition of the practice of all forms of female circumcision. Under the impetus of an international women's health movement, female circumcision has been targeted to be "eradicated" as though it were a disease. One outspoken advocate has written:

> "Genital mutilation should be treated as a public health problem and recognized as an impediment to development that can be prevented and eradicated much like any disease."
> (Hosken, 1978: 155)

Paradoxically, those who emphasize female "circumcision" as a public health issue at the same time oppose any medical intervention designed to minimize health risks and pain for women being cut. Medical interventions have been attempted in various forms. In some regions local health workers promote precautionary steps, such as the use of clean sterile razors on each woman, and dispensing prophylactic antibiotics and anti-tetanus injections. Other regions have incorporated training on genital cutting as part of traditional birth attendant programs (Van Der Kwaak, 1992). Traditional circumcisers are encouraged to perform the milder forms of genital cutting, and are given training on anatomy on septic procedures. In addition, in some areas, genital operations take place in clinics or hospitals by trained nurses and physicians (see, for example, Orubuloye et al., 2000, for a discussion of this trend in Nigeria). The impact of these programs on the health of women in rural communities has received surprisingly little attention. Without consideration of health improvements resulting from these training programs, this approach has been strongly criticized. Opponents argue that incorporation of female genital cutting procedures in the biomedical healthcare system institutionalizes the custom, and counteracts efforts to eliminate the practice of female "circumcision" (Gordon, 1991; Toubia, 1993). The position of the World

Health Organization is that all forms of female "circumcision" should be abolished without intermediate stages (WHO, 1982). The purely medical arguments for abolition of even the mildest forms of female "circumcision" lose weight when health risks of female circumcision are minimized.

A less controversial approach has been to promote change in values and attitudes toward female "circumcision" as part of a larger process of social change. This approach, termed "development and modernization" (van der Kwaak, 1992), suggests that improvements in socioeconomic status and education, particularly for women, will have far-reaching social effects, including a decline in the demand for female circumcision.[2] By analogy, such changes have been shown to have an important effect on morbidity and mortality patterns (van der Kwaak, 1992). Recognizing that circumcision is in many African societies a prerequisite for marriage, and that marriage and children are vital aspects of women's roles and economic survival, helps contextualize this practice in a larger social setting. Hence, it is believed that effective change can occur only in the context of a women's movement directed at the social inequality of women, particularly economic dependency, educational disadvantages, and limited employment opportunities (Gruenbaum, 1982). It has been argued, however, that changing social conditions will not automatically change strongly held beliefs and values on female "circumcision"; it is still important to convince men and women that female "circumcision" has an overall negative effect on their lives (van der Kwaak, 1992). It is also recognized that changing traditions and values is possible, but will not occur quickly.[3] Lessons can be learned from family planning initiatives in Kenya and other parts of sub-Saharan Africa, demonstrating that change in deep seated values, such as high fertility and large family size, can occur. However, the process of change is slow. Family planning programs were underway in Kenya for more than 20 years before taking hold (Brass et al., 1993). There is no reason to assume that attitudes toward female "circumcision" would change more quickly. Therefore, we believe that short-term "intermediate" solutions, including medical support for female "circumcision," need to be at least considered and evaluated.

2. FEMALE "CIRCUMCISION" AMONG THE RENDILLE OF NORTHERN KENYA

The purpose of this study is to examine the context and consequences of female "circumcision" among a non-Islamic African group in northern Kenya, the Rendille. Traditionally, the Rendille subsist in the Kaisut Desert of northern Kenya as nomadic herders of mainly camels, and to a lesser degree cattle and small stock (goats and sheep). The settlement pattern of the Rendille began to change when a series of droughts in the 1970s and 1980s diminished large portions of Rendille livestock (Fratkin, 1991). The Rendille were forced to settle alongside mechanized waterholes which were also sites of famine relief food distribution (Fratkin, 1991). Today more than one-half of the estimated 22,000 Rendille people are settled in towns forming around these permanent waterholes (Fratkin, 1991). The lifestyles in newly formed towns vary considerably, representing a broad range of development in terms of infrastructure, market integration, and educational and economic opportunities.

This setting provides a unique opportunity to examine the manner in which the practice of female "circumcision" and surrounding attitudes change as a society experiences the transformations brought about by settlement and development. The people here share a common language and common cultural beliefs and practices, yet participate in widely varying social and production systems (Nathan et al., 1996). Hence, it is possible to examine whether and how the practice and perceived value of female "circumcision" has altered in segments of this society that have experienced dramatic change. Noteworthy, however, is the fact that even in the most "urbanized" location, Marsabit Town, which is the capital of Marsabit District, the level of development lags far behind that of central Kenya. Because of the rugged terrain, lack of any paved roads, and threat of banditry, this northern region is quite isolated from the rest of Kenya. Additionally, because this region is sparsely populated as well as remote, it is commonly overlooked in national surveys and given low priority in national development efforts.

Another interesting feature of this study setting is that it allows us to preliminarily evaluate the effect of medical support on the health consequences of female circumcision. In the settled sector of the Rendille community health workers have in recent years been promoting the use of sterile razors, antibiotics and anti-tetanus injections. A comparison of the long- and short-term health consequences following medically assisted versus unassisted operations will provide insight to the degree to which minimal medicalization can alter the health risks of female genital cutting.

2.1. The Rendille Maternal and Child Demographic and Health Study

Data on female circumcision were gathered as part of a larger demographic and health survey of women ages 15 and over in five distinct Rendille communities. A questionnaire was constructed to obtain information on socioeconomic characteristics, reproductive history, obstetrical outcomes, and female "circumcision." This questionnaire was pretested once in 1994, and again in 1995. Interviewers were trained by Walter Obiero, and data collection was completed in 1996. We first interviewed women in the traditional nomadic pastoralist sector of the population, and obtained information that could serve as a baseline frame of reference for comparison with various settled communities. We then randomly sampled women in four different settled communities:

1. The District Capital, Marsabit Town, which is located atop an isolated mountain in the Kaisut Desert. Many women here are involved in wage labor or small businesses, such as working at the post office or in local markets.
2. The town of Songa, which is also atop Mt. Marsabit. It is inhabited by approximately 2,000 people who practice irrigation agriculture in a mission-sponsored project in Marsabit National Forest. Women here grow and sell vegetables at the Marsabit market. We expected to see the greatest amount of difference in beliefs, values and practices of female circumcision in these two communities in comparison to the nomadic community because their lifestyles and subsistence strategies differ the most.
3. The third community we surveyed is Karare, a town on the slopes of Mt. Marsabit where people subsist by either sedentary cattle keeping and milk marketing, or by dryland horticulture.
4. Finally, we visited a lowland desert town called Korr which emerged when the Catholic Mission dug permanent mechanized waterholes and began distributing famine relief food in the 1970s. Some people here are involved in market trade, but

as is the case with Karare, many people still rely on animal production for subsistence, and seemingly have a greater degree of connection with the nomadic sector of the population. We therefore expected changes in the practice of female circumcision to be intermediate in these locations.

Characteristics of survey respondents are shown in Table 1. A total of 920 Rendille women ages 15 to 76 years were surveyed across the five communities in Marsabit District. Respondents from each community shared similar socioeconomic characteristics. Overall 16.1% of women were from poor families.[4] The majority of respondents have never attended school and are illiterate. Of those who attended school, less than 1% continued their education beyond primary school. The majority of women practice the traditional Rendille religion, although in the towns of Korr, Karare and Marsabit where Catholic missions are present, a large number of Catholics are found. The majority of respondents are ever married women (91.6%) and have given birth to at least one child (95.4%). As a follow-up to the survey we returned to Marsabit District in 1997 to conduct in-depth interviews and focus group discussion on the cultural context of female "circumcision." Focus group discussions were stratified according to marital status and gender as follows: unmarried teenage boys and girls, men and women married for less than ten years, men and women married for more than ten years. Each group was split by gender to facilitate open discussions. Interviews were also conducted among traditional circumcisers (usually in the course of observing the circumcision ritual), and local healthcare personnel. This qualitative information allowed us to develop an understanding of gender and intergenerational differences in perspective, and factors promoting change.

2.2. Cultural Context of Female "Circumcision" among the Rendille

In the Rendille language, the term *khandi* is used to refer to the circumcision of both men and women. Today excision is the form widely practiced among Rendille women. In contrast to other regions of Kenya, discussion about this practice are not shrouded in secrecy. In 1994 before conducting the first pre-test of the survey questionnaire, the senior author (BSD) inquired about the sensitivity of raising questions on female "circumcision." My assistant assured me that women would speak openly on the topic, and shortly

Table 1. Socioeconomic Characteristics of Female Respondents, Ages 15–76, in Five Rendille Communities, 1996.

| | Community | | | | | |
	Nomad	Korr	Karare	Songa	Marsabit	Total
Number sampled	95	246	223	219	137	920
% ever married	92.5	93.5	89.6	88.8	94.9	91.6
% ever given birth	93.7	95.1	92.8	97.2	98.5	95.4
% poor	12.6	16.7	19.8	13.3	16.1	16.1
% ever attended school	11.6	17.5	17.6	13.2	8.8	14.6
Religion:						
Catholic	15.9	51.4	39.6	41.6	50.0	42.4
Protestant	3.2	3.7	18.0	14.4	2.2	9.4
Muslim	0.0	1.2	1.8	0.9	1.5	1.1
Traditional	80.8	42.5	58.1	45.0	46.3	48.6

afterward I was invited to attend a wedding celebration the following morning. It was there that I discovered that excision is performed as part of the wedding ceremony. Being unprepared for what I was to witness, I watched with amazement the bravery of the newly cut bride, and the joyous celebration that ensued. Although I was asked to not take photographs, the circumciser, the bride, and attending women all spoke freely with me about the importance of this custom for Rendille women. And as so often happens to anthropologist, as many questions were asked of me regarding the customs in my country. Women politely tried to conceal their disgust as they learned that I, a married woman and mother, was not excised. One woman replied: "In your place this might be fine, but for Rendille women, circumcision is the only thing that separates us from animals."

Among the Rendille, excision is traditionally performed during the marriage ceremony. Marriages are typically arranged by a woman's family without her involvement. Marriage negotiations are conducted between the elders of the bride's and groom's clans, and once completed, the girl's mother is informed about the decision and begins preparations for the celebration. Marriage is one of the most important rites of passage for Rendille men and women, and the celebration extends over several days. All preparations are kept secret from the bride-to-be if the family suspects that she may object to the chosen husband, and run away, disgracing the family.

Two days before the wedding, the groom and his family form a ritual procession (*guro*) from their home to the bride's home, and outside the hut women sing praises of the groom and the fortuitous union of the two families (Beaman, 1981). A ram is slaughtered, and presented to the bride's mother for a feast among married women, and the skin of the ram is prepared for the bride to sit upon while being cut. The procession arrives again on the eve of the wedding, and girls bring red ochre to smear on the bride's beads.

On the wedding day, before dawn the circumciser (*kamaratan*), who was selected by the bride's mother, is brought to the hut, and at first light the young bride is circumcised. The groom's family brings animals to pay the *kamaratan* for her service, and one sheep is given as a gift to the bride's mother, and tied to the outside of the hut. It is said that the sheep, as it bleats in the midday heat, forewarns the bride of the burden and responsibilities accompanying marriage.

Among the Rendille, excision is not performed by a traditional birth attendant (*tagan*), but rather by a woman who is a specialized circumciser (*kamaratan*). A *kamaratan* is usually an elderly woman with a reputation for being careful and observant who acquires her skills by watching many ritual circumcisions. A skillful circumciser is considered to be one who cuts quickly and accurately. In the past the cutting was done with a knife, although now razor blades are more commonly used. Excision typically takes place inside a traditional hut in the early morning hours, lighted by a low fire. Males and uncircumcised girls are required to leave the hut, the latter for fear of elevating their anxiety over their future fate. Often the mother of the bride waits outside as well, to avoid the torment of watching her daughter suffer. While being cut, the girl is held still by two women, one holding her legs, and the other holding her torso. The *kamaratan* kneels between the girl's legs, and quickly removes the clitoris and labia minora, and the cut tissue is buried beneath a hearthstone in the center of the hut. The fresh wound is rinsed with milky water steeped with herbs (*lkiroriti*), and examined for mistakes. The bride is then advised to sit quietly with her legs together, and periodically, women tending her check to see if her blood has clotted, and remove the formed clot. If the bleeding continues, several remedies may be tried. In some cases the girl is told to drink plain milk or milk mixed with blood until the bleeding stops. Alternatively, she is given water boiled with small bitter seeds known locally as *khankho*, which are said to make the blood clot. For the next several days the bride will be nursed, resting and rinsing the wound with water boiled with herbs several times a day. Additionally, it is believed that a good diet, one of milk, blood and meat, will help the bride replace lost blood and regain strength.

Once the operation is complete, the bride's mother and mother-in-law are allowed to enter the hut to see that the bride has fared well. When the news is announced, a collective sense of anxiety vanishes, and a festive celebration begins. The *guro* procession is repeated, with relatives singing praises of the bride and groom in echoing choruses. This time, the procession also delivers the initial payment of brideweath.[5,7] Amid songs of praise, visitors come to congratulate and bless the new couple, bringing gifts of sugar, tea and small livestock, and the families feast on a sacrificed sheep.

Later in the day the families select a site for building a new hut for the couple. The Rendille term for marriage, *min discho*, literally means "house building," and no unmarried man can have his own house (Beaman, 1981). Women in the bride's family walk long distances to gather new green branches for building a new hut, and rebuilding the bride's mother's hut. Dried sticks from the mother's hut are removed and divided, half remaining and half used to build a new hut for the bride and groom, symbolizing that this house will join the two families. Later that evening, in a ceremonial procession, the bride and her new husband walk to their new hut. The bride carries her possessions on her back in a manner that she will carry an infant, symbolizing the anticipation of children. As Beaman (1981 : 336) notes, "Children are the most important reason a man marries, and are the central focus of a woman's life, assuring her place in her husband's clan, her allotment of milch camels, and her social acceptance throughout society." The groom makes a fire anew from sticks, rather than from burning embers, to symbolically mark the beginning of their new life together. The marriage is consummated after the bride's wounds have been allowed to heal, usually one week.

Circumcision symbolically marks that the bride is no longer a girl (*inam*), but now a woman (*aronto*). To mark the transformation, the bride discards her father's name and takes her husband's name, and leaves her clothes and ornaments of a girl behind. She is given new skins, a large woven necklace (*bukhurcha*) made of sisal or giraffe hair, and a strand of red and white marriage beads (*irtitior*) that she will wear until she bears her first child. Women stress the importance of circumcision as a central part of their rite of passage into womanhood, marking that they are now mature and worthy of respect. The bravery and self control displayed by enduring the pain of the operation is seen as central to the transformation from childhood to womanhood. By withstanding the pain of being cut, a woman demonstrates her maturity and readiness to endure the pain of childbirth and hardships of married life. One issue raised in focus group discussions was the acceptability of using anesthesia on women when excised. This suggestion stemmed from the fact that when the most recent age set of warriors were initiated, circumcision was performed by a nurse who used a local anesthesia. When asked if this same practice would be acceptable for girls, women, particularly recently married women, strongly objected, viewing the proposed change as a trespass on women's affairs. Enduring the pain, they explained, is a constitutive experience in preparing a woman for her role as wife and mother, her central role in Rendille society.

Concomitant with circumcision is improved social status. A woman, once circumcised, is recognized as the female head of her new household. She is allocated livestock, and most importantly, she is now allowed to bear children. Prior to marriage, it is acceptable for Rendille girls to be sexually active. Adolescents often spend weeks or months tending animals in the bush, where they are free to have sexual partners. It is taboo, however, for unmarried girls to become pregnant. Young warriors are counseled by elders on how to avoid impregnating their girlfriends by practicing withdrawal. If an unmarried girl becomes pregnant, both she and her boyfriend are disgraced, and the girl is faced with the danger of induced abortion. One young married man explained,

> "Children of girls are outcasts or can be killed. You must have a circumcised mother to be accepted. If an unmarried woman is pregnant, we must jump on her stomach until she aborts. The mother and child may die … If we bear children with an uncircumcised girl, we lose all respect in the community."

Once married and circumcised, a woman is allowed only one partner, her husband, and is allowed to bear children. Marriage cannot be disentangled from excision in legitimating reproduction.

For men, the importance of circumcising a bride is that it marks the woman, socially signifying that she is sexually exclusive and allowed to bear his children. As one informant described, "Circumcision is a brand. If a girl is not circumcised, she can stay with her family, and can have sex with boyfriends. We get a brand to show that she is mine, and can only be with me, and will bear my children … Branding makes her mine."

When discussing the decision to circumcise their daughters, men also stress the economic ramifications of circumcision. An uncircumcised girl is not socially sanctioned to marry or bear children. Excision is central in preparing a girl for marriage, and entitling her family to receive large bridewealth payments. Therefore failing to circumcise a daughter would have significant social and economic consequences for the girl and her entire family.[6,7]

When asked to describe the advantages of female "circumcision", women in particular emphasized that excision reduces a wife's sexual desire, and helps her remain abstinent during the often long absences of her husband, who may leave for months at a time to take animals to the *fora* (bush) or to work in larger towns. One elderly woman explained, "If you are circumcised, your emotions (sexual desires) are reduced, and you don't have to sleep around and lose respect."

It has been debated, however, whether excised women do, in all cases, experience reduced sexual pleasure, particularly since the extent of cutting can vary considerably (see Hernlund, 2000, for a more detailed discussion of this issue). It is often difficult to determine whether excision alters sexual response since many African women are circumcised long before becoming sexually active. Rendille women, by contrast, usually become sexually active prior to being excised. They therefore provide a unique perspective in evaluating the effect of excision on sexual pleasure. When asked if they experienced diminished sexual pleasure after being circumcised, responses were divided. Some women claimed to have lost all enjoyment: "Only when I was a girl did I feel pleasure. All pleasure was reduced when I was circumcised. There is now enough for one purpose—bearing my husband's children." Another informant concurred, "One disadvantage (of excision) for men is that the sexual desire of the wife is reduced. In a way it is good because you don't go to other men. In a way it is bad because you are not interested in him." Other women claimed to experience fulfilling sexual pleasure after marrying. In response to a question about differences in sexual enjoyment before and after excision, several women insisted that sex was actually better *after* being circumcised, a finding also reported by anthropologist Anne Beaman (1981 : 321–22). As one informant told anthropologist Elliot Fratkin (1991 : 70) in what he described as consciously contradictory terms, "But you know, the men like to see us circumcised. They think we won't see other men if we are circumcised. But they are wrong," she laughed.

While women certainly play a central role in promoting the practice of excision, the interests of men are not irrelevant. The importance of engenderment for women cannot be separated from the fact that they are oppressed by men throughout their life. Rendille women have very low status in terms of control over economic resources or decision making about events in their life or household affairs. Although they play an important role in economic production, women own no independent property, and are thus dependent upon their husbands or, when widowed, upon their sons. Male dominance over women expresses itself in ideology, economic control, and behavior. One elderly female informant explained, "When a man has married, he takes control of you. You must follow his rules ... If you make a mistake—if he thinks that you have gone to another man, or that you did not care for the goats—he can leave you without animals, and without food and water, and he can beat you." Circumcision transforms a girl into a bride, and she becomes property that is transferred upon payment of bridewealth. Women receive their status of full wife upon bearing a child, emphasizing the importance of fertility in defining a woman's role. Without being excised a woman can never bear legitimate children or achieve the full status of womanhood in Rendille society.

2.3. Recent Changes in Attitudes and Practice

As noted by Leonard (2000), descriptions of the cultural context of female "circumcision" are shaded with an image that the tradition is deep-seated and unchanging. While it is important to recognize that the practice of female "circumcision" is a response to complex social concerns, it is also necessary to realize that social conditions are dynamic, and consequently so is the nature of this practice.

By conducting questionnaires among women and focus groups and in-depth interviews with men and women across a broad range of ages, we were able to identify intergenerational differences in the practice and attitudes toward female circumcision. These changes are not the product of anticircumcision campaigns since none have been implemented as yet in Marsabit District.[7] Rather changing attitudes and practices reflect the rapid social transformations accompanying settlement and development in Rendille communities.

According to the "development and modernization" approach to eliminating female "circumcision," as societies experience dramatic social change in the process of development, traditional values erode, and adherence to the practice of "circumcision" declines (see van der Kwaak, 1992). What is not clear is how great these social changes need be, or whether educational campaigns are required to initiate a public discourse questioning the value of female "circumcision."

Throughout the Rendille community, the manner in which female "circumcision" is performed departs little from tradition (Table 2). In the vast majority of cases excision is performed by a *kamaratan* at the mother's home. Across communities, the prevalence of excision among women ages 15 and over is very high (97.8%), and is universal among ever-married women in all communities except Marsabit (Table 3). While we hypothesized that the practice of female circumcision would decline in settled communities, particularly in Marsabit town and Songa, the practice shows no sign of diminishing in Songa, even in the younger cohort. In Marsabit, however, 5 percent ($n = 7$) of married women were not circumcised. All these were Rendille women married to men from a tribe that does not circumcise either males or females. Abandonment of female "circumcision" in cases of inter-ethnic marriage has been reported elsewhere (see, for example, Orubuloye et al., 2000), and demonstrates the importance of cultural constructs of marriageability in

Table 2. Characteristics of Female Genital Cutting Among the Rendille ($n = 900$).

		%
Type of genital cutting performed	a. clitoridectomy	0
	b. excision	99.6
	c. infibulation	0.4
Location of the operation	a. clinic	0.7
	b. bush	0.7
	c. mother's home	98.6
Training of the circumciser	a. traditional circumciser (kamaratan)	86.6
	b. traditional birth attendant	12.2
	c. trained nurse	0.2
	d. doctor	0

Table 3. Prevalence of Excision Among Rendille Women.

	Community					
	Nomad	Korr	Karare	Songa	Marsabit	Total
All Women Ages 15–17 ($n = 920$)	88.8%	97.5%	98.5%	99.1%	95.4%	97.8%
Ever-Married Women ($n = 843$)	100%	100%	100%	100%	95%	99.2%

determining whether or not the practice is retained. In communities in which lifestyle changes are so recent, changes in attitude would be expected to herald changes in behavior. Therefore, a more telling indicator might be to consider women's intentions for their daughters. Of women with unmarried daughters, we found that 98 percent intend to circumcise their daughters, and that the preferred type of circumcision is still excision. Together, these findings indicate that, except in cases of inter-ethnic marriage, there appears to be no decline in the demand for traditional excision in settled Rendille communities.

Although demand for excision remains high, the circumstances surrounding the practice are changing. In settled Rendille communities the tradition of arranged marriage is eroding. As it becomes increasingly common for girls to attend school, a rising number of girls elope without the consent of their families, some without being excised. While male elders express disapproval of this departure from arranged marriage, women most often tacitly approved, recalling their own unhappiness with this tradition. Commonly recounted were feelings of distress over being forced to marry an old man, and becoming separated from their friends. However, men and women alike expressed deep concern over a girl initiating childbearing without being excised. A woman of any age is regarded as a "girl" if uncircumcised, and unable to legitimately bear children. The solution to avoiding childbearing by uncircumcised girls has been to uncouple circumcision and marriage. One informant explained: "We now circumcise schoolgirls before they marry so they can go on with their education ... At school, a girl may decide to marry without parents' permission. It is bad to go with a husband if you are not circumcised. This mother will bring shame to her family. Somalis circumcise early, and our schoolgirls want to do this also. A girl's blood is shed in her mother's house so she can go on with her education."

Our survey results confirm this trend. Overall, the mean age of circumcision is 17 years, and in most cases corresponds to the age of marriage. However, in the most recent generation, circumcision is preceding marriage by an average of two years. Additionally, among never-married women, 88 percent ($n = 77$) were already excised. These unmarried circumcised women were all attending school.

The most obvious recent change in the practice of female circumcision has been the gradual adoption of medical treatment to prevent infection. Each of the settled communities are served by mission-sponsored dispensaries staffed with trained nurses.[8] These health practitioners have encouraged families to bring girls and boys to the clinic to receive an anti-tetanus injection and antibiotics the day before being circumcised, and families are instructed to purchase a new sterile razor blade. In focus groups among both married and unmarried men and women we found a consensus that the health hazards arising from circumcision, particularly infection and hemorrhage, are considered to be serious problems. Among informants there was a widespread sense that medical interventions have been effective, and that further education and certain efforts to minimize the health risks would be welcomed. While men expressed willingness, even eagerness, to adopt changes that would decrease the pain and health risks associated with excision, women were divided. Most women supported the idea of providing training for traditional circumcisers, but there was considerable debate over other suggested innovations: the use of anesthesia, decreasing the extent of cutting, employing a trained nurse to perform the excision, and performing the excision at a clinic. The suggestion to stop practicing excision was adamantly opposed by all women. In a discussion among recently married men, one informant was open to this idea, but, received very little support from other participants:

Husband 1: "We know that women bleed and die. I think we should stop this practice if everybody agrees."

Husband 2: "No, if it killed too many women, we would have stopped long ago. Nobody complains much about it, so it is good. To some communities it is bad, but to us it is good."

Husband 3: "There is no way to stop female circumcision, but we can help it. It is our custom, and we cannot abandon it. We can try training circumcisers, or going to the clinic, just like with traditional birth attendants and delivering babies."

The rest of the group agreed with this view.

3. HEALTH CONSEQUENCES OF EXCISION

A clear understanding of the extent and nature of health consequences of female genital cutting is warranted since health outcomes form the basis of anticircumcision educational campaigns, and inform policy decision making. An enormous body of literature has been devoted to describing the medical consequences of female genital cutting, yet serious gaps in knowledge exist. After extensively reviewing the literature on health outcomes, Obermayer (1999:92) reported that, in actuality, only eight studies systematically assessed health complications. Attempts have been made to quantify the range and frequency of circumcision-related complications from clinic and hospital records (e.g., Aziz, 1980; Rushwan, 1980; De Silva, 1989). However, because these data suffer from selection bias, they need to be interpreted with caution. Women are often reluctant to seek medical attention because of cost, modesty, and in rural settings, inaccessibility of health services. Consequently, complications tend to be reported only if they are severe and prolonged. Furthermore, in some regions, such as the Sudan, certain types of genital cutting are illegal, and women hide medical complications for fear of legal repercussions (El Dareer, 1982; Toubia, 1993).

The best information available on the incidence of various complications comes from a few case-control studies, as well as several large-scale population-based surveys, the first of which was conducted by El Dareer (1982) in northern Sudan. However, self-reported retrospective survey data also suffer from a number of limitations. Recall error on details surrounding events that occurred many years ago is inevitable. Moreover, since many women may have been cut as infants or very young children, it may be impossible to remember immediate adverse health outcomes. Questionnaire design must take into account that informants' concept of illness or abnormality may be different from that of medical researchers. For example, Lightfoot-Klein (1989:59) reported that women who claimed to have no difficulty with urination also indicated that it took up to 15 minutes to empty their bladder. This condition was considered normal in a community where all adult women were infibulated. Self-reported retrospective survey data also suffer from selection bias in that they are limited to women who survived genital cutting.

Despite these limitations, survey data currently provide the best information on health risks associated with genital cutting. Unfortunately, the many reports fail to distinguish between variation in medical complications associated with different types of genital cutting. Furthermore, the conditions under which the procedures are conducted and the medical support available are not considered.

Data from the Rendille Maternal and Child Demographic and Health Survey allow us to examine the range and frequency of complications associated with excision. Additionally, we can evaluate the effect of medicalization in reducing these risks. In order to estimate the prevalence of immediate short-term post-circumcision complications, women were asked whether they experienced any symptoms within one month of being cut. Overall 16 percent of women reported experiencing at least one complication.[9] As set out in Table 4, the most commonly reported condition was general infection (9.9%), followed by extreme pain and hemorrhage. In addition, 4.2 percent of respondents reported specifically contracting tetanus following the procedure.

Women were also asked to report chronic long-term health problems. While a small fraction of women in our study suffered from primary or secondary sterility, it is unclear that the cause is excision, rather than other factors such as sexually transmitted disease.

Table 4. Percent of Excised Women Reporting Specific Short-term Complications.

Type of complication	%
Infection	9.9
Pain	9.1
Hemorrhage	8.1
Tetanus	4.2

Table 5. Percent of Procedures Performed with Medical Assistance* in each Community.

	Community				
Nomad	Korr	Karare	Songa	Marsabit	Total
0	8.9	16.1	17.4	13.3	12.2

* Sterile razor, anti-tetanus injection, and prophylactic antibiotics.

Table 6. Logistic Regression Analysis of Use of Medical Support as a Predictor of Complications within One Month of Excision.

Outcome Variable: Complication[1]

Predictor variable[2]	Coefficient	Odds Ratio	95% CI for Odds Ratio	Excess risk
Medical assistance[3]	−1.367	0.321	(−2.149, −0.585)	3.1%

[1] No complication used as reference category.
[2] Location of the operation and community were not significant covariates, and were removed from the final model.
[3] Assistance used as reference category.

None of our respondents reported dermoid cysts, or chronic problems with urination or menstruation.

Among all excised women 12.2 percent received medical assistance in the form of receiving an anti-tetanus injection, prophylactic antibiotics, and use of a new, sterile razor (Table 5). This occurred only in settled communities, all of which are served by dispensaries. To evaluate the effect of medical assistance on the risk of short-term complications, several variables were examined as predictors of complications arising within one month of being excised. These include: the use of medical assistance, location of the operation, and community. In a logistic regression analysis (Table 6), the use of medical support was found to have a significant effect on the outcome of the operation. Women who received no medical assistance experienced a 3.1 times higher risk of developing complications following the procedure. Stated the other way, women who received medical assistance experienced a nearly 70 percent lower risk of short-term complications.

For decades debate has continued as to whether excised women experience elevated risk of obstetrical complications (Epelboin and Epelboin, 1981). It has been suggested that delayed healing due to infection results in the formation of fibrous scar tissue that is non-elastic, contributing to obstructed labor. Tearing of the scar tissue is also reported to result in an elevated risk of hemorrhage (Mustafa, 1966; Sami, 1986). Among respondents who had at least one live birth, we examined the association between delayed healing due to short-term complications and problems during the last labor and delivery. Overall, 20.4% of women reported experiencing excessive bleeding following their last labor and delivery, while nearly 40% reported prolonged labor. A history of complications following excision was examined as a predictor of prolonged labor (labor lasting more than 24 hours), while controlling for other potential causes and confounding factors: maternal age at delivery, parity, short stature (<150 cm), training of assistant during labor and delivery, location of delivery and presentation of the baby (breech or normal) (Table 7). None of these factors are significant predictors of prolonged labor among Rendille women.

Table 7. Variables Examined as Predictors of Obstetrical Complications.

Variables examined as predictors of:	Prolonged labor	Excessive bleeding
Maternal age at delivery	n.s.	n.s.
Parity	n.s.	n.s.
Short stature	n.s.	n.s.
Who assisted in delivery (doctor, nurse, trained midwife, traditional midwife, relative)	n.s.	n.s.
Location of delivery (clinic, home, bush)	n.s.	n.s.
Presentation of the baby (normal, breech)	n.s.	$p < .05$
Complications following excision	n.s.	n.s.

Complications during healing from excision was also examined as a predictor of excessive bleeding, while controlling for other potential causes and confounding. The results of a logistic regression analysis (not shown) reveal that breech births are associated with a 6.1 times greater risk of excessive bleeding, while short-term post-circumcision complications does not have a significant effect.

These results demonstrate that delayed healing and scarring from excision in this study sample is not associated with the types of complications during labor and delivery that are commonly found among infibulated women. It may be the case that obstetrical complications, such as tearing and hemorrhage, are more common during first births. Since our study is retrospective, there is likely to be underreporting of complications because of recall bias. Therefore, this issue would be better studied through a prospective study of first pregnancies. Our survey did, however, reveal that excision is associated with significant risk of short term complications, including infection and hemorrhage, particularly when the procedure is performed without preventive medical support. When the excision procedure was performed with sterile cutting instruments, antibiotics and anti-tetanus injections, the risk of subsequent complications was substantially reduced.

4. SUMMARY AND POLICY IMPLICATIONS

The Rendille case illuminates several key issues that arise in debates surrounding the practice of female genital cutting. First, development and social change, at least on the level experienced by the Rendille, does not in and of itself lessen the demand for excision. While the nature and meaning of the practice has changed, the new context of excision does not undermine its importance for Rendille people. Secondly, excision is not associated with the types of long-term and obstetrical complications reported among infibulated women, although it is associated with serious, potentially life-threatening short-term complications. Finally, even minor levels of medical intervention dramatically reduce the risk of developing immediate complications. Given these findings, we must at least consider medicalization as an intermediate solution to improving the health and welfare of women.

Opposition to all forms of medicalization is central in international efforts to eliminate female genital cutting. In 1982, the World Health Organization issued a statement declaring it unethical for female genital cutting to be performed by "any health officials in any setting – including hospitals or other health establishments" (WHO, 1982). In 1994, the

International Federation of Gynecology and Obstetrics passed a resolution calling on all doctors to refuse to perform "FGM," and were joined by many other major organizations such as the American College of Obstetricians and Gynecologists, the United Nations International Children's Emergency Fund, and the American Medical Association (ACOG Committee Opinion, 1995). Additionally, in response to threats of withholding international aid and World Bank loans, ministries of health in many African countries have issued similar statements.

This staunch opposition to medical intervention rests on one central assumption: that medicalization will counteract efforts to eliminate female genital cutting. This assumption is, however, not based on empirical evidence, and deserves critical examination. Our role as scholars is to assess whether (and if so, how) medicalization influences anticircumcision efforts (see Mandera, 2000; Shell-Duncan, 2001 for an extensive discussion of the medicalization debate). This need is underscored by the fact that, regardless of legislation or international opinion, female genital cutting is, and will continue to become, increasingly medicalized throughout sub-Saharan Africa. It is therefore necessary to determine whether medicalization is best viewed as one in a series of steps in improving women's health, or an impediment to change efforts.

Among the Rendille, change efforts must take into account the sociocultural situation in which the people are enmeshed, and the socioeconomic hardships that families that try to change the practice would experience. Excision is an integral part of marriage and gender identity. It is performed as part of the marriage ceremony, a step in preparing a bride for her future husband, transforming her from a girl to a woman who is ready to assume the rights and responsibilities of marriage and childbearing. Awareness of the fact that female "circumcision" is associated with adverse health consequences is widespread, yet the Rendille view the risks as worth taking in light of the implications for marriageability. Marriage and children are a central part of a woman's role in society, and not being circumcised would define a woman as an outcast, and render her unmarriageable. Although women are economically active, participating in livestock production, agriculture, milk marketing, and more recently, wage employment, their cultural identity and economic well-being is embedded in large family production units, with their role as a wife and mother. Men also stress the economic consequences of female "circumcision." A daughter who is not excised is not marriageable, and the family risks losing the large bridewealth and social ties gained through marriage. To outsiders, one surprising finding is that genital cutting is not simply imposed on women by men, but also perpetuated by women themselves, despite their awareness of the pain and serious health risks involved. However, the motivation for this practice is best understood in terms of its social and economic ramifications for men and women, and in terms of its wider cultural meaning. In this case ritual "circumcision" entwines the issues of gender identity, cultural heritage, marriageablity, and socially sanctioned childbearing. Efforts to discourage female "circumcision" will need to consider these issues, and help seek alternative solutions to these social concerns.

Until acceptable alternative solutions to female circumcision are found, Rendille women are women without choices. A woman who is not excised cannot marry or bear children and is ostracized. Yet Rendille society is rapidly undergoing change, and one can hope that the demand for female "circumcision" will decline as educational and economic opportunities improve, particularly for women, and as educational campaigns highlight the disadvantages of excision and encourage members of the community to seek alternative practices. Yet this process is slow, and interim solutions for improving women's health need to be at least considered. This study has shown that even small steps in improving the septic conditions for excision significantly reduce medical risks for the women involved. If improvement in women's health is truly targeted as a priority, intermediate solutions, including the improvement of medical conditions for female "circumcision," merit careful consideration.

ACKNOWLEDGEMENTS. This research was supported by funding from the Andrew W. Mellon Foundation and the John D. Rockefeller Foundation. Helpful comments on an earlier draft of this paper were given by Ylva Hernlund and Dr. Lynn Thomas. Permission for this research was granted by the Office of the President of Kenya, permit number OP 13/001/25C282/5. Logistical support in planning this research was kindly provided by Drs. Elliot Fratkin and Eric Roth. We wish to thank our team of research assistants, particularly the field managers, Larion and Anna Marie Aliayaro, for their dedication, patience and humor during data collection. We also thank the many Rendille people who generously shared their time, insight, and opinions.

NOTES

1. As this manuscript was going to press, a paper by Obermeyer (1999) appeared, making a similar point.
2. See Obermeyer (1999) for a recent critical review of empirical data supporting or contradicting the assertion that the prevalence of FGC declines with improvements in economic status and female education.
3. Mackie (2000) describes a case in which, upon taking a pledge to not circumcise daughters, the practice has been eradicated in several Senegalese community virtually overnight. We would like to stress that efforts preceding the pledge, that is, introducing the concept to a community and obtaining support, will take varying amounts of time in different social settings. In southeast Nigeria (describe by Orubuloye et al., 2000) this process may be rapid since marriageability is the main factor blocking elimination of the practice of female "circumcision". Among the Rendille (described later in this chapter), the notion of abandoning the practice was deemed entirely unacceptable by most informants. Clearly, change efforts could take much more time in this community.
4. Classification of economic status (poor vs. not poor) is a relative measure in each community, based on livestock holdings (total livestock units), garden size, and other income sources.
5. Traditionally, bridewealth among the Rendille is fixed at eight camels, but this can be paid over many years. According to Beaman (1981), usually no more than half of the bridewealth is paid at one time. There is more variability in the amount and form of bridewealth payments in the settlements, particularly when the bride elopes.
6. Mackie (2000) emphasizes the importance of recognizing the link between female genital cutting and marriageability, and argues that eradication efforts must assist in developing a forum in which uncircumcised women retain marriageabilty.
7. Although central Kenya has a long history of anticircumcision campaigns (see Thomas, 2000), the remote Rendille community has not yet been targeted in these campaigns. According to Kratz (1994:342), "In Kenya … debates about circumcision began as soon as missionaries arrived, and were framed within the question of whether (and which) local customs violated standards of Christian behavior and had to be condemned and eliminated." In the 1920's and 1930's, moral arguments gave way to medical arguments, and a health approach is still emphasized in current campaigns. For example, the Kenyan women's group Maendeleo Ya Wanawake Organization is distributing pamphlets to schoolgirls entitled "The Dangers of Female Circumcision: Female Circumcision is Harmful to Women's Health." Therefore, it is of interest to determine whether the health message delivered will ring true when it eventually reaches Rendille communities.
8. Although central Kenya has a long history of anticircumcision campaigns (see Thomas, 2000), the remote Rendille community has not yet been targeted in these campaigns. According to Kratz (1994:342), "In Kenya … debates about circumcision began almost as soon as missionaries arrived, and were framed within the question of whether (and which) local customs violated standards of Christian behavior and had to be condemned and eliminated." In the 1920's and 1930's, moral arguments gave way to medical arguments, and a health approach is still emphasized in current campaigns. For example, the Kenyan women's group Maendeleo Ya Wanawake Organization is distributing pamphlets to schoolgirls entitled

"The Dangers of Female Circumcision: Female Circumcision is Harmful to Women's Health." Therefore, it is of interest to determine whether the health message delivered will ring true when it eventually reaches Rendille communities.
9.   In an attempt to estimate the degree of recall error in reporting complications, we compared the frequency of reported complications among women "circumcised" within the past 10, 15 and 20 years. Not surprisingly, the frequency of reported complications declines with time lapsed since "circumcision." Therefore, these figures reflect an estimated 6–10% underreporting.

REFERENCES

ACOG [American College of Obstetricians] Committee Opinion, 1995, Female Genital Mutilation. *International Journal of Gynaecology and Obstetrics* 49:209.

Aziz, F.A., 1980, Gynecologic and Obstetric Complications of Female Circumcision. *International Journal of Gynaecology and Obstetrics* 17:560–563.

Beaman, A.W., 1981, *The Rendille Age-set System in Ethnographic Context.* Ph.D. Dissertation, Boston University.

Brass, William and Carole L. Jolly (eds.), 1993. *Population Dynamics of Kenya.* Washington DC: National Academy Press.

De Silva, S., 1989, Obstetric Sequelae of Female Circumcision. *European Journal of Obstetrics, Gynecology, and Reproductive Biology* 32:233–240.

El Dareer, A., 1982, *Women, Why Do You Weep? Circumcision and its Consequences.* London: Zed Press.

Eliah, E., 1996, REACHing for a Healthier Future. *Populi* (May): 12–16.

Epelboin, S. and A. Epelboin, 1981, Excision: Traditional Mutilation or Cultural Value? *African Environment* 16:177–188.

Fratkin, E., 1991, *Surviving Drought and Development: Ariaal Pastoralists of Northern Kenya.* Boulder: Westview Press.

Gordon, D., 1991. Female Circumcision and Genital Operations in Egypt and the Sudan: A Dilemma for Medical Anthropology. *Medical Anthropology Quarterly* 5(1):3–14.

Gruenbaum, E., 1982, The Movement Against Clitoridectomy and Infibulation in Sudan: Public Health Policy and the Women's Movement. *Medical Anthropology Newsletter* 13:4–12.

Hosken, F., 1978, Female Circumcision and Fertility in Africa. *Women and Health* 1:3–11.

Hernlund, Y., 2000, Cutting Without Ritual, and Ritual Without Cutting: Female "Circumcision" and the Re-ritualization of Initiation in the Gambia. In *Female 'Circumcision' in Africa: Culture, Controversy and Change*, edited by B. Shell-Duncan and Y. Hernlund, pp. 235–252. Boulder: Lynne Rienner Publishers.

Leonard, L., 2000, Female Circumcision in Southern Chad: Origins, Meaning, and Current Practice. *Social Science and Medicine* 43:255–263.

Lightfoot-Klein, H., 1989, *Prisoners of Ritual: An Odyssey into Female Circumcision in Africa.* New York: Harrington Park Press.

Mandera, M.U., 2000, Female Genital Gutting in Nigeria: Views of Nigerian Doctors on the Medicalization Debate. In *Female 'Circumcision' in Africa: Culture, Controversy and Change*, edited by B. Shell-Duncan and Y. Hernlund, pp. 95–108. Boulder: Lynne Rienner Publishers.

Mustafa, A.Z., 1966, Female Circumcision and Infibulation in the Sudan. *Journal of Obstetrics and Gynecology, British Commonwealth* 73:302–306.

Nathan, M.A., E. Fratkin, and E.A. Roth, 1996, Sedentism and Child Health among Rendille Pastoralists of Northern Kenya. *Social Science and Medicine* 43(4):503–515.

Obermayer, C.M., 1999, Female Genital Surgeries. *Medical Anthropology Quarterly* 13:79–106.

Orubuloye, I.O., P. Caldwell, and J.C. Caldwell, 2000, Female 'Circumcision' Among the Yoruba of Southwestern Nigeria: The Beginning of Change. In *Female 'Circumcision' in Africa: Culture, Controversy and Change*, edited by B. Shell-Duncan and Y. Hernlund, pp. 73–94.

Rushwan, H., 1980, Etiological Factors in Pelvic Inflammatory Disease in Sudanese Women. *American Journal of Obstetrics and Gynecology* 138:877–879.

Sami, I.R., 1986, Female Circumcision with Special Reference to the Sudan. *Annals of Tropical Medicine* 6:99–115.

Shell-Duncan, B., 2001, The Medicalization of Female 'Circumcision': Harm Reduction or Promotion of a Dangerous Practice? *Social Science and Medicine* 52(7):1013–1028.

Shell-Duncan, B. and Y. Hernlund, 2000, Female "Circumcision" in Africa: Dimensions of the Practice and Debates. In *'Female Circumcision' in Africa: Culture, Controversy and Change*, edited by B. Shell-Duncan and Y. Hernlund, pp. 1–40. Boulder, CO: Lynne Rienner Publishers, Inc.

Talle, A., 1993, Transforming Women into "Pure" Agnates: Aspects of Female Infibulation in Somalia. In *Carved Flesh, Cast Selves: Gender Symbols and Social Practices*, edited by V. Broch-Due, I. Rudie, and T. Bleie. Oxford: Berg.

Thomas, L., 2000, Ngaitana (I Will Circumcise Myself). In B. Shell-Duncan and Y. Hernlund, *Female Circumcision in Africa*, pp. 129–150. Boulder: Lynne Rienner Publishers, Inc.

Toubia, N., 1993, *Female Genital Mutilation: A Call for Global Action*. New York: Rainbow/Women Ink.

Toubia, N., 1994, Female Circumcision as a Public Health Issue. *New England Journal of Medicine* 331:712–716.

Toubia, N. and S. Izette, 1998, *Female Genital Mutilation: An Overview*. Geneva: World Health Organization.

Van Der Kwaak, A., 1992, Female Circumcision and Gender Identity: A Questionable Alliance? *Social Science and Medicine* 35:777–787.

WHO (World Health Organization), 1982, Female Circumcision. *European Journal of Obstetrics, Gynecology, and Reproductive Biology* 45:153–154.

Chapter 13

Female Education in a Sedentary Ariaal Rendille Community
Paternal Decision-Making and Biosocial Pathways

ERIC ABELLA ROTH AND ELIZABETH N. NGUGI

1. INTRODUCTION

The noted Australian demographer John Caldwell's (1979) seminal analysis of Nigerian survey data revealed that maternal education influenced offspring mortality more than father's income. This finding sparked over twenty years of demographic concern with female education in Third World populations. Caldwell (1980, 1982) went on to make female education a cornerstone of his emerging "Wealth Flows Theory", while subsequent World Fertility Survey and Demographic Health Survey Data consistently included female education as an important descriptive and analytical variable. Both sets of international surveys revealed strong negative associations between maternal education and high levels of infant/child mortality and fertility across a set of world-wide cultures (for reviews see Bledsoe et al., 1999; Cleland and Kaufmann, 1998; Cleland and Van Ginnekan, 1989; United Nations, 1995). Largely as a result of these findings, the International Conference on Population and Development meeting in Cairo in 1994 strongly called for universal female access to education because:

> increase in the education of women and girls contributes to greater empowerment
> of women, to a postponement of the age of marriage, and to a reduction in the size of
> families (United Nations, 1994: Ch. XI, Para. 11.3)

ERIC ABELLA ROTH • Department of Anthropology, University of Victoria, P.O. Box 3050, Victoria British Columbia, Canada V8W 3P5   ELIZABETH N. NGUGI • Strengthening STDs/HIV Control Unit, Department of Community Health, University of Nairobi, Nairobi, Kenya.

Despite this long and intensive interest in the strong causal relationship between female education and demographic change, problems remain in delineating actual pathways connecting girls' schooling and demographic parameters. Levine et al. (1994: 304) honestly describe the current situation by saying:

> Among the least understood processes are those that mediate the effects of school experience during childhood on the behaviour of adult women as mothers and reproductive decision-makers in Third World countries. This is the "black box" in survey data linking years spent in school with demographic and health outcomes; it is often covered by speculative assumptions and interpretations, but rarely investigated.

Since 1990, one of us (Roth) has investigated two related aspects of female education among Rendille and Ariaal populations of Marsabit District, northern Kenya. The first is how parents choose which children to send to school (Roth, 1991). The second concerns the search for pathways linking female education to fertility and morbidity (Roth et al., 2001). In this chapter we update and link these two research questions by focusing on recent work in the Ariaal Rendille village of Karare. Since this volume already contains broad cultural, economic and ecological descriptions of both Rendille and Ariaal populations we begin by focusing on our two specific research problems. An overview of the study setting, an introduction to our methodological approaches and analyses of recently collected data follow this.

2. RESEARCH ISSUES

2.1. Parental Decision-Making and Female Education

In a 1991 paper entitled, "Education, tradition and household labor among Rendille pastoralists", Roth examined survey data from the sedentary lowland community of Korr to delineate possible Rendille parental decision-making patterns of child selection to attend school. Despite Caldwell's initial recognition of the demographic importance of female education, the topic of parental decision-making is strangely lacking in the formulation of his subsequent Wealth Flows Theory. Regarding this subject he says only,

> One other factor little mentioned in research reports but frequently mentioned in the villages is the view that not all children are sufficiently gifted to give an adequate return on educational investment or to achieve sufficient success to provide their parents with a channel to the modern world. It is very much of a lucky dip. (Caldwell, 1982: 44)

Rather than constituting just "a lucky dip" Roth proposed that cultural factors molded parental decision-making. Noting that Rendille society features patrilineality, patrilocality, and primogeniture, that wealth is measured in livestock, and the primary unit of production is the household labor force, he predicted that latter-born boys from poor families are most likely chosen for schooling. In contrast, first-born sons are too valuable to remove from the family labor pool, since they inherit the entire household herd upon the death of the male household head, while in wealthy households large family herds necessitates labor from all sons. Daughters would not be selected for schooling often; because of patrilocality and patrilineality daughters leave the natal household labor pool upon marriage and their subsequent labor, and that of their children, belongs to their husband's lineage. Under these conditions, and given the early entry into and substantial contribution of females to the Rendille household labor force (Fratkin and Smith, 1995; Smith, 1999), it would make little economic sense to educate daughters.

Roth (1991) tested these predictions upon data collected in a 1987 demographic survey of Korr that included a question asking respondents if any children enumerated in maternal histories ever went to school. For children old enough to begin public education (n. children = 931) "Yes/No" responses to the question, "Did this child ever attend school" formed a dichotomous categorical dependent variable for logistic regression analysis. Independent variables included child's sex, parity, household wealth, and family size. As predicted, results showed that the most likely candidates for schooling were latter-born sons of poor families. Children from wealthy families were approximately one-third (Odds Ratio = 0.368) as likely to receive schooling than offspring from poor families, while latter-born boys were almost three times (Odds Ratio = 2.71) more likely to ever have gone to school than first-born sons. Most importantly for this chapter, girls were less than one-half as likely (Odds Ratio = 0.457) as likely as boys to be selected for schooling.

Repeating this exercise in Korr ten years later (n. of children = 546) (Fratkin et al., 1999) revealed the same bias against educating daughters, with resulting Odds Ratio indicating that boys were almost two and one-half times (2.46) more likely than girls to be selected for school. Rather than constituting merely "a lucky dip" as Caldwell predicted, these results revealed that parental decision-making strategies followed specific cultural patterns, in the Rendille case including primogeniture, patrilineality, patrilocality, and notions of wealth embedded in livestock ownership. Results also showed a consistent bias spanning ten years against educating females, negating the role of mass education as an engine of demographic and social change proposed by Caldwell (1982) in his Wealth Flows Theory as well as minimizing the potential demographic and public health benefits arising from female education (e.g., United Nations, 1994, 1995).

In subsequent sections we again test for parental decision-making patterns and their links to cultural features, now based on data from the upland Ariaal town of Karare, which features important cultural and socio-economic differences with the lowland Rendille community of Korr. These include partible, instead of impartible inheritance, reliance upon cattle, rather than camel pastoralism, dryland agriculture, higher levels of polygyny, and greater access to the large regional market center of Marsabit Town.

2.2. Female Education and the Risk of Sexually Transmitted Infection

While the recent United Nations position emphasizes the potential positive benefits of female education within sub-Saharan Africa there may also be a potentially negative relationship between female education and sexually acquired infections (STIs), particularly HIV/AIDS. Specifically adolescent females' need to pay their schooling fees may make them susceptible to older men, or "Sugar Daddies" (cf. Meekers and Calves, 1997). Heise and Elias (1995:939) succinctly state this relationship as:

> Subsidizing the uniform and school fees of adolescent girls in Africa might actually do more to reduce HIV transmission—by eliminating the need for Sugar Daddies—than the most sophisticated "peer education" campaign. It would also reduce unwanted pregnancy, raise the age of marriage and decrease infant mortality, not to mention promoting gender equality.

For Rendille and Ariaal communities one possible link between female education and STIs arose during the course of a 1996 survey designed to obtain baseline levels of knowledge, attitudes and practices concerning towards HIV/AIDS (n. respondents = 282 reproductively aged men and women) undertaken in Karare. This survey (Roth et al., 1999) recorded a strong dichotomy in female responses to the question "Should a woman be a virgin at the time of her marriage", with a large minority (40%) of women answered affirmatively. In subsequent interviews both men and women suggested that the positive responses came from educated women, and reflected changes in Ariaal Rendille sexual culture. To clarify this point it is necessary to briefly describe Ariaal sexual culture, following earlier outlines by Fratkin (1998) and Roth et al. (2001) and link this to larger sub-Saharan sexual patterns, as outlined by Caldwell (1999).

Ariaal culture is maintained and regulated by an elaborate age-set system identical in length, but not in precise calendar years, to the Rendille age-set system described in detail by Spencer (1965, 1973), Beaman (1981), Fratkin (1991, 1998) and Roth (1993, 1999, 2001). Ariaal age-sets open with male circumcision ceremonies, marking the transition from boys to warriors for males. Warriors (*lmurran*) provide both labor and defense of livestock for an eleven-year period before they marry *en masse*. As soon as they marry, men become elders, and take on a managerial role in their households, making decisions about the movement and care of livestock, which are increasingly left to younger men, including their sons, to carry out.

One consequence of this system is to effectively remove warriors from the marriage pool for the duration of their warriorhood, during which time elders practice polygyny, taking sequentially younger wives from the pool of unmarried women (Spencer, 1965). The control elders maintain over warriors is alleviated partially by the institutionalization of pre-marital sexual relationships. Within Ariaal and Rendille culture there are severe proscriptions for births out of wedlock (Fratkin, 1996; Spencer, 1973), in part arising from fear of inheritance conflicts between children born within, and outside, marriage. At the same time neither Ariaal nor Rendille think that men should be celibate. This was evident from the very high level of positive responses from both men and women to the 1996 survey question, "Should a man be a virgin at marriage?" (women = 89%, men = 97%). Later interviews with Ariaal men in Karare revealed the belief that male sexuality was a powerful biological drive that must be satisfied, and that failure to have frequent sexual intercourse damages a man's health.

This view is widespread across sub-Saharan Africa; with multiple authors (Caldwell et al., 1992, 1999; Orubuloye et al., 1997; Varga, 1999) recording the belief that the male biological sex drive is innately strong and "programmed" to need multiple partners. In interviews conducted with young males in Nigeria, Caldwell and colleagues found the need for multiple, sometimes concurrent, sexual partners was linked not to sexual pleasure, but to a biological drive (Caldwell et al., 1999). This is further linked to an indigenous model of overall personal health in which failure to have multiple sexual partners results in general ill health.

Given this belief, Rendille and Ariaal men are not expected to abstain from sex during the eleven years that they are warriors. One alternative is to enter into a socially institutionalized pre-marital sexual tradition known as *nykeri*. In this tradition warriors present an unmarried girl's mother a series of beaded necklaces, which the mother puts around her daughter's neck. Once completed, warriors and their chosen young women, known as *nkeryi*, initiate sexual relationships. Warriors and *nkeryi* usually do not marry, instead taking spouses chosen by their parents upon the successful completion of bridewealth negotiations. Thus while warriors and *nkeryi* are expected to have full sexual intercourse, procreation is severely discouraged because it raises biological questions of paternity and socio-economic issues concerning inheritance.

In Ariaal culture, very young girls (the 1996 survey denoted girls as young as 10–12) are beaded to and begin sexual relationships with much older warriors (Roth et al., 2001).

In addition both men and women stated that warriors frequently have sex with their age-mates' *nkeryi*, and that such sexual sharing is culturally condoned. The *nykeri* tradition therefore represents high risk sexual behavior, with both age at sexual debut and rate of partner change important parameters in the risk of contracting sexually transmitted infections (Aral et al., 1992; Korenromp, 2000).

In post-survey interviews, Ariaal repeatedly stated that uneducated women are usually beaded, often at a very early age. In contrast, we were told educated women, particularly if they stay in school, are infrequently sought as *nkeryi*. These latter women are accorded more autonomy and respect as a result of both their own learning and their parents' economic investment. Educated women take an active role in selecting their marital partner, and usually marry later than same-aged uneducated women. This distinction suggested an explanation to the highly divided female responses to the survey question, "Should a woman be a virgin before marriage", with educated women answering positively, and uneducated women responding in the negative.

This explanation also suggests a model for the effect of female education upon the rates of transmission of STIs. In this model educated women are less likely to be beaded, thereby delaying their age of sexual debut, and avoiding sexual sharing among age-mates, reducing the rate of partner exchange. Combining survey data with post-survey interview information we hypothesized links between female education, beading and attitudes towards pre-marital virginity, as shown in Figure 1. We expected strong negative associations between female education and beading, as well as between beading and the attitude that a woman should be a virgin at marriage. In contrast, we predicted a positive association between female education and virginity.

To test this model Roth and Ngugi returned to Karare in January–February 1998 and interviewed 127 unmarried, adolescent Ariaal women, using an abbreviated version of the 1996 survey instrument. We employed log-linear modeling (Stokes et al., 1995:126) to analyze these data. Key categorical variables included whether or not the interviewee ever attended school (yes/no) or been beaded (yes/no), and whether they thought a woman should be a virgin before marriage (yes/no). Results, shown in Figure 2, supported our hypothesized model. There was a strong positive interaction between the EDUCATION variable, denoting that a women went to school, and the VIRGINITY variable representing attitudes favoring female pre-marital virginity (EDUCATION*VIRGINITY, maximum likelihood coefficient = +0.4426, p = 0. 0154). Also as predicted, there was a strong negative interaction between EDUCATION and the variable denoting beading (EDUCATION*BEADED, maximum likelihood coefficient = −0.4293, p = 0.0190). By far the strongest association is that between female education and beading (EDUCATION*BEADED maximum

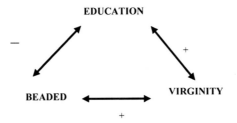

Figure 1. Proposed relationships between female education, beading status and opinion about pre-marital virginity (After Roth et al., 2001: 41).

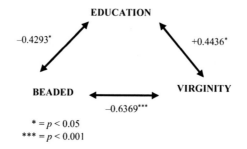

Figure 2. Maximum likelihood coefficients for variable interaction in log-linear model of female education variables (After Roth et al., 2001: 43).

likelihood coefficient = -0.6369, $p = 0.0005$), indicating that educated girls are far less likely to be beaded than uneducated girls.

These results delineated culture-specific pathways by which female education potentially can affect the risk of acquiring STIs, as well as linking female education with increased female autonomy as originally proposed by Caldwell (1979, 1982). However, to fully quantify risks over the adolescent period entails application of a life history approach. In the following section we apply such an approach via survival analysis to new data from Karare.

3. METHODS AND MATERIALS

3.1. Delineation of Parental Decision-Making Patterns

Located on the main north–south road from Isiolo to Marsabit Town (see earlier maps, this volume), Karare actually represents two adjoining yet distinctive communities. Settlements on the west side differ significantly from those on the east side in that the former do not practice dryland agriculture, instead remaining almost totally involved in the pastoral economy. On the eastern side is the agricultural settlement of Nasakakwe, founded by the African Inland Church in the mid-1970s, and whose inhabitants rely almost entirely upon agriculture, raising maize, beans and sorghum for consumption and sale in Marsabit Town. Having two different subsistence patterns in the same locale provides a unique opportunity to see if the transition from pastoralism to agriculture within a sedentary context alters parental decision-making with respect to childhood education.

To realize this opportunity we conducted a survey relating to childhood education in 1996 consisting of 256 houses, divided amongst eighteen settlements within Karare/ Nasakakwe. As with earlier surveys in Korr, questions pertaining to childhood education were nested within sections reconstructing maternal birth histories. These yielded a total of 1071 children of school age, including over 300 children from the Nasakakwe agricultural scheme. In addition to previously utilized independent variables denoting sex and parity we added 1) subsistence pattern (pastoralism versus agriculture), 2) whether the child came from the union of a first or latter-married wife to account for the possible effects of polygyny, and 3) whether sibs were also selected for schooling (yes/no). To measure household wealth, we used the sample (median value) for annual income (11,000 Kenyan shillings) to distinguish between wealthy and poor households. To determine if parental decision-making patterns change over time, we separated children of fathers from age-sets married before

1951 (*Il-Kiliako*, marriage year = 1935, *Il-Mekuri*, marriage year = 1950) from those of more recent age-sets (*Il-Kaminiki*, marriage year = 1964, *Il-Kachili*, marriage year = 1978, and *Il-Kororo*, marriage year = 1992). The resulting data formed part of Giles' (1999) unpublished MA thesis. These data are analyzed here using the SAS® Version 6.12 PROC LOGISTIC procedure with the same dichotomous dependent variable pertaining to childhood education as in earlier logistic analyses.

3.2. Female Education, Survival Analysis, and STI Risk

To examine the effects of female education upon beading and age at sexual debut Roth interviewed 160 married and unmarried men and women in Karare/Nasakakwe in May–June, 2001. Interview questions asked about age at sexual debut and number of partners in the past three years. Men were asked if they had ever beaded *nykeri*, and if so how many *nykeri* they beaded. Women were asked if they were ever beaded and if so, at what age, and how many times. Results were analyzed using the SAS® 8.0 survival analysis procedures LIFETEST and PHREG, which modeled the time to sexual debut as "failure event" (Allison, 1996) in relation to covariates concerning education and beading history.

4. ANALYSIS AND RESULTS

4.1. Logistic Analysis of Parental Decision-Making

For the 1996 survey data, Table 1 presents the results of multivariate logistic regression analysis, employing the SAS® PROC LOGISTIC backward elimination procedure, which removes non-significant ($p > 0.05$) variables from the full main effect model to yield a more parsimonious fit between model and data. As shown in Table 1, analysis featured the respective elimination of the WIFE (monogamous vs. polygynous households), HOUSEHOLD WEALTH (poor vs. wealth), and the AGE-SET (recent vs. old) variables. All remaining variables were highly significant ($p < 0.001$). The Odds Ratio associated with the variable SUBSISTENCE (pastoral vs. agricultural) showed that children in the agricultural sample were almost two and one-half times (2.489) more likely to have ever gone to school relative to children from the pastoral sample. A negative coefficient was generated for the BIRTH (first vs. latter-born) school variable, indicating that latter born children were less than one-half (0.442) as likely as first-born children to ever go to school. This result reverses previous findings among Rendille children (Roth, 1991), where latter-born children were far more likely to have ever attended school, a finding attributed to Rendille practicing primogeniture. The negative coefficient recorded for the SIBS (sibs in school yes/no) variable indicated that families sent multiple children for schooling, rather than just selecting one. As in all previous analyses, the SEX variable showed a continuing distinct bias against choosing girls for schooling, with boys were almost one and three-quarters (1.727) times more likely to have ever gone to school than girls.

Regardless of algebraic sign, the largest coefficient was associated with the SUBSISTENCE variable. Therefore we subdivided the entire data set by subsistence base and performed separate analyses, as shown in Tables 2A and 2B for the pastoral and agricultural samples. These subsistence-specific runs yielded very different results. For the pastoral sample the remaining significant variables included SEX, SIBS and AGESET, with the

Table 1. Results of Logistic Regression Analysis, Total 1996 Karare Sample,
Backward Selection Option.

Step	Effect removed	Degrees of freedom	Number In	Chi-square	Pr > Chi-square
Summary of Backward Elimination					
1	WIFE	1	6	0.0102	0.9195
2	HHWEALTH	1	5	0.0107	0.9178
3	AGESET	1	4	3.2744	0.0704

Parameter	Degrees of freedom	Parameter estimate	Standard error	Chi-square	Prob.	Odds Ratio
Analysis of Maximum Likelihood Estimates						
INTERCEPT	1	−0.6942	0.5057	1.884	0.1698	
ECONOMY	1	0.9120	1.1467	38.640	<0.0001	2.489
BIRTH	1	−0.8163	0.1620	25.383	<0.0001	0.442
SEX	1	0.5465	0.1454	14.121	0.0002	1.727
SIBS	1	−0.7264	0.1634	19.753	<0.0001	0.666

most highly significant variable SEX ($p < 0.001$) showing the continued bias against educating daughters. Overall, this constellation of significant independent variables indicates that households headed by men from more recent age-sets sent multiple children to school, but in doing so still favored boys at the expense of girls. In contrast to the pastoral analysis, and indeed notably different from all previous analyses, the SEX variable is non-significant for the agricultural sample, which retains only BIRTH and SIBS as significant independent variables. Negative coefficients associated with both these variables indicate that parents are adopting an "all or none" strategy, with families who choose to educate children sending multiple children, regardless of parity and, most importantly, sex.

We interpret the results for the sex variable as partly reflecting differences in the labor requirements for males in the two subsistence bases. For pastoral families sons' labor is still required. Although pastoral families now maintain sedentary residences in Karare this does not alter the demands of male labor in herding and guarding livestock in remote *fora* or animal camps, where household herds still move in search of water and graze. In contrast, agricultural families do not have such far-flung responsibilities and their sons can attend school in the day and still return home to help with household and agricultural chores. However, changing labor requirements can not explain sending girls for schooling, because time allocation studies of sedentary agricultural Ariaal populations (Fratkin and Smith, 1995; Smith, 1999) show early and continued substantial labor contributions from girls, *nykeri*, and women. Equally, if not more important, there has been little or no change in the cultural factors of patrilocality and patrilineality that remove daughters' labor from her natal home upon her marriage.

The past ten years of research and analysis always showed a distinct parental bias against educating daughters. The lack of such a bias in the agricultural sample constitutes true innovative behavior on the part of their parents. Educating daughters in a social context featuring the continuation of some previous social structures, specifically patrilocality and patrilineality, combined with rapid changes in others, notably settlement and subsistence patterns, may be an empirical example of what Cleland (2001) termed the *blended theory* of innovation diffusion. This hypothesizes that structural change is accompanied and aided

Table 2A. Results of Logistic Regression Analysis, Children from Pastoral Households, Backward Selection Option ($n = 679$).

Step	Effect removed	Degrees of freedom	Number In	Chi-square	Pr > Chi-square
Summary of Backward Elimination					
1	WIFE	1	5	0.0023	0.9615
2	HHWEALTH	1	4	0.2012	0.6537
3	SIBS	1	3	1.0086	0.3153

Parameter	Degrees of freedom	Parameter estimate	Standard error	Chi-square	Prob.	Odds Ratio
Analysis of Maximum Likelihood Estimates						
INTERCEPT	1	−0.7343	0.5648	1.6901	0.1963	
BIRTH	1	−0.6277	0.2136	8.6393	< 0.0033	0.534
SEX	1	0.9086	0.2054	19.5694	< 0.0001	3.710
AGESET	1	−0.7653	0.2329	10.7983	0.0010	0.734

Table 2B. Results of Logistic Regression Analysis, Children from Agricultural Households, Backward Selection Option ($n = 392$).

Step	Effect removed	Degrees of freedom	Number In	Chi-square	Pr > Chi-square
Summary of Backward Elimination					
1	SEX	1	5	0.5398	0.4625
2	WIFE	1	4	0.7408	0.3894
3	HHWEALTH	1	3	0.7593	0.3836
4	AGESET	1	2	1.3216	0.2503

Parameter	Freedom	Estimate	Error	Square	Prob.	Ratio
Analysis of Maximum Likelihood Estimates						
INTERCEPT	1	3.0318	0.6295	23.1930	<0.0001	
BIRTH	1	−0.9215	0.2545	13.1135	<0.0003	0.655
SIBS	1	−1.4694	0.2993	24.0988	<0.0001	0.414

by the diffusion of ideas, so that a blend of structural and ideational changes results in behavioral innovation. Specifically developed with reference to fertility change in Third World populations (see Retherford, 1985; Retherford and Palmore, 1983), Cleland (2001: 45) described the relationship between the two major variables as: "Under the blended theory the engine of demographic change is the structural transformation of societies, and diffusion is the lubricant". In this case, rather than the idea of family planning and/or modern contraceptive technology, we suggest that the idea of female education is diffusing throughout the agricultural sample. While this idea itself would be a powerful mechanism for changing parental behavior, we adopt the logic of the blended theory and argue the acceptance of female education was only possible following the dramatic social structure change from nomadic pastoralism to sedentary agriculture. In our concluding section we return to this point, and discuss possible approaches to test this claim.

4.2. Survival Analysis and the Risk of Sexually Transmitted Infections

We now turn our attention to delineating the ramifications of parental decisions to educate daughters, focusing on the higher risks of transmitting sexual infections because of early age of sexual debut. To understand the overall age-specific pattern of sexual initiation regardless of sex and/or beading, we first compared the age distribution of sexual debut for all women in our sample, relative to men. Figure 3 presents survival distribution curves derived from the SAS® PROC LIFETEST procedure. These show a significant (Log-Rank Chi-Squared = 6.90, p = 0.009) difference between the two curves, with males on average featuring a sexual debut two years later than females (males n. = 97, median = 18.0 years, mean = 18.2 years, females n. = 84, median = 16.0 years, mean = 16.6 years). Next we considered the age distribution for women only, differentiating the sample into beaded versus never-beaded women. The resulting difference in curves was even greater than that distinguishing males from females (Log-Rank Chi-Squared = 29.6, p = 0.0001). As shown in Figure 4, never-beaded women featured average values a full four years (never beaded n. = 64, median = 17.0 years, mean = 17.6 years) above *nykeri* (n. = 20, median = 13.0 years, mean = 13.7 years), supporting the hypothesis that beaded women feature a significantly earlier age at sexual debut than do never-beaded women.

We next used the (SAS, 1999) PROC PHREG procedure to analyze the same data. This routine uses the Cox (1972) proportional hazard model that permits calculation of covariate effects on failure time distribution. Covariates included categorical variables for education and beading, respectively dichotomized as never versus ever-educated in the covariate EDUCATION and ever versus neverbeaded for the variable BEADED. As shown in Table 3, results featured a negative Risk Ratio for EDUCATION, indicating that educated females have approximately one-third (0.340) the hazard of the failure event (sexual debut) as uneducated women. The same measure for the BEADED covariate yielded a

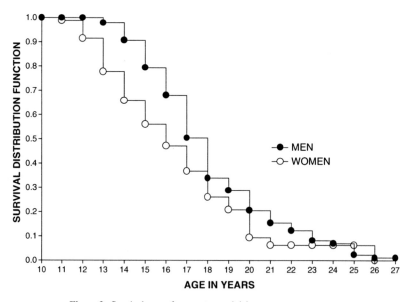

Figure 3. Survival curve for age at sexual debut, men versus women.

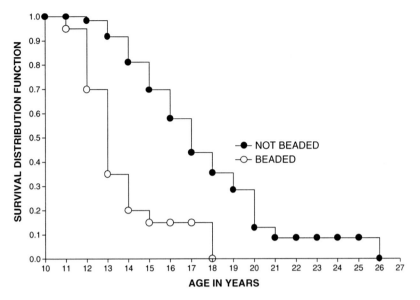

Figure 4. Survival curve for female age at sexual debut, beaded versus non-beaded women.

Table 3. Results of Proportional Hazards Analysis for Female Sample.

parameter	df	Estimate	Standard error	Chi sq.	Prob.	Risk ratio
Analysis of Maximum Likelihood Estimate						
BEADED	1	0.9907	0.3607	7.54	0.006	2.693
EDUCATE	1	−1.0795	0.3532	9.33	0.002	0.340

value of 2.693, indicating that beaded girls feature more than two and one-half times the hazard relative to non-beaded girls.

Overall both the LIFETEST and PHREG procedures results strongly support the earlier Roth et al. (2001) model hypothesizing that educated daughters would feature a later age at sexual debut because they would be protected from the Ariaal Rendille beading tradition. Combining the age distribution generated by the LIFETEST procedure with the covariate analysis provided by PROC PHREG provides another independent verification of the original model shown in Figures 1 and 2.

5. SUMMARY AND DISCUSSION

In this chapter, we considered female education in a sedentary Ariaal community in terms of two different, but related research issues. The first concerned parental decision-making with respect to which children to send for schooling. Rather than representing sheer serendipity, as previously suggested (Caldwell, 1982), our analyses repeatedly revealed strong, predictable linkages with cultural and/or socio-economic patterns. When cultural factors changed, represented by the differences between Korr and Karare with respect to subsistence bases, inheritance patterns and the economic value of children, so too did parental decision-patterns. In addition to between community differences our analyses revealed substantial within community variation, represented by the highly distinctive sets of statistically significant independent variables for the pastoral versus agricultural Karare samples. To us the most exciting find of this last analysis was the absence of the bias against educating daughters we invariably found throughout ten years of research and in two different communities. While the numbers in our sample were small, we hope that we are seeing an incipient transition away from son preference and towards educating both sons and daughters. One important point gleaned from these results is that the blanket comparison of sedentism versus nomadism is a facile dichotomy with little explanatory power. Rather, we argued that it is a combination of both structural and ideational innovations that determines change in parental decision-making patterns.

The second stage of our study examined concomitants of female education with respect to the transmission of sexual infections. Although the demographic consequences of female education with respect to fertility and offspring mortality have been studied for over twenty years, there are still large lacunae in our knowledge of the precise pathways by which female education affects vital rates. These gaps are most evident at the community level, and should provide impetus for future anthropological fieldwork. For sub-Saharan Africa, one underdeveloped avenue of research is the delineation of the effects of female education upon the risk of sexual transmitted infections. In the present study survival analysis strengthened our previous model's predictions that educated women have a lower risk of STIs because of a later age of sexual debut.

While delighted that our quantitative *etic*, or externally imposed, analysis supported an originally qualitative *emic*, or internally derived, model of Ariaal female adolescent sexuality, we recognize the limitations of this study. First, our samples are small, even for anthropological research, and we are now making efforts to increase our data sets. We also need increased time depth to accurately assess how the ramifications of female education echo beyond adolescence into the reproductive period. If the information given to us in interviews is true, then whether or not a daughter is educated can result in two different life-course trajectories, with educated women featuring a higher age at sexual debut, a lower rate of partner change, an increased age at marriage, higher contraceptive usage rates and finally lower realized fertility levels, than their uneducated counterparts. While certainly plausible, this extended model, shown in Figure 5 remains untested.

Results presented here suggest future paths for our two research topics. First, our argument that agricultural Ariaal parents now do not distinguish between sons and daughters with respect to schooling, as predicted by the Blended Theory of Diffusion, certainly needs testing. Specifically, we need to apply the social network methodology recently developed by demographers for the diffusion of contraceptive usage (cf. Montgomery and Casterline, 1998; Montgomery et al., 2001) to Ariaal communities. Secondly it may be fruitful to reverse the direction of work already completed and begin by asking parents to describe their household strategies for educating, or not educating, their children in their own words. The resulting emic models could then be compared with our quantitative, etic results to highlight conformance or deviation from what people say and what they really do. For the concomitants of female education on STIs research we need further quantitative tests to see if the emic model shown in Figure 5 is correct with respect to the effect of female education on changes in the age at marriage, the timing of the onset of fertility and the adoption of modern contraceptives. We also need to incorporate biological testing for STIs along with further research on sexual behavior to firmly determine relationships between socially condoned sexual behaviour and sexual disease transmission.

BIRTH

↓

EDUCATION

NO   **YES**

↓   ↓

AGE AT SEXUAL DEBUT

–   +

↓

RATE OF PARTNER CHANGE

+   –

↓

AGE AT MARRIAGE

–   +

↓

FAMILY PLANNING

–   +

↓

FERTILITY

+   –

↓

OFFSPRING MORTALITY

+   –

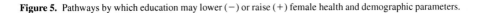

Figure 5. Pathways by which education may lower (−) or raise (+) female health and demographic parameters.

These avenues of future research highlight the importance of considering female education as a potentially powerful engine for demographic and social change in now sedentary northern Kenyan communities. We need to continue these lines of research enquiry to determine if in newly sedentarized villages of northern Kenya female education actually achieves the potential outlined in Caldwell's (1982) Wealth Flow Theory and in the United Nations' (1994, 1995) policy statements.

ACKNOWLEDGEMENTS. The authors are grateful for funding support provided by the Social Sciences and Humanities Research Council of Canada, the Association of Universities and Colleges of Canada, and the Canadian International Development Agency. We wish to express our heartfelt appreciate to all the people of Karare/Nasakakwe, and in particular to our hosts, the Leala family.

REFERENCES

Allison, P., 1998, *Survival Analysis Using the SAS® System: A Practical Guide.* SAS Press, Cary, NC.

Aral, S., 1992, Sexual Behaviour as a Risk Factor for Sexually Transmitted Diseases. In *Reproductive Tract Infections: Global Impact and Priorities for Women's Reproductive Health,* edited by A. Germain K. Holmes, P. Piot, and J. Wasserman, pp. 185–198, Plenum Press: New York.

Beaman, A., 1981, *Rendille Age-Set System in Ethnographic Context: Adaptation and Integration in a Nomadic Society,* Unpublished Ph.D. Dissertation, Department of Anthropology, Boston University.

Bledsoe, C.J.C., J. Johnson-Kuhn, and J. Haaga, 1999, *Critical Perspectives on Schooling and Fertility in the Developing World,* National Academy Press: Washington, D.C.

Caldwell, J., 1979, Education as a Factor in Mortality Decline: An Examination of Nigerian Data. *Population Studies* 33:395–413.

Caldwell, J., 1980, Mass Education as a Determinant in the Timing of Fertility Decline. *Population and Development Review* 6:225–255.

Caldwell, J., 1998, *Theory of Fertility Decline.* New York: Academic Press.

Caldwell, J., 1999, Reasons for Limited Sexual Behavioural Change in the sub-Saharan African AIDS Epidemic and Possible Future Intervention Strategies. In *Resistances to Behavioural Change to Reduce HIV/AIDS Infection in Predominantly Heterosexual Epidemics in Third World Countries,* edited by J. Caldwell, P. Caldwell, J. Anarfi, K. Awusabo-Asare, J. Ntozi, I.O. Orubuloye, J. Marck, W. Cosford, R. Colombo, and E. Hollings, pp. 241–256, Health Transition Centre, Australia National University, Canberra.

Caldwell, J., I.O. Orubuloye, and P. Caldwell, 1992, Underreaction to AIDS in sub-Saharan Africa. *Social Science and Medicine* 34(11):1169–1182.

Caldwell, J., I.O. Orubuloye, and P. Caldwell, 1999, Obstacles to Behavioural Change to Lessen the Risk of HIV Infection in the African AIDS Epidemic: Nigerian Research. In *Resistances to Behavioural Change to Reduce HIV/AIDS Infection in Predominantly Heterosexual Epidemics in Third World Countries,* edited by J. Caldwell, P. Caldwell, J. Anarfi, K. Awusabo-Asare, J. Ntozi, I.O. Orubuloye, J. Marck, W. Cosford, R. Colombo, and E. Hollings, pp. 113–124, Health Transition Centre, Australia National University, Canberra.

Cleland, J., 2001, Potatoes and Pills: An Overview of Innovation-Diffusion Contributions to Explanations of Fertility Decline. In *Diffusion Processes and Fertility Transition: Selected Perspectives,* edited by J. Casterline, pp. 39–65, National Academy Press: Washington, D.C.

Cleland, J. and G. Kaufmann, 1998, Education, Fertility and Child Survival: Unraveling the Links. In *The Methods and Uses of Anthropological Demography,* edited by A. Basu and P. Aaby, pp. 128–152, Clarendon: Oxford.

Cleland, J. and J. Van Ginnekan, 1988, Maternal Education and Child Survival in Developing Countries: The Search for Pathways of Influence, *Social Science and Medicine* 27:1357–1368.

Cox, D.R., 1972, Regression Models and Life Tables, *Journal of the Royal Statistical Society, Series B,* 34:187–224.

Fratkin, E., 1991, *Surviving Drought and Development,* Westview Press: Boulder, CO.

Fratkin, E., 1996, Traditional Medicine and Concepts of Healing among Samburu Pastoralists of Northern Kenya. *Journal of Ethnobiology* 16:53–100.

Fratkin, E. 1998, *Ariaal Pastoralists of Northern Kenya: Surviving Drought and Development in Africa's Arid Lands.* Allyn and Bacon: Boston.

Fratkin, E. and E. Roth, 1996, Who Survives Drought: Measuring Winners and Losers among the Ariaal Rendille Pastoralists of Kenya. In *Case Studies in Human Ecology,* edited by D. Bates and S. Lees, pp. 159–174. Plenum Press: New York.

Fratkin, E. and K. Smith, 1995, Women's Changing Economic Roles with Pastoral Sedentarization: Varying Strategies in Four Rendille Communities. *Human Ecology* 23(4):433–454

Fratkin, E., E. Roth, and M. Nathan, 1999, When Nomads Settle: Commoditization, Nutrition and Child Education among Rendille Pastoralists. *Current Anthropology* 40(5):729–735

Giles, J., 1999, *Who Gets to Go To School: Parental Schooling Choices Among the Ariaal Rendille of Northern Kenya*. Unpublished MA Thesis, Department of Anthropology, University of Victoria, Victoria, BC.

Heise, L. and Elias, C., 1995, Transforming AIDS Prevention to Meet Women's Needs: A Focus on Developing Countries. *Social Science and Medicine* 40:931–943.

Korenromp, E., C. Van Vliet, C. Baker, S. De Vlas, and J.D.F. Habbema, 2000, HIV Spread and Partnership Reduction for Different Patterns of Sexual Behaviour—A Study with the Microsimulation Model *STDSIM*. *Mathematical Population Studies* 8(2):135–173.

LeVine, R., S. LeVine, A. Richman, F. Tapia Uribe, and C. Correa, 1994, Schooling and Survival: The Impact of Maternal Education on Health and Reproduction the Third World. In *Health and Social Change in International Perspective*, edited by L. Chen, A. Kleinman, and N. Ware, pp. 303–338, Harvard University Press: Boston.

Meekers, D. and A.-E. Calves, 1997, Main Girlfriends, Girlfriends, Marriage and Money: The Social Context of HIV Risk Behavior in sub-Saharan Africa. *Health Transition Review* 7(supplement):361–375.

Montgomery, M. and J. Casterline, 1998, Social networks and the diffusion of fertility control. *Working Papers of the Population Council*, No. 119.

Montgomery, M., G.-E. Kiros, D. Aryman, J. Casterline, P. Agolobitse, and P. Hewett, 2001, Social Networks and Contraceptive Dynamics in Southern Ghana. *Working Papers of the Population Council*, No. 153.

Orubuloye, I.O., J. Caldwell, and P. Caldwell, 1997, Perceived Sexual Needs and Male Sexual Behaviour in Southwest Nigeria. *Social Science and Medicine* 44:1195–1207.

Retherford, R., 1995, A Theory of Marital Fertility Transition. *Population Studies* 39(2):703–740.

Retherford, R. and J. Palmore, 1983, Diffusion Processes Affecting Fertility Regulation. In *Determinants of Fertility in Developing Countries*, Volume 2, edited by R. Bulatao and R. Lee, pp. 295–339. Academic Press: New York.

Roth, E., 1991, Education, Tradition and Household Labour among Rendille Pastoralists of Northern Kenya. *Human Organization* 50:136–141.

Roth, E., 1993, A Re-examination of Rendille Population Regulation. *American Anthropologist* 95:597–611.

Roth, E., 1999, Proximate and Distal Demographic Variables among Rendille pastoralists. *Human Ecology* 27:517–536.

Roth, E., 2001, The Demise of the Sepaade Tradition: Cultural and Biological Explanations. *American Anthropologist* 103:1014–1024.

Roth, E., E. Fratkin, A. Eastman, and L. Nathan, 1999, Knowledge of AIDS Among Ariaal Pastoralists of Northern Kenya. *Nomadic Peoples (NS)* 3(2):161–175.

Roth, E., E. Ngugi, E. Fratkin, and B. Glickman, 2001, Female Education, Adolescent Sexuality and the Risk of Sexually Transmitted Infection in Ariaal Rendille Culture. *Culture, Health and Sexuality* 3:35–48.

SAS, 1999 *SAS/STAT Users' Guide, Version 8*, SAS Institute, Inc., Cary, NC.

Smith, K., 1999, Sedentarization and Market Integration: New Opportunities for Rendille and Ariaal women of Northern Kenya. *Human Organization* 57:469–480.

Spencer, P., 1965, *The Samburu: A Study of Gerontocracy in a Nomadic Tribe*, University of California Press: Berkeley.

Spencer, P., 1973, *Nomads in Alliance: Symbiosis and among the Rendille and Samburu in Kenya*. Oxford University Press: Oxford.

Stokes, M., C. Davis, and G. Koch, 1995, *Categorical Data Analysis Using the SAS System*. SAS Institute, Inc., Cary, NC.

United Nations, 1994, Population and Development: Programme of Action Adopted at the International Conference on Population and Development. Cairo, 5–14, September, 1994.

United Nations, 1995, *Women's Education and Fertility Behaviour*, United Nations Department of Social and Economic Affairs, New York.

Varga, C., 1999, South African Young People's Sexual Dynamics: Implications for Behavioural Responses to HIV/AIDS. In *Resistances to Behavioural Change to Reduce HIV/AIDS Infection in Predominantly Heterosexual Epidemics in Third World Countries*, edited by J. Caldwell, P. Caldwell, J. Anarfi, K. Awusabo-Asare, J. Ntozi, I.O. Orubuloye, J. Marck, W. Cosford, R. Colombo, and E. Hollings, Health Transition Center, pp. 13–34. Australian National University Press, Canberra.

Index